Calculus
and the
Computer

CONTENTS

Chapter 20. Multiple Integration

Chapter 21. Vector Calculus

Chapter 22. Differential Equations

Calculus
and the
Computer

INTRODUCTION

The calculus is a powerful tool for solving problems, especially those dealing with motion. In fact, it was invented in the seventeenth century specifically for that purpose. Since that time the advent of the computer has allowed the opportunity to provide numerical solutions to calculus problems which were difficult or impossible to obtain by other methods. It is this alliance of the calculus and the computer that this text addresses.

The organization of the text provides for both the novice and the student familiar with BASIC. Thus, the first three chapters provide an introduction to BASIC and can be skipped by the student familiar with the language; a brief review of BASIC can be found in Appendix A. Chapter four is a discussion of computer error, and chapter five addresses devising an algorithm for a computer program. The following chapters develop the calculus through computer programs and their analysis. It should be noted that different computers can produce different output, errors, and BASIC statements themselves. Thus, computer results may, at times, be quite different from those in this text. The exercise sets are graded in difficulty with zero, one, two, or three asterisks (*) and, together with the challenge activities, check comprehension and extend the material covered. Thus, it is hoped that this text will not only deepen your understanding and knowledge of calculus, but will also motivate an appreciation of the value and difficulties of computer solutions.

CHAPTER 1
Programming In BASIC Part I:
Input, Output, And The Assignment Statement

A computer program is a sequence of instructions telling a computer to perform specific tasks. In BASIC each instruction is generally written on a separate line and consists of a line number and a statement. Each statement contains a keyword, a command to the computer, and the information required by that keyword. The order in which instructions are performed is established by the line numbers, with the computer starting with the instruction with the lowest line number, then proceeding to the next higher line number, and ending with the highest line number.

From the beginning it is necessary to realize that writing a computer program requires two vastly different skills. First, we must know the programming language's syntax, that is, its vocabulary and the rules governing that vocabulary. In this book that language will be BASIC. The second skill is the ability to create an algorithm, a list of steps, which can be performed by a computer to solve a problem. This ability to fashion an algorithm is a fundamental, yet the most difficult, step in the process of writing a computer program.

In this and the next four chapters we introduce that portion of BASIC and related concepts necessary for writing and understanding the majority of programs in this book. The remaining chapters are devoted to the development of algorithms and computer programs related to the calculus.

1.1 OUTPUT

Before we can start writing programs in BASIC we must learn certain commands which allow us to communicate with the computer. Since these commands do vary among computers, the reference manual for the computer being used should be consulted for the correct variation if one of the commands listed does not function properly.

1. Entering Information.

     To enter a statement or command into the computer the RETURN key must be pressed after the statement or command is typed. Pressing this key not only transmits information to the computer but also acts in a manner similar to a carriage return on a typewriter causing the next entry to be typed at the beginning of the next line. On some computers the corresponding key is labeled CR or ENTER.

2. Clearing The Computer's Memory.

     Typing NEW followed by pressing the RETURN key erases any program in the computer's memory. It is a good habit to always clear the memory before entering a new program to insure that no program lines from a previous program will affect the new program being entered. In place of NEW some computers use CLEAR or SCRATCH.

3. Displaying A Program.

     Typing LIST followed by pressing the RETURN key will display whatever program is currently in memory. Most computers have several LIST commands.

     a. LIST   without a line number displays the entire
                   program.
     b. LIST 100   will display the line numbered 100 only.
     c. LIST 100,200   will display all lines numbered between
                   100 and 200 inclusive.

4. Running A Program.

     Typing RUN and pressing RETURN starts execution of the program currently in memory. Execution will halt when an error, an END statement, or the highest line number is encountered. Although most computers do not require that the last statement of a BASIC program be the END statement, it is a good programming habit to do so.

The Print Statement

     A computer can manipulate two different types of data--numerical data or numbers, and alphanumeric data, a sequence

of letters, numbers, and/or special characters enclosed in
quotation marks("). Alphanumeric data is also referred to by the
term "character string". The PRINT statement can be used to
output both numeric and alphanumeric data as illustrated by the
the following program.

Example 1:

```
NEW
10 PRINT "*THE PRINT STATEMENT*"
20 PRINT "1+2+3"
30 PRINT 1+2+3
40 PRINT "7/2"
50 PRINT 7/2                          using / for division
60 END


RUN(press the RETURN key)
*THE PRINT STATEMENT*
1+2+3
6
7/2
3.5
```

Notice that alphanumeric data, enclosed in quotes, is printed
verbatim--that is, everything entered between the quotation marks,
including punctuation marks and blank spaces, but not the quote
marks, is printed. Data not enclosed in quotes is numeric data
and its value is printed.

The PRINT statement can also be used to print several
different data items on a single line. This is accomplished by
listing in a single PRINT statement all the data items separated
by either a comma or semicolon. On most computers, items
separated by a comma(,) will be printed spread across the page,
while items separated by a semicolon(;) will be printed next to
each other.

When the computer encounters commas within a PRINT statement,
it usually divides the output line into predefined print zones
each having the same width. The exact number of zones per line
and width is dependent upon the computer being used. However,
most computers have three to five zones per line with about 15
print positions in each zone. Data items are printed in
successive zones with one data item per zone. If there are more
data items than zones on a line the computer prints the remaining
items on the next line. The effect of using commas in a PRINT
statement is illustrated by the following program.

Example 2:

```
NEW
10 PRINT 1,-22,333,-4444
20 END


RUN
1               -22             333
-4444
```

Notice that the fourth data item is printed in the first print zone of the second line. Also note that a minus sign takes up a full space in the print zone.

Character strings also normally start in the first position of each zone as shown below.

Example 3:

```
NEW
10 PRINT "HELLO","THERE"
20 END


RUN
HELLO           THERE
```

When a computer prints all data starting in the first position of the print zones we say that the output is "left justified".

The use of semicolons packs the output closely together as the following program illustrates.

Example 4:

```
NEW
10 PRINT 1;-22;333;-4444
20 END


RUN
1-22333-4444
```

Computers vary as to the exact location of successive data items when semicolons are used. Some computers leave one, two, or possibly three spaces between printed items while others leave none as illustrated above. In a manner similar to using commas, the computer will advance to the next line and print any data which cannot fit on the current output line. However, since there are no predefined print zones when using semicolons, it is possible that data items will be "split", that is, printed partly on two different lines.

As we have seen commas and semicolons can be used to control, to a degree, the location of data items within a line of output. BASIC also provides a method for controlling spacing between successive lines of output. A PRINT statement followed by no data items will cause the computer to skip a line each time it is encountered. On the other hand, a comma or semicolon encountered at the end of a PRINT statement will suppress line feed. The different effects of ending a PRINT statement with a comma or semicolon are demonstrated by the following program.

Example 5:

```
NEW
10 PRINT "TO SKIP A LINE USE"
20 PRINT
30 PRINT "A PRINT BY ITSELF"
40 PRINT
50 PRINT "2+3+5 =",
60 PRINT 10
70 PRINT
80 PRINT "2+3+5 =";
90 PRINT 10
100 END


RUN
TO SKIP A LINE USE

A PRINT BY ITSELF
2+3+5 =             10

2+3+5 =10
```

As always, a different computer may generate a different looking output.

The TAB Function

The TAB function provides greater flexibility in the horizontal spacing of data items by overriding print zones and skipping to a specific column to display an item. When the computer encounters TAB(n), where n is a numeric value, it will begin printing in the n-th column. The columns are numbered from the left 1,2,3 etc. Finally, the TAB function should only be used in a PRINT statement with semicolons as demonstrated in the program below. The numbers across the top are included to indicate the print positions.

Example 6:

```
NEW
10 PRINT "1234567890123456789012
   345678901234567890"
20 PRINT TAB(5);"HI";TAB(21)
   ;"THERE"
30 PRINT
40 PRINT TAB(5),"HI",TAB(21)
   ,"THERE"
50 END
```

```
RUN
123456789012345678901234567890
    HI                THERE

                  HI
THERE
```

Notice that using commas in line 40 has resulted in the improper functioning of the TAB function.

Editing

Eventually we are going to want to change a line in a program. BASIC provides several editing features by which this and other operations can be accomplished without having to re-enter the entire program.

1. To add a new line: Simply type it.

```
NEW
10 PRINT "HI THERE"
20 END
```

7

```
RUN
HI THERE


15 PRINT "BILL"

LIST
10 PRINT "HI THERE"
15 PRINT "BILL"
20 END


RUN
HI THERE
BILL
```

2.   To change a line: Type it over.

```
LIST
10 PRINT "HI THERE"
15 PRINT "BILL"
20 END

15 PRINT "WILLIAM"

LIST
10 PRINT "HI THERE"
15 PRINT "WILLIAM"
20 END


RUN
HI THERE
WILLIAM
```

3.   To delete a line: Type the line number and press the RETURN
     key.

```
LIST
10 PRINT "HI THERE"
15 PRINT "WILLIAM"
20 END

15

LIST
10 PRINT "HI THERE"
20 END
```

```
RUN
HI THERE
```

    Besides these three editing features some computers do provide additional features which are described in the reference manual for the computer being used.

Exercise Set 1.1

1. Run the program in Example 2 to determine the number of predefined print zones per line and the number of print positions in each zone.

2. Run the program in Example 4 to determine the number of spaces left between two data items separated by a semicolon.

*
3. Write and run a program to print your name in a box of asterisks. A sample output is shown below.

```
******************
*                *
* WILLIAM OBERLE *
*                *
******************
```

4. Write and run a program to print the sum of the first seven odd whole numbers (1+3+5+7+9+11+13). Include sufficient information in the output so that anyone reading only the output would be able to tell what is happening. Can you see a relation between the sum and the number of terms in the sum? If so, what is the relation.

**
5. Write and run a program which prints 1/2 in the first print zone on a line, 1/3 and 2/3 on the next line in the first two print zones, and on the third line 1/4, 2/4, and 3/4 in the first three print zones. The numbers printed will not be fractions but decimals. Are the exact decimal values of the numbers printed? Could this cause a problem in doing numerical calculations? Explain.

1.2 INPUT AND THE ASSIGNMENT STATEMENT

Variables

For most programming problems we can think of a variable as a symbol which represents a data value. In this respect it is similar to what we think of as a variable in algebra. However, there is an important difference between variables in algebra and BASIC. In algebra, we generally manipulate variables without actually knowing their value. For example, we factor the expression $3x^2+16x+5$ and solve the equation $2x+7=13-x$ without knowledge of the value of x. When dealing with BASIC this is not possible; a variable must always have a value before it can be manipulated.

A variable in BASIC is a name, assigned by the programmer, which refers to a location in the computer's memory. Each time a variable is referenced in a program the computer retrieves the value stored in the appropriate memory location and performs whatever operations are specified on that value. Thus, to a computer the only thing it can do to $3x^2+16+5$ is evaluate it for the current value of x. Similarly, the computer cannot solve the equation $2x+7=13-x$, but only decide if it is a true or false statement for the current value of x.

In BASIC two different types of variables are used to distinguish between numeric and alphanumeric data. A numeric variable is denoted by a single letter or a single letter followed by a single digit. A string variable is distinguished from numeric variables by the use of the dollar sign($) following a single letter. The following are examples of valid and invalid variables.

```
Numeric Variables
Valid:    X      and    X1
Invalid: 2X (must begin with a letter) and
              X; (no special symbol)

Alphanumeric Variables
Valid:    A$     and    T$
Invalid:  B (must end in a $) and
              $ ( must begin with a letter)
```

While the above rules for naming variables hold for all computers, most do allow for more descriptive variable names. But, for all computers, variable names must start with a letter and string variables must end with a dollar sign. The exact number of characters allowed varies from one machine to the next and even when more characters are allowed they may not all be recognized.

For example, on some computers the variables AB1 and AB2 are treated as being the same. Many microcomputers recognize only the first two characters of both numeric and alphanumeric variables.

BASIC provides two methods for assigning data values to variables, the INPUT statement and the LET or assignment statement.

The INPUT Statement

Note: If you are not using an interactive system(teletype, microcomputer, etc.) you will not be able to use the INPUT statement but must use a READ and DATA statement. See APPENDIX B.

The INPUT statement is used to assign data values to variables from the keyboard while the program is running. The general format of the INPUT statement is:

line# INPUT (variable list)

For example,

10 INPUT A

or

20 INPUT A,B$

There is no restriction as to the number of variables or type, numeric or alphanumeric, which may appear in the variable list.

When the computer encounters an INPUT statement it prints a question mark(?), then stops and waits for the user to enter the appropriate data. Once the data is entered and the RETURN key pressed the computer continues running the program.

Examples illustrating the use of the INPUT statement are given below.

Example 1:

```
NEW
10 INPUT A,B,C
20 PRINT
30 PRINT A+B-C
40 END


RUN
?4,6,2 (? printed by computer; 4,6,2 entered by user)

8
```

The order in which the values are assigned to the variables is determined by the order in which they appear in the INPUT statement. Therefore, the variable A is assigned 4, B is assigned 6, and C is assigned 2. Also, remember that when a variable occurs in a program, other than in an INPUT statement or LET statement, the computer automatically replaces it with its current value. Thus, to the computer line 30 is PRINT 4+6-2.

Example 2:

```
NEW
10 INPUT A$
20 PRINT
30 PRINT A$
40 END


RUN
?HI THERE

HI THERE
```

Since a string variable appears in the INPUT statement the computer automatically makes the data entered a character string. Note the absence of the quote marks about the character string entered after the ?.

Example 3:	**Example 4:**	**Example 5:**
NEW	NEW	NEW
10 INPUT A,B,C	10 INPUT A,B,C	10 INPUT A,B$
20 PRINT	20 PRINT	20 PRINT
30 PRINT A+B-C	30 PRINT A+B-C	30 PRINT A,B$
40 END	40 END	40 END
RUN	RUN	RUN
?1,2	?5,3,2,1	?Y,4
??1,3	?EXTRA IGNORED	?REENTER
EXTRA IGNORED		?4,Y
	6	
2		4 Y

In examples 3, 4, and 5 we have illustrated what happens when we
enter too few, too many, or the wrong type of data values in
response to the INPUT statement. Again different computers will
respond to these kinds of errors in different ways.

One way to minimize these types of errors is to precede each INPUT
statement with a PRINT statement providing the user   with
information about the data required by the INPUT statement. This
PRINT statement is referred to as a prompt and is illustrated in
the following program.

Example 6:

```
NEW
10 PRINT "ENTER TWO NUMBERS"
20 INPUT X,Y
30 PRINT "X-Y=";X-Y
40 END


RUN
ENTER TWO NUMBERS
?10.578,7.29
X-Y=3.288
```

Finally, it is a good programming habit to display or "echo"
the value assigned to each variable when using the INPUT statement

Such a printout provides a record of the data used in the program and can be quite helpful in detection of program errors and interpreting program output. An example of "echoing" data is given below.

Example 7:

```
NEW
10 PRINT "ENTER TWO NUMBERS"
20 INPUT X,Y
30 PRINT
40 PRINT "THE TWO NUMBERS ARE ";X;" AND ";Y;"."
50 PRINT "X-Y=";X-Y
60 END


RUN
ENTER TWO NUMBERS
?12.5,15.6

THE TWO NUMBERS ARE 12.5 AND 15.6.
X-Y=-3.1
```

The LET Statement

In addition to the INPUT statement, data values can also be assigned to variables through the use of the LET or assignment statement. This statement not only allows direct assignment of values to numeric and string variables but also the assignment of a numerical calculation to a numeric variable. The general format of the LET statement is:

line# LET variable=(expression or value)

The expression may be a constant, a numeric expression, a character string, or another variable. The following are examples of valid assignment statements.

```
10 LET A=5.6

15 LET A$="HI THERE"

20 LET X=Y

30 LET X=4+Y+Z

40 LET N=N+1
```

When the LET statement is encountered during the running of the program the computer will first compute the value of the expression, whatever is to the right of the equal sign, and then assign the computed value to the variable on the left of the equal sign. Thus, only a single variable may appear to the left of the equal sign in an assignment statement. Finally, many computers do not require the use of the keyword LET. However, we will always use it in this book.

Although we have already used numeric expressions several times, we have never exactly defined them. Basically, a numeric expression is just what we think it is--an expression composed of constants, numeric variables, and arithmetic operations. In BASIC the permissible numeric operations are summarized in the following table.

Operation	Symbol	Example
Addition	+	A+B
Subtraction	−	B-D
Multiplication	*	F1*F2
Division	/	PI/T
Exponentiation	^	B^2

When more than one operation is to be performed the computer will follow the same hierarchy of operations as in algebra. Namely, in descending order:

1. Expressions within parentheses.
2. Exponentiation.
3. Multiplication and division.
4. Addition and subtraction.

In the event of ties, operations of the same priority are performed from left to right. Thus, in BASIC the following expressions are equivalent

$$4*5/3+2^3^4 \qquad \text{and} \qquad ((4*5)/3)+((2^3)^4) \ .$$

The following example illustrates not only the assignment statement but also the use of prompts and data "echo" to enhance program clarity.

Example 8:

The expression ax^3+bx^2+cx+d can be written as $x(x(ax+b)+c)+d$. Write a program which has as input the coefficients and a value for x and prints out the evaluation of each expression. Run the program for $4x^3-2x^2+x-3$ with $x=-2$.

```
NEW
100 PRINT "ENTER COEFF.  OF X CUBED."
110 INPUT A
120 PRINT "ENTER COEFF.  OF X SQUARED."
130 INPUT B
140 PRINT "ENTER COEFF.  OF X."
150 INPUT C
160 PRINT "ENTER CONSTANT."
170 INPUT D
180 PRINT "ENTER VALUE OF X."
190 INPUT X
200 PRINT
210 LET V1=A*X^3+B*X^2+C*X+D
220 LET V2=X*(X*(A*X+B)+C)+D
230 PRINT A;"X^3+";B;"X^2+";C;"X+";D;"=";V1
240 PRINT "X(X(";A;"X+";B;")+";C;")+";D;"=";V2
250 END


RUN
ENTER COEFF. OF X CUBED.
?4
ENTER COEFF. OF X SQUARED.
?-2
ENTER COEFF. OF X.
?1
ENTER CONSTANT.
?-3
ENTER VALUE OF X.
?-2

4X^3+-2X^2+1X+-3=-45
X(X(4X+-2)+1)+-3=-45
```

As programs become longer and more complicated it is difficult to keep track of what is happening. This is especially true if we come back to a program written weeks ago or have someone else try to follow the program. To avoid this problem as much as possible we should annotate or document our programs. In BASIC this is accomplished by a REM statement, which appears when a program is listed but has no effect when the program is run. Adding REM or remark statements to the program in Example 8 would result in the program below.

Example 9:

```
NEW
100 REM ** EVALUATING A CUBIC **
110 REM ** INPUT **
120 PRINT "ENTER THE COEFF.  OF X CUBED."
130 INPUT A
140 PRINT "ENTER THE COEFF.  OF X SQUARED."
150 INPUT B
160 PRINT "ENTER THE COEFF.  OF X."
170 INPUT C
180 PRINT "ENTER THE VALUE OF X.
190 INPUT X
200 PRINT
210 REM ** EVALUATION OF EXPRESSIONS **
220 LET V1=A*X^3+B*X^2+C*X+D
230 LET V2=X*(X*(A*X+B)+C)+D
240 REM ** OUTPUT **
250 PRINT A;"X^3+";B;"X^2+";C;"X+";D;"=";V1
260 PRINT "X(X(";A;"X+";B;")+";C;")+";D;"=";V2
270 END
```

Exercise Set 1.2

1. For each variable name, decide if it is a valid numeric variable or alphanumeric variable name. If it is not valid, explain why.
    a. A   e. X$
    b. SUM   f. X+Y
    c. 5A   g. AB
    d. $A   h. A;

2. Which of the following are valid assignment statements? For those which are not valid explain why.
    a.   100 LET A$=3*B+C
    b.   100 LET A=(B+C)*(B+C)
    c.   100 LET A+B=C
    d.   100 LET A=B=0
    e.   100 LET A=B$+C
    f.   100 LET A=(X*(X+6)-5)+8

3. Write the following algebraic expressions as valid expressions in BASIC.

    a. $\dfrac{n^3+3n+1}{6}$

    b. $\dfrac{x+y}{z+t}\,^4$

    c. $\dfrac{2(x/y)^{k-1}}{(a-3b)^{(1/m)}}$

    d. $\dfrac{ax^3+bx^2+cx+d}{ex^2+fx+g}$

4. When are quote marks not needed around a character string?

5. Decide if your computer requires the keyword LET by running the program below. If no error occurs then you do not need to use the keyword LET.

```
100 REM ** TESTING FOR LET **
110 LET X=5
120 PRINT X
130 Y=50
140 PRINT Y
150 END
```

*

6.   Write and run a program which accepts three character strings as input and prints out the three strings with four blank spaces between them. Test your program with the strings "CALCULUS", "IS", "MATHEMATICS".

7.   Write and run a program which asks for the name of an object and its weight in pounds and outputs the name of the object and its weight in pounds and kilograms. One pound=.454 kilograms. Test your program with a car weighing 2800 pounds which is 1271.2 kilograms.

8.   Write and run a program to compute the hypervolume of a four dimensional figure. As input enter the four dimensions. The output should "echo" the dimensions and give the hypervolume. Test your program with a figure of dimensions 8'x7'x6'x5' which has a hypervolume of 1680 ft.^{4}.

**

9.   One of VanDerWaal's equations for a gas is

$$P= \frac{NRT}{V-NB} \; \frac{-AN^2}{V^2}$$

where   P=pressure in atmospheres
V=volume of gas in cubic centimeters
T=temperature in degrees Kelvin
N=number of moles
R,A,B   constants dependent on the gas

Write and run a program to compute P. Your input should ask for values of N,R,T,V,A, and B. Your output should "echo" the data and give the value of P. Test your program with the following values: N=2, R=82, T=315, V=26000, A=3600000, and B=43 which produces a pressure of about 1.97 atmospheres.

***

10. Write and run a program to interchange the values of the variables A and B. That is, after the program is run the value assigned to A should be stored in B and vice versa.

CHALLENGE ACTIVITY

Given that the sides of a triangle have lengths a,b, and c then the area of the triangle by Hero's formula is

$$\sqrt{s(s-a)(s-b)(s-c)}$$

where s=(a+b+c)/2. In addition, the radius of the inscribed circle to the triangle is given by (area of triangle)/s and the radius of the circumscribed circle of the triangle is given by (abc)/(4 times the area of the triangle). Write and run a program which accepts as input the lengths of the three sides of a triangle and outputs the lengths of the three sides, the area of the triangle, and the area of both the inscribed and circumscribed circles. Test your program for a triangle with sides of length 32,18, and 25. The area of the triangle is about 224.22, the area of the inscribed circle is about 112.31, and the area of the circumscribed circle about 809.87 The square root in BASIC can be found using A^.5 Use 3.1415927 for pi.

CHAPTER 2
Programming In BASIC Part II: Control Statements

So far all of our programs have contained instructions which were executed one after another starting with the lowest line number and ending with the highest. However, sometimes it is desirable to alter the flow of program execution. We may want to repeat a set of instructions several times, or we may want to perform a set of instructions only when some condition is satisfied.

In this chapter we introduce several BASIC statements which allow us to transfer program control to statements out of sequence. When the statement sends the computer back to a line that has already been executed, we say that the program has a loop. On the other hand, if the statement causes the computer to jump ahead, skipping over lines in the program, we say that a selection has been made. These two ideas, looping and selection, are fundamental concepts and techiques in computer programming which will be used throughout this book.

2.1 GO TO, IF/THEN, And ON/GOTO Statements

The GO TO Statement

The simplest method for transferring program control is to use the GO TO statement. The general format of the GO TO statement is:

                    line# GO TO (line#)

Valid examples of GO TO statements are:

50 GO TO 100   and   100 GO TO 20

When encountered during the program the GO TO causes the computer to branch to the line whose line number follows the GO TO. Execution of the program then continues from that line. The GO TO statement is often referred to as an unconditional branch since the flow of the program is always changed when it is executed.

Example 1:

```
100 REM ** A BAD PROGRAM **
110 PRINT "ENTER A VALUE FOR X"
120 INPUT X
130 GO TO 160
140 LET A=X*X
150 PRINT "X*X=";A
160 LET B=X+1
170 PRINT "X+1=";B
180 END


RUN
ENTER A VALUE FOR X
?28
X+1=29
```

In this example the GO TO caused the program to branch forward in the program, thus a selection has been made. However this is a bad program since lines 140 and 150 can never be executed.

Example 2:

```
100 REM ** A SECOND BAD PROGRAM **
110 PRINT "ENTER A VALUE FOR X"
120 INPUT X
130 LET A=X*X
140 PRINT X;"*";X;"=";A
150 GO TO 110
160 END
```

23

```
RUN
ENTER A VALUE FOR X
?5
5*5=25
ENTER A VALUE FOR X
?10
10=100
ENTER A VALUE FOR X
?
BREAK IN LINE 120
```

Since the GO TO in line 150 sends the computer back to line 110, we are branching back to execute lines which have already been performed, thus we are creating a loop. Unfortunately, there is no way to terminate this program. We are caught in what is called an infinite loop, and we can stop the program only by manually interrupting it. This can be done by turning the computer off or by any method specified for the computer being used.

As these two examples have illustrated the GO TO statement used by itself normally results in programs which contain infinite loops or statements which can never be executed. However, the GO TO statement can be quite effective in achieving the desired program flow when used in combination with other control statements.

The IF/THEN Statement

The IF/THEN statement is used when it is necessary to transfer control only when a specific condition is satisfied. Also referred to as a conditional branch, the general format of the IF/THEN statement is:

line# IF (condition) THEN (statement)

The condition in the IF/THEN statement must be an expression which is either true or false. Determining the truth value of the condition involves testing a relationship between two expressions using the format:

expression (relational operator) expression

Expressions may contain either numeric or alphanumeric data and variables and any legal arithmetic operation symbols. The relational operators which can be used in BASIC are summarized below.

For Comparing Alphanumeric Expressions

Operator	Meaning	Example
=	equals, identical	A$="HI"
<>	not equal, not the exact same thing	A$<>"HI" "HI"<>"HELLO"

For Comparing Numeric Expressions

Operator	Meaning	Example
=	equals	A=4*B
<>	not equal	A<>B
<	less than	A<10/B
>	greater than	A>B*B
<=	less than or equal	A<=B+3
>=	greater than or equal	A>=B+C

The statement following the THEN may be any executable BASIC statement. Examples of valid IF/THEN statements are:

    IF A>=B+5 THEN LET A=A+1

    IF A$<>"YES"A THEN GO TO 100

    IF A+B<=C*C*C THEN PRINT "A=";A;" B=";B

As always, different computers may implement the IF/THEN statement in different ways. Generally, this will involve the statement following the THEN. However, all computers implement the IF/THEN statement with the format

    line# IF (condition) THEN (line#)

and thus we will use only this form of the IF/THEN. An example of this form of the IF/THEN is:

    IF A*A<=50 THEN 210

When an IF/THEN statement is encountered during the execution of the program, the computer will first determine if the condition is true or false. Any variables in the condition will be assigned their current values.   If the truth value of the condition is false, the computer will proceed to the next line in the program. If the truth value of the condition is true, the computer will execute the statement following the THEN. When this statement is a GO TO or a line number the computer will unconditionally branch to the indicated line and continue execution from that point in the program.   Otherwise the computer will execute the statement following the THEN and then proceed to the line following the IF/THEN statement.

Example 3:

```
100 REM ** ABSOLUTE VALUE **
110 PRINT "ENTER THE VALUE OF X"
120 INPUT X
130 PRINT "THE ABSOLUTE VALUE OF ";X;" IS";
140 IF X>=0 THEN 160
150 LET X=-1*X
160 PRINT X;"."
170 END
```

```
RUN
ENTER A VALUE FOR X
?23
THE ABSOLUTE VALUE OF 23 IS 23.

RUN
ENTER A VALUE FOR X
?-34
THE ABSOLUTE VALUE OF -34 IS 34.
```

An alternate method for determining the absolute value of a number is illustrated in the following program.

Example 4:

```
100 REM ** ABSOLUTE VALUE #2 **
110 PRINT "ENTER A VALUE FOR X"
120 INPUT X
130 IF X>=0 THEN 170
140 LET A=-1*X
150 PRINT "THE ABSOLUTE VALUE OF ";X;" IS";A;"."
160 GO TO 180
170 PRINT "THE ABSOLUTE VALUE OF ";X;" IS";X;"."
180 END
```

```
RUN
ENTER A VALUE FOR X
?23
THE ABSOLUTE VALUE OF 23 IS 23.

RUN
ENTER A VALUE FOR X
?-34
THE ABSOLUTE VALUE OF -34 IS 34.
```

In examples 3 and 4 the IF/THEN statement caused the computer to jump ahead in the program. Thus, in both programs a selection has been made. However, there is a difference in the type of selection made in the two programs. In example 3 the IF/THEN statement is used to decide whether to do something, multiply by -1, or not to do it. Contrast that with example 4 where the IF/THEN(and a GO TO) statement are used to select one of two alternatives. This second use of the IF/THEN is often referred to as an IF/THEN/ELSE selection. If the condition is true, THEN the computer does one thing, ELSE(otherwise) the computer does something different. Many computers implement an IF/THEN/ELSE statement.

Example 5:

```
100 REM ** EVALUATING 3X+5 **
110 PRINT "ENTER A VALUE FOR X"
120 INPUT X
130 LET V=3*X+5
140 PRINT "3X+5=";V;" FOR X=";X;"."
150 PRINT "ANOTHER VALUE FOR X?(Y/N)"
160 INPUT A$
170 IF A$="Y" THEN 110
180 END
```

```
RUN
ENTER A VALUE FOR X
?4
3X+5=17 FOR X=4.
ANOTHER VALUE FOR X?(Y/N)
?Y
ENTER A VALUE FOR X
?-3
3X+5=-4 FOR X=-3.
ANOTHER VALUE FOR X?(Y/N)
?N
```

In this example we have used the IF/THEN statement to branch back
to an instruction which has already been executed. Thus, we have
created a loop.   Unlike example 2, however, this loop can be
terminated since we can "fall through" the IF/THEN statement by
making the condition false.   Also, notice that the body of the
loop, lines 110 to 170, must be executed at least once but the
total number of times the loop will be executed is unspecified,
dependent upon the user.

     The next example illustrates creating a loop when the number
of times the loop is to be executed is known beforehand.

Example 6:

```
100 REM ** SUMMING THE INTEGERS 1-50 **
110 LET S=0
120 LET N=0
130 LET N=N+1
140 LET S=S+N
150 IF N<50 THEN 130
160 PRINT "THE SUM OF THE FIRST 50 INTEGERS IS ";S;"."
170 END
```

```
RUN
```
THE SUM OF THE FIRST 50 INTEGERS IS 1275.

     The loop is controlled by the three statements in lines 120,
130, and 150. The assignment statement in line 120 initializes
the counter which in this program is the variable N.   The
assignment statement in line 130 increases or increments the
counter.   Finally, the IF/THEN statement in line 150 tests the
counter to determine if the loop has been executed the required
number of times.

     To understand the importance of initializing the variables as
done in line 120, suppose that lines 110 and 120 were deleted from
the program. Then the first two lines would be:

```
100 REM ** SUMMING THE INTEGERS 1-50 **
130 LET N=N+1
```

When the computer comes to line 130 it will add 1 to the current
value of the variable N. But the first time through the loop we
do not know what value is assigned to N. The computer may assign
zero to N or it may use some other value. To avoid the latter
situation we should assign the desired starting value to N as we

did in line 120. Likewise, the variable S is initialized in line 110. Fortunately, not all variables in a program need to be initialized. Only those variables which appear for the first time in an assignment to the right of the equal sign need to be initialized .

The ON/GO TO Statement

The ON/GO TO statement is used when it is necessary to transfer control to more than two alternatives. Also known as a computed GO TO, this statement transfers control based upon the evaluation of a numeric expression. Its general format is:

line# ON (expression) GO TO (line#1,line#2,...,line#n)

The expression is always evaluated as an integer. Valid examples of ON/GO TO statements are:

ON X+3*Y GO TO 110,150,160

and

ON X*X GO TO 50,60,70,80,90

When encountered during the execution of the program, the computer will first determine the value of the numeric expression. If the value is not an integer, it will be rounded or truncated(digits to the right or the decimal point are dropped) depending on the computer being used to an integer value. Then if the value of the expression is 1, control will be transferred to the statement whose line number appears first after the GO TO. A value of 2 transfers control to the statement whose line number appears second. Other values for the numeric expression transfer control in a similar manner. If the value of the numeric expression is less than 1 or larger than the number of line numbers after the GO TO the program will terminate with an error message or ignore the ON/GO TO. The specific action taken by the computer in these cases depends on the computer being used.

The following example illustrates the operation of the ON/GO TO statement.

Example 7: Consider the ON/GO TO statement below.

ON X/3 GO TO 400,500,600

Value of x	Action
6	6/3=2, so control is passed to line 500.
8	8/3=2.6666..., if truncation is used the value of the expression is 2 and control is passed to line 500. If rounding is used the value of the expression is 3 and control is passed to line 600.
15	15/3=5, the value of the expression exceeds the number of line numbers. On most computers the ON/GO TO would be ignored and control would pass to the line following the ON/GO TO. However some computers give an error message in this situation.

One of the most common uses of the ON/GO TO statement is in making a selection from a list of operations which can be performed by a program. Such a listing is called a menu. The program below illustrates using an ON/GO TO statement in this manner.

Example 8:

```
100 REM ** POWERS OF NUMBERS **
110 PRINT "ENTER THE VALUE OF THE NUMBER"
120 INPUT X
130 PRINT "******* MENU *******"
140 PRINT "1.   SQUARE THE NUMBER."
150 PRINT "2.   CUBE THE NUMBER."
160 PRINT "3.   FOURTH POWER OF THE NUMBER."
170 PRINT "ENTER 1, 2, OR 3 FOR WHAT IS TO BE DONE."
180 INPUT N
190 ON N GO TO 200,220,240
200 PRINT X;" SQUARED IS ";X*X;"."
210 GO TO 250
220 PRINT X;" CUBED IS ";X*X*X;"."
230 GO TO 250
240 PRINT X;" TO THE FOURTH POWER IS ";X*X*X*X;"."
250 END
```

```
RUN
ENTER THE VALUE OF THE NUMBER
?4
******* MENU *******
1. SQUARE THE NUMBER.
2. CUBE THE NUMBER.
3. FOURTH POWER OF THE NUMBER.
ENTER 1, 2, OR 3 FOR WHAT IS TO BE DONE.
?2
4 CUBED IS 64.
```

Ch. 2 PROGRAMMING IN BASIC PART II

Exercise Set 2.1

1. Identify the valid BASIC statements. For those which are not valid explain why not.
a. IF A+B>3*B THEN 100
b. IF A="NO" THEN 200
c. ON 3*N+1 GO TO 100,110,120.5
d. IF X=X*X THEN 100
e. IF A$="YES" THEN 50
f. GO TO 50,100
g. IF A>5 THEN B>7
h. IF X*2 THEN 100

2. How do each of the following BASIC statements differ as to the number of alternatives offered in transferring control?
a. GO TO
b. IF/THEN
c. ON/GO TO

3. What happens when the value of the numeric expression in the ON/GO TO contains a decimal. Give an example.

4. Write an ON/GO TO statement which transfers control to line 50 if X is an integer between 3 and 6, transfers to line 100 if X is a 1 or a 2, and transfers control to line 150 if X is a 7.

5. Run the following program on your computer to determine what happens when the numeric expression in an ON/GO TO statement is larger or smaller than the number of line numbers listed after the GO TO. To do this first run the program and enter the value of -1 to test for a value which is too small. Then run the program and enter a value of 3, which is too large. Also, determine if the computer being used rounds or truncates by running the program for a value of 1.832. Rounding will give selection two while truncation will give selection one.

```
100 REM ** TEST OF ON/GO TO **
110 PRINT "ENTER THE VALUE OF THE"
120 PRINT "NUMERIC EXPRESSION."
130 INPUT A
140 ON A GO TO 170,190
150 PRINT "ON/GO TO SKIPPED."
160 GO TO 200
170 PRINT "THE FIRST SELECTION WAS MADE."
180 GO TO 200
190 PRINT "THE SECOND SELECTION WAS MADE."
200 PRINT "ANOTHER VALUE?(Y/N)"
210 INPUT B$
220 IF B$="Y" THEN 110
230 END
```

*
6. Write and run a program to determine the first positive integer whose square is greater than 92176.

**
7. A sequence of IF/THEN statements can be used to produce the same program flow as an ON/GO TO statement. Write the necessary IF/THEN statements to replace the statement

ON (A+2) GO TO 100,200,300

8. In example 5 the loop had to be executed at least one time since the test to end the loop occurred after the loop was executed. This type of loop is called a posttest loop. However, it may not always be that we want the loop executed at least once. This can be performed by having the test to end the loop at the beginning of the loop. Such a loop is called a pretest loop. Modify the program in example 5 so that it is a pretest loop. Run the modified program to verify that it is actually executing a pretest loop.

9. One of the properties of Hero's formula is that if $s(s-a)(s-b)(s-c)$ is negative then line segments of lengths a, b, and c cannot form a triangle($s=(a+b+c)/2$). Write and run a program which has as input the lengths of three line segments. Then have the program determine if the lengths could be the lengths of three sides of a triangle. If yes print an appropriate message and the area of the triangle(See Challenge Activity Chapter 1). If no, print an appropriate message. Test your program with the following sets of data:

a=20, b=16, and c=30

a=10, b=15, and c=27

2.2 FOR/NEXT STATEMENTS

In example 6 of the previous section we created a loop which would be executed exactly 50 times. This was accomplished through the use of a counter and an IF/THEN statement. Loops of this nature occur so often in programming that BASIC provides a seperate control statement for creating them. Known as a FOR/NEXT loop this statement has the the general format:

        line# FOR (index)=(initial value) TO (terminal value)
                          STEP (step size)

The index, also called the control variable, must be a numeric variable, but the initial value, terminal value, and step size may be numeric constants, integer and decimal; numeric variables; or numeric expressions. If the step size is +1, the keyword STEP and the step size may be omitted. Examples of valid FOR/NEXT loops are:

```
10 FOR I=1 TO 15 STEP 2        10 LET Y=6.2
20 PRINT I,I*4                 20 FOR X=13*Y TO 9.7 STEP -.4
30 NEXT I                      30 PRINT X+Y
                               40 NEXT X
```

Although the FOR and NEXT are seperate keywords, every FOR must be matched with a corresponding NEXT. We can think of the FOR and NEXT keywords as forming the boundaries of the loop.

When a FOR statement is encountered in a program, the computer will assign the initial value to the control variable or index. Then the value of the index is compared to the terminal value.   If the index exceeds the terminal value and the step size is positive, the statements in the loop are skipped and control is transferred to the statement following the matching NEXT. When the step size is negative, the loop is skipped if the index becomes less than the terminal value. If the index has not passed the terminal value, the computer executes the statements between the FOR and NEXT statements. Upon reaching the NEXT statement the value of the step size is added to the index and control is transferred back to the FOR statement where the comparison with the terminal value is repeated. The workings of the FOR/NEXT loop is illustrated in the following program.

Example 1:

```
100 REM ** ILLUSTRATING THE FOR/NEXT STATEMENT **
110 PRINT "X","2X+2"
120 FOR X=1 TO 9 STEP 2
130 LET Y=2*X+2
140 PRINT X,Y
150 NEXT X
160 END


RUN
X                  2X+2
1                  4
3                  8
5                  12
7                  16
9                  20
```

Notice that the index, X, started with the initial value and increased by 2, the step size, each time through the loop.

If implemented in the manner described the FOR/NEXT loop creates a pretest loop. This means that the comparison of the index to the terminal value, the test, is performed before the loop is executed. Thus, if the initial value exceeded the terminal value and the step size is positive the loop would not be executed at all. However, many microcomputers implement the FOR/NEXT loop as a posttest loop. Meaning that the comparison of the index and terminal value takes place after the loop has been executed. Think of the comparison being performed at the NEXT statement before the step size is added. In this case, the loop will always be carried out at least one time, even if the initial value is already past the terminal value as shown in the program below.

Example 2:

```
100 REM ** DETERMINING TYPE OF FOR/NEXT LOOP **
110 FOR N=10 TO 5
120 PRINT "THIS IS A POSTTEST LOOP"
130 NEXT N
140 PRINT "IF THIS IS THE ONLY LINE PRINTED"
150 PRINT "THEN WE HAVE A PRETEST LOOP."
160 END
```

```
RUN
```
THIS IS A POSTTEST LOOP
IF THIS IS THE ONLY LINE PRINTED
THEN WE HAVE A PRETEST LOOP.

An important feature of the FOR/NEXT loop, which we used in example 1, is the ability to use the index in statements within the loop. However, care must be taken so as not to change its value within the loop, since this would cause the loop to be executed a wrong number of times as illustrated in the following progaram.

Example 3:

```
100 REM ** CHANGING THE INDEX IN THE LOOP **
110 FOR I=1 TO 50
120 PRINT I
130 LET I=I*I
140 NEXT I
150 PRINT "THIS LOOP SHOULD BE EXECUTED 50 TIMES."
160 END
```

```
RUN
1
2
5
26
```
THIS LOOP SHOULD BE EXECUTED 50 TIMES.

In both this and the previous section all of our loops have had a single exit point. That is, there has been only one statement in the program that the computer transfers control to when the loop was finished. This type of loop is referred to as a process loop. However, there are many problems which are more easily solved by using a multiple exit loop. These problems are characterized by wanting to perform a sequence of steps repeatedly until a specific condition is satisfied. A multiple exit loop is also referred to as a search loop.

To illustrate a search loop, we will decide if a positive integer is a perfect square in a rather unusual way. Instead of using exponentiation we use the fact that if a positive integer is a perfect square, say n squared, then it is the sum of the first n odd positive integers.

Example 4:

```
100 REM ** TESTING FOR PERFECT SQUARES **
110 PRINT "ENTER A NUMBER TO BE TESTED."
120 INPUT X
130 LET Y=X
140 LET C=0
150 FOR N=1 TO 1000 STEP 2
160 LET X=X-N
170 LET C=C+1
180 IF X=0 THEN 220
190 NEXT N
200 PRINT Y;" IS NOT A PERFECT SQUARE."
210 GO TO 230
220 PRINT Y;" IS THE SQUARE OF ";C
230 END


RUN
ENTER A NUMBER TO BE TESTED.
?121
121 IS THE SQUARE OF 11

RUN
ENTER A NUMBER TO BE TESTED.
?1567
1567 IS NOT A PERFECT SQUARE.
```

Notice how the IF/THEN statement in line 180 is used to decide when we should transfer out of the loop.

Nested FOR/NEXT Statements

Nested FOR/NEXT statements occur when one FOR/NEXT loop is contained in the body of a second FOR/NEXT loop. An example of a valid nesting of FOR/NEXT loops is:

```
FOR I=1 TO 100
FOR J=1 TO 10
    .
    .                    } BASIC Statements
    .
NEXT J
NEXT I
```

We may nest as many loops and position them as desired as long as they are positioned so that any nested loop is totally contained within any outer loop and the control variables on nested loop are distinct. The following diagrams depict both valid and invalid ways of nesting FOR/NEXT loops.

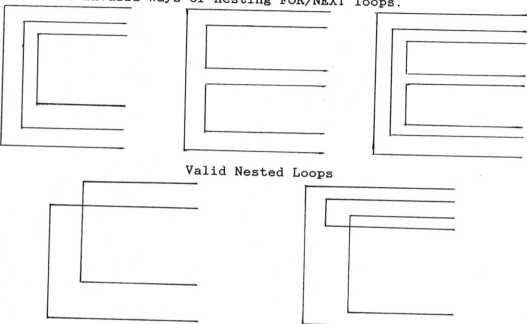

Valid Nested Loops

Invalid nested loops

To illustrate the mechanics of nested FOR/NEXT loops consider the following program segment.

```
         ⎡ FOR I=1 TO 2
         ⎢ FOR J=1 TO 3 ⎤
Outer Loop ⎢ PRINT I,J    ⎥    Inner Loop
         ⎢ NEXT J       ⎦
         ⎣ NEXT I
```

The outer loop will be executed twice and the inner loop three times. Now the index on the inner loop will go through its complete cycle before the index on the outer loop is incremented. Thus, the PRINT statement within the inner loop will be executed six times(2x3) with the values of I and J as indicated below.

```
          I  J

          1  1
          1  2
          1  3
          2  1
          2  2
          2  3
```

We finish this section by illustrating the use of nested FOR/NEXT loops to determine all positive integers which have the property that the number is equal to the sum of the cube of its digits. For example, $153=1^3+5^3+3^3=1+125+27$.

Example 5:

```
100 REM ** SPECIAL NUMBERS **
110 PRINT "THE FOLLOWING NUMBERS"
120 PRINT "EQUAL THE SUM OF THE"
130 PRINT "CUBE OF THEIR DIGITS."
140 FOR I=1 TO 9              first digit cannot be 0
150 FOR J=0 TO 9             second digit
160 FOR K=0 TO 9              third digit
170 LET N=100*I+10*J+K        forming the number
180 LET S=I*I*I+J*J*J+K*K*K   summing the digits
190 IF N<>S THEN 210
200 PRINT N
210 NEXT K
220 NEXT J
230 NEXT I
240 END


RUN
THE FOLLOWING NUMBERS
EQUAL THE SUM OF THE
CUBE OF THEIR DIGITS.
153
370
371
407
```

Ch. 2 PROGRAMMING IN BASIC PART II

Exercise Set 2.2

1. When the initial value is greater than the terminal value
should the step size be positive or negative if the loop is to be
executed?

2. Will a FOR/NEXT loop terminate when the index is less than or
greater than the terminal value if the step value is negative?

3. If no step size is given what value is assumed?

4. Identify any errors in the following program segments.

a.100 NEXT I
     .
     .
  200 FOR I=1 TO 10

b.100 FOR K=1 TO 100
     .
     .
  200 NEXT J

c.100 FOR J>10 TO 50
     .
  300 NEXT J

d. 100 LET L=10
    110 FOR L=1 TO 20
    .
    .
    200 NEXT L

e. 100 FOR I=1 TO 100
    110 FOR J=1 TO 50
    .
    .
    150 NEXT J
    .
    200 NEXT I

f. 100 FOR I=2 TO 15 STEP 4
    .
    150 FOR J=1 TO 15
    .
    200 NEXT J
    .
    300 FOR K=1 TO 50
    .
    400 NEXT K
    .
    500 NEXT I

5. Run the program in example 2 of this section to determine if the computer being used implements the FOR/NEXT loop as a pretest or posttest loop.

*

6. Run the program below a sufficient number of times to determine the value of the index variable when the FOR/NEXT loop has completed execution. <u>Hint</u>: Try the triples (1,10,1), (1,10,2), (1,10,3), (10,2,-1), and (10,5,1) for I, J, and K.

```
100 PRINT "ENTER INITIAL, TERMINAL, AND STEP SIZE VALUES."
110 INPUT I,J,K
120 FOR N=I TO J STEP K
130 PRINT "INDEX VALUE IN LOOP ";N
140 NEXT N
150 PRINT "INDEX VALUE AFTER LOOP IS EXECUTED ";N
160 END
```


7. A fixed point of a function, f(x), is a value "a" such that f(a)=a. For example, 4 is a fixed point of f(x)=3+.25x. One technique which often works to determine a fixed point is an interative method(interation means the same as repetition). We start by choosing any value we like for x, this value is termed the seed value. Next the function is evaluated at the seed value. Then this computed value is used as the new seed value and the process is repeated. Hopefully, after a sufficient number of repetitions the fixed point will be determined. To illustrate, consider the function f(x)=3+.25X. Starting with a seed value of 0, we obtain the sequence of values

```
f(0)=3
f(3)=3.75
f(3.75)=3.9375
f(3.9375)=3.984375
f(3.984375)=3.9960937
etc.
```

which appear to be getting close to the value 4. Run the porgram below to determine any fixed point of f(x)=1+(1/x). Start with a seed value of 2.

```
100 REM ** FINDING FIXED POINTS **
110 LET X=2
120 FOR I=1 TO 20
130 LET X=1+1/X
140 PRINT X,
150 NEXT I
160 END
```

Use a calculator to show that the fixed point determined has a value close to $(1+\sqrt{5})/2$. This value was of interest to both Greek and Roman mathematics. Many historians believe that this was the first irrational number to be discovered . The Pythagorean society used the pentagram shown below as its badge.

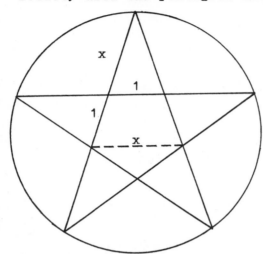

By similar triangles we have,

$$\frac{x}{1}=\frac{x+1}{x} \quad \text{or } x=1+(1/x)$$

Thus, $x=(1+\sqrt{5})/2$.

To the Romans this value was known as the golden mean and was used by architects in the design of structures. A structure in which the ratio of the length to the width equalled the golden mean was felt to be the most pleasing aesthetically.

**

8. The method for determining a fixed point of a function in exercise 7 can also be adapted for approximating roots of functions. Suppose we want to determine a root of the function $f(x)$, that is the values of x for which $f(x)=0$. We change the problem by adding x to both sides of the equation to obtain the new equation

   $x+f(x)=x.$

Thus a fixed point of the function $g(x)=x+f(x)$ will be a root of $f(x)$. For example, finding a root of $f(x)=0.0007x^2-1.05x-.5$ is the same as finding a fixed point of $g(x)=0.0007x^2-1.05x-.5+x=0.0007x^2-.05x-.5$ Modify the program in exercise 7 to find a fixed point of $g(x)=0.0007x^2-.05x-.5$ and hence a root of $0.0007x^2-1.05x-.5$

***
9. We know that the FOR statement

        FOR I=1 TO 10

will create a loop which will be executed 10 times. But how many
times will the loop

        FOR I=2 TO 7 STEP3

be executed? Or in general the loop

        FOR N=I TO J STEP K ?

Run the program below for different values of I, J, and K to
determine a formula for the number of times the loop

        FOR N=I TO J STEP K

will be executed. Assume I<J and K is positive and all three are
integers.

```
100 REM ** TESTING FOR/NEXT LOOP **
110 PRINT "ENTER INITIAL, TERMINAL, AND STEP SIZE VALUES."
120 INPUT I,J,K
130 LET S=0
140 FOR N=I TO J STEP K
150 LET S=S+1
160 NEXT N
170 PRINT "THE LOOP WAS EXECUTED ";S;" TIMES."
180 END
```

One of the most widely studied sequences of numbers is the
Fibonacci sequence, 1, 1, 2, 3, 5, 8, 13, 21, 34, 55 ... in which
each number in the sequence, starting with the third, is the sum
of the preceding two numbers. The program segment below generates
the first sixty Fibonacci numbers.

```
200 LET f1=1
210 LET f2=1
220 PRINT F1,F2, 230 FOR I=2 TO 60
240 LET F3=F1+F2
250 PRINT F3,
260 LET F1=F2
270 LET F2=F3
280 NEXT I
```

10.   Write and run a program to print the first sixty numbers in the Fibonacci sequence.

11.   Let the i-th Fibonacci number be denoted by F_i.   Modify the program of exercise 10 to print the   sequence F_{i+1}/F_i for   i equal 1 to 59.   Compare your results to exercise 7.   If possible, explain algebraically why the results are the same.

12.   If F_{n-1}, F_n, and F_{n+1} are   three   consecutive   Fibonacci numbers   define $D_n = F_{n-1}*F_{n+1}-F_n^2$.   For   example, $D_4 =$ $F_3F_5-F_4^2$.   Find D_2, D_3, D_4, D_5, D_6,   and D_7. Try   and   discover   a   pattern.   If you can, state the pattern and write and run a program to verify the pattern for n=2 to 30.

13.   Same as exercise 12, except let $D_n = 1+F_1+F_2+...+F_n$.

14.   What value does the index have when   control   is   transferred out of a FOR/NEXT loop before the loop is finished?

CHALLENGE ACTIVITY

The roots of a quadratic function $f(x)=ax^2+bx+c$ can be determined by using the quadratic formula $(-b\pm\sqrt{b^2-4ac})/2a$. If $b^2-4ac >0$ then the quadratic function has two distinct real roots, $b^2-4ac=0$ results in a double root, while $b^2-4ac <0$ implies that the quadratic has two complex roots, $(-b/2a)\pm i\sqrt{4ac-b^2}/2a$. Write a program which accepts as input the coefficients of a quadratic function and then determines and prints out the roots. Be sure to take into account the three possible cases for the roots. Test your program with the following quadratic functions which have the indicated roots.

$f(x)=x^2-5x+4$; roots:1,4

$f(x)=4x^2-28x+49$; roots:+7/2 double root

$f(x)=x^2-6x+13$; roots 3+2i,3-2i

CHAPTER 3
Programming In BASIC Part III: Functions

In the first two chapters we formed numeric expressions using variables and the basic arithmetic operations; addition, subtraction, multiplication, division, and exponentiation. However, BASIC provides numerous "built-in" functions which perform a variety of mathematical operations, such as extracting square roots or evaluating trigonometric functions. Such functions are useful since they spare the programmer the necessity of writing the sequence of statements needed to perform the operation. In addition, BASIC also provides a method by which the programmer can define a function to meet a particular application.

In this chapter we introduce both of these useful programming tools: built-in functions (known also as predefined, or library functions) and user-defined functions.

3.1 Library Functions

Almost all computer systems provide twelve library functions in BASIC. These twelve library functions are summarized in the following table.

FUNCTION	PURPOSE
ABS(X)	Absolute value of x.
ATN(X)	Trigonometric Arctangent.
COS(X)	Trigonometric Cosine Function.
EXP(X)	The exponential function, e raised to the power x.
INT(X)	Greatest Integer Function.
LOG(X)	Natural Logarithm Function.
RND	Random Number Generator.
SGN(N)	Signum Function.
SIN(X)	Trigonometric Sine Function.
SQR(X)	Square Root Function.
TAB(X)	Tab Function-Used In Printing.
TAN(X)	Trigonometric Tangent Function.

As can be seen from the table the general format for referencing a library function is a three letter name designating the operation to be performed, followed by a number in parenthesis, known as the argument, to which the operation is to be applied. The argument of a function can be any legal BASIC expression, a numeric variable, constant, expression, or even another function. With the exception of the TAB function, which was discussed in Chapter 1, and the RND function which is not used in this book, the remaining functions can be grouped into three categories: Trigonometric Functions, Exponentiation Functions, and Special Mathematical Functions.

Trigonometric Functions

The four library functions ATN(X), COS(X), SIN(X), and TAN(X) are trigonometric functions. For most computers it is important to remember that the argument for the functions must be expressed in radians. However, we are more familiar with degree measure. To convert between degree and radian measure we can make use of the fact that 180 degrees is equal to π radians. Thus, to change degrees to radians we use

1 degree= π/180 radians = .017453292 radians

N degrees= N* .017453292 radians

and to change radians to degree measure

1 radian= 180/π degrees= 57.295779 degrees

N radians= N * 57.295779 degrees.

In the following example we illustrate using the COS(X), SIN(X), and TAN(X) functions as well as converting from degrees to radians by printing a table of trigonometric values.

Example 1:

```
100 REM ** TABLE OF TRIG VALUES **
110 PRINT "DEGREES";TAB(10);"SIN(X)";TAB(24);
120 PRINT "COS(X)";TAB(39);"TAN(X)"
130 FOR X=0 TO 180 STEP 10
140 LET R=X*.017453292              converting to
150 PRINT X;TAB(10);SIN(R);TAB(24);COS(R);   radians
160 PRINT TAB(39);TAN(R)
170 NEXT X
180 END
```

```
RUN
DEGREES   SIN(X)          COS(X)           TAN(X)
0         0               1                0
10        .173648173      .984807754       .176326975
20        .342020134      .939692624       .363970223
30        .499999986      .866025412       .577350248
40        .642787594      .766044456       .839099596
50        .766044426      .64278763        1.19175353
60        .866025388      .500000027       1.73205068
70        .939692608      .342020178       2.74747711
80        .984807746      .173648219       5.67128044
90        1               4.6567654E-08    21474133.1
100       .984807762      -.173648126      -5.67128355
110       .93969264       -.34202009       -2.74747791
120       .866025435      -.499999946      -1.73205106
130       .766044487      -.642787558      -1.19175376
140       .642787665      -.766044396      -.839099755
150       .500000067      -.866025365      -.577350373
160       .342020221      -.939692592      -.363970328
170       .173648264      -.984807738      -.176327072
180       9.3135308E-08   -1               -9.3135308E-08
```

Looking closely at the output notice that the cosine at 90 degrees is not identically zero. The value 4.6567654E-08 is a number in exponential form meaning .000000046567654. This inaccuracy is the result of several factors. One is that in converting to radians we expressed irrational numbers, such as $\pi/2$ and π, as terminating decimals. Other factors will be discussed in Chapter 4. However, this output does illustrate a problem, loss of accuracy, which can affect any numerical calculation.

The last trigonometric function, ATN(X), the arctangent function, yields a value in radian measure between $-\pi/2$ and $\pi/2$ such that the tangent of the value is X. For example,

$$\text{ATN}(1)=.78539816 \text{ radians}$$

which when converted to degrees gives about 45. This is the correct value since tan(45)=1. One common use of the ATN(X) function is to generate the value of π. Since ATN(X) corresponds to 45 degrees, the expression 4*ATN(1) would correspond to 180 degrees or π radians.

Exponentiation Functions

The EXP(X), LOG(X), and SQR(X) functions deal with raising a number to a power and thus are referred to as the exponentiation functions. The SQR(X) function determines the square root of its argument and on most computers has the same effect as X^.5 The exponential function, EXP(X), computes e^X, where e is an irrational number known as Euler's number. To nine decimal places e is approximated by 2.71828183 The LOG(X) or natural logarithm function is the inverse of the EXP(X) function. That is, if Y=EXP(X) then X=LOG(Y), and the two identities EXP(LOG(X))=X and LOG(EXP(X))=X hold as illustrated in the program below.

Example 2:

```
100 REM ** EXPONENTIATION FUNCTIONS **
110 FOR X=1 TO 5
120 PRINT "FOR X=";X;"."
130 PRINT
140 PRINT "EXP(X)=";EXP(X)
150 PRINT "LOG(X)=";LOG(X)
160 PRINT "EXP(LOG(X))=";EXP(LOG(X))
170 PRINT "LOG(EXP(X))=";LOG(EXP(X))
180 PRINT
190 NEXT X
200 END
```

```
RUN
FOR X=1.

EXP(X)=2.71828183
LOG(X)=0
EXP(LOG(X))=1
LOG(EXP(X))=1

FOR X=2.

EXP(X)=7.3890561
LOG(X)=.693147181
EXP(LOG(X))=2
LOG(EXP(X))=2

FOR X=3.

EXP(X)=20.0855369
LOG(X)=1.09861229
EXP(LOG(X))=3
LOG(EXP(X))=3

FOR X=4.

EXP(X)=54.5981501
LOG(X)=1.38629436
EXP(LOG(X))=4
LOG(EXP(X))=4

FOR X=5.

EXP(X)=148.413159
LOG(X)=1.60943791
EXP(LOG(X))=5
LOG(EXP(X))=5
```

For most computers there is no built-in function for the common logarithm, that is, the base 10 logarithm. However, the base 10 logarithm of X can be found using the expression LOG(X)/LOG(10).

Special Mathematical Functions

The special mathematical functions include the ABS(X), INT(X), and SGN(X) functions. The ABS(X) or absolute value function gives the absolute value of the argument. If we think of the absolute value function as stripping away the sign of a number then the signum function, SGN(X), has the opposite affect, stripping away the number and leaving the sign. The function returns a 1 if the argument is positive, a 0 if the argument is zero, and a -1 in the argument is negative.

Example 3:

```
100 REM ** SPECIAL MATHEMATICAL FUNCTIONS **
110 PRINT "ENTER A VALUE FOR X."
120 INPUT X
130 PRINT "FOR THE VALUE X=";X
140 PRINT "ABS(X)=";ABS(X)
150 PRINT "SQR(ABS(X+3))=";SQR(ABS(X+3))
160 PRINT "SGN(X)=";SGN(X)
170 PRINT
180 PRINT "ANOTHER VALUE FOR X?(Y/N)"
190 INPUT A$
200 IF A$="Y" THE 110
210 END


RUN
```

ENTER A VALUE FOR X.
?-7.8
FOR THE VALUE -7.8
ABS(X)=7.8
SQR(ABS(X+3))=2.19089023
SGN(X)=-1

ANOTHER VALUE FOR X?(Y/N)
?Y
ENTER A VALUE FOR X.
?0
FOR THE VALUE X=0
ABS(X)=0
SQR(ABS(X+3))1.73205081
SGN(X)=0

ANOTHER VALUE FOR X?(Y/N)
?Y
ENTER A VALUE FOR X.
?25.25
FOR THE VALUE X=25.25
ABS(X)=25.25
SQR(ABS(X+3))=5.31507291
SGN(X)=1

ANOTHER VALUE FOR X?(Y/N)
?N

Note the nesting of functions in computing SQR(ABS(X+3)) in line 150 of the previous example.

The INT(X) or greatest integer function determines the largest integer which is less than or equal to the argument. Thus, for positive arguments, the INT function simply truncates or drops the fractional portion of the argument. However, for negative arguments care must be exercised. The greatest integer less than or equal to -6.7 is -7 not -6.

Example 4:

```
100 REM ** GREATEST INTEGER FUNCTION **
110 PRINT "ENTER A VALUE FOR X."
120 INPUT X
130 PRINT
140 PRINT "FOR THE VALUE ";X;", INT(";X;")=";INT(X);"."
150 PRINT
160 PRINT "ANOTHER VALUE FOR X?(Y/N)"
170 INPUT A$
180 IF A$="Y" THEN 110
190 END
```

```
RUN
ENTER A VALUE FOR X.
?-6.7

FOR THE VALUE -6.7, INT(-6.7)=-7.

ANOTHER VALUE FOR X?(Y/N)
?Y
ENTER A VALUE FOR X.
?5.67

FOR THE VALUE 5.67, INT(5.67)=5.

ANOTHER VALUE FOR X?(Y/N)
?Y
ENTER A VALUE FOR X.
?0.89

FOR THE VALUE .89, INT(.89)=0.

ANOTHER VALUE FOR X?(Y/N)
?N
```

The INT(X) function is especially useful in programs when it is necessary to determine if one number is divisible by another. In the following program we test a positive integer to determine if it is a prime by checking to see if it is divisible by any integer less than itself and greater than or equal to 2. The main step in the program is line 140 where the INT(X) function is used to decide about the divisibility.

Example 5:

```
100 REM ** PRIME NUMBERS **
110 PRINT "ENTER AN INTEGER TO BE TESTED."
120 INPUT X
130 FOR N=2 TO (X-1)
140 IF (X/N)=INT(X/N) THEN 180
150 NEXT N
160 PRINT "THE INTEGER ";X;" IS A PRIME NUMBER."
170 GO TO 200
180 PRINT "THE INTEGER ";X;" IS NOT A PRIME NUMBER."
190 PRINT "ONE FACTOR IS ";N;"."
200 PRINT
210 PRINT "TEST ANOTHER INTEGER?(Y/N)"
220 INPUT A$
230 IF A$="Y" THEN 110
240 END
```

```
RUN
ENTER AN INTEGER TO BE TESTED.
?37
THE INTEGER 37 IS A PRIME NUMBER.

TEST ANOTHER INTEGER?(Y/N)
?Y
ENTER AN INTEGER TO BE TESTED.
?1517
THE INTEGER 1517 IS NOT A PRIME NUMBER.
ONE FACTOR IS 37.

TEST ANOTHER INTEGER?(Y/N)
?N
```

Exercise Set 3.1

1. What type of expressions are valid arguments for a library function?

2. Explain the operation of the INT(X) function. What does INT(-15.6) equal?

3. What values can the SGN(X) function return?

4. What are the values of the following expressions?

a. INT(75/14)
b. ABS(INT(-7/5))
c. SQR(SQR(625))

d. 75+INT(15/4)*3
e. INT(3.795+.5)/10
f. SQR(SQR(49)-2*SGN(-4))

*
Exercises 5 and 6 refer to the triangle below where a, b, and c are the lengths of the sides and A, B, and C are the measures of the angles. In Chapter 1, Hero's formula for the area of a triangle given the lengths of the three sides was introduced. Formulas for finding the area of a triangle given the length of two sides and the included angle or two angles and the length of the included side also exist. In exercises 5 and 6 write and run a program to compute the area of a triangle for the given information and the given formula.

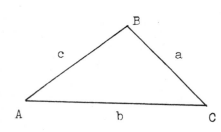

5. Given two sides a and b and the included angle C the area of the triangle is given by the formula:

$$a*b*\sin(C/2)*\cos(C/2).$$

Test your program using a=3, b=4, and C=90 which forms a right triangle with area 6 square units.

6. Given two angles A and B and the length of the included side,
c, the area is given by the formula:

$$(c^2*\sin(A)*\sin(B))/(2*\sin(C)).$$

Test your program using the values A=B=60 and c=6 which forms a
triangle with area approximately 15.588457 square units.

7. In exercise 6 could the length of any side of the triangle be
given in addition to the measure of the two angles in order to
compute the area? Explain your answer.

**
8. For two positive integers m and n the modulus, written
mod(m,n), is defined to be the remainder when the first number is
divided by the second. For example mod(15,4)=3 since 15/4=3R3 and
mod(27,9)=0 since 27/9=3R0. Using the INT function write a single
BASIC expression to determine the modulus of two positive integers
m and n.

9. To determine the month and date of Easter Sunday for any
year, Y, we can use the following steps.

Let A=mod(Y,19)
Let B=mod(Y,4)
Let C=mod(Y,7)
Let D=mod(19A+24,30)
Let E=mod(2B+4C+6D+5,7)
Easter Sunday has the date March(22+D+E), which is a date in April
if 22+D+E is larger than 31.

Write and run a program to determine the date of Easter Sunday for
a given year. Test your program for the year 1984 in which Easter
Sunday was April 22.

10. Write and run a program which truncates any number. For a
value of -7.3 the program should give an answer of -7 and for 6.3
the program should give 6. Test your program for both positive
and negative values.

11.   In Example 5, we gave a program which determined if a positive integer was a prime number. However, that program can be made more efficient by using the following facts about integers.

1. If an integer is not divisible by 2, it is not divisible by any even integer.

2. A divisor of an integer must be less than or equal to the square root of the number being tested.

Modify the program in Example 5 to incorporate these facts so as to improve the efficiency of the program. Test your program with the integers 37 and 1517. (Hint: Treat divisibility by 2 outside the loop.)

***
12.   Escape velocity is the speed at which an object must travel if it is to escape from the gravitational pull of another object. Escape velocity is given by the formula:

$$v = \sqrt{(2*G*M)/R}$$

where G is the gravitational constant, M the mass of the object from which the escape is being made, and R is the radius of the object from which we are escaping. If the escape velocity for an object exceeds the speed of light the object is referred to as a **black hole**. Thus, the values of R and M for a black hole must satisfy the inequality:

$$c \leq \sqrt{(2*G*M)/R}$$

where c is the speed of light (about 3×10^8 meters per second).

   a. Write and run a program which has as input the radius (in meters) of an object and prints out the weight (in kilograms) which a black hole of that radius must have. Use 6.67×10^{-11} for G. Test your program with a radius of one centimeter (.01 meters) for which the weight must exceed about 7×10^{24}.

   b. Write and run a program which has as input the weight of an object (in kilograms and outputs the radius of a black hole of that weight. Test your program using the weight of the earth, about 6×10^{24} kilograms. The black hole will have a radius of about .00889 meters while the radius of the earth is about 6377830 meters.

3.2 User Defined Functions

In the previous section we introduced several numeric functions which were referenced by three letter names such as SIN or ABS. The purpose of those functions was to save the programmer the necessity of writing the sequence of statements needed to perform the particular operation called for by the function. Unfortunately, the list of functions available was limited. To compensate, BASIC provides a means by which we can write our own functions which can then be referenced throughout a program in the same manner as built-in functions. Unfortunately, these functions, which are referred to as user-defined functions, do have limitations. the principal one being that the function must be defined on a single line.

To define our own function we use a DEF statement, which unlike other statements in BASIC has two slightly different formats. These general formats are:

line# DEF FN(letter)= numeric expression

or

line# DEF FN(letter)(argument) = numeric expression

As can be seen, both forms start with a line number followed by the keyword DEF, which abbreviates "define". Following the keyword is the function definition. This consists of the function name, an optional argument in parentheses and an equal sign followed by a numeric expression. The function name must be three letters in length and start with the letters "FN". thus, we are limited to a maximum of twenty-six, FNA, FNB,...,FNZ, user-defined functions in a program. As for the optional argument, it must be a numeric variable name. This argument is also referred to as a "dummy argument", since its only purpose is to be a place-holder for the value which will be operated on when the function is referenced during the execution of the program. The numeric expression to the right of the equal sign is constructed by the programmmer. This expression may be any legal numeric expression containing the dummy argument, numeric constants and variables, library functions, and other user-defined functions. In general, the only restriction is that the expression may not reference the function it is defining. That is,

100 DEF FNA(X)=FNA(X-1)+FNA(X-2)

is not allowed. Below we give several valid examples of the DEF
statement.

      100 DEF FNR(X)=INT(X * 100+.5)/100   rounding function

      100 DEF FNP=4*ATN(1)   pi function

      100 DEF FNF(X)=A*X*X+B*X+C   quadratic function

    Once defined in the DEF statement the function is referenced
or called in the program by using its three letter name just as we
did for library functions. When called, the dummy argument is
replaced by whatever value is to be operated on. All other
variables which occur in the numeric expression are replaced by
their current values. If the function has no argument then it is
referenced by its name alone. The position of DEF statements
within the program depends upon the computer being used. On some
the user-defined functions must be defined before they are
referenced in the program. On other computers they may be placed
anywhere in the porgram. Normally, however, all DEF statements
are located together at the beginning of the program. In the
following examples we illustrate using the DEF statement.

Example 1:

```
100 REM ** EVALUATING A FUNCTION **
110 DEF FNF(X)=5*COS(X)+3*SIN(X)
120 PRINT "ENTER AN ANGLE IN DEGREES."
130 INPUT X
140 LET R=X*.017453292
150 PRINT "FOR ";X;" DEGREES, F(X)=";FNF(R)
160 PRINT
170 PRINT "ANOTHER ANGLE?(Y/N)"
180 INPUT A$
190 IF A$="Y" THEN 120
200 END
```

```
RUN
ENTER AN ANGLE IN DEGREES.
?0
FOR 0 DEGREES, F(X)=5

ANOTHER ANGLE?(Y/N)
?Y
ENTER AN ANGLE IN DEGREES.
?90
FOR 90 DEGREES, F(X)=3.00000019

ANOTHER ANGLE?(Y/N)
?Y
ENTER AN ANGLE IN DEGREES.
?45
FOR 45 DEGREES, F(X)=5.65685429

ANOTHER ANGLE?(Y/N)
?N
```

Using the variable X in line 130 and as the dummy argument in the function is valid. A dummy argument has no effect in a program except in the DEF statement. For this program, the function is referenced only one time in the PRINT statement of line 150 and is evaluated at the current value of the variable R not the variable X.

Example 2:

```
100 REM ** FUNCTION CALLING A FUNCTION **
110 DEF FNF(X)=INT(X)+3
120 DEF FNG(X)=ABS(X*X-100)+10*X
130 PRINT "ENTER A VALUE FOR X."
140 INPUT X
150 LET Z=FNF(FNG(X)+SQR(X*X*X*X))
160 PRINT "FOR X=";X;" THE VALUE OF Z IS ";Z;"."
170 END


RUN
ENTER A VALUE FOR X.
?17.89
FOR X=17.89 THE VALUE OF Z IS 722.
```

This program illustrates that although the argument of a user-defined function must be a numeric variable in the DEF statement, when referenced or called the function argument can be any valid numeric expression, even including other user-defined functions and library functions.

Example 3:

```
100 REM ** DIFFERENCE QUOTIENT **
110 DEF FNF(X)=X*X-9
120 LET X=5
130 PRINT "X+H","DIFFERENCE QUOTIENT"
140 FOR I=1 TO 5
150 LET H=.5^I
160 LET D=(FNF(X+H)-FNF(X))/H
170 PRINT H,D
180 NEXT I
190 END
```

```
RUN
```

X+H	DIFFERENCE QUOTIENT
5.5	10.5
5.25	10.25
5.125	10.125
5.0625	10.0625
5.03125	10.03125

As this program illustrates, it is valid to call the same function several times with different arguments in the same expression.

Since the calculus deals mostly with functions the majority of our programs will involve user-defined functions. In fact, we will want to use the same program for several different functions. However, BASIC provides no means by which a DEF statement can be changed except by typing the DEF statement over before running the program. For example, to run the program in Example 3 for the function f(x)=6x+17, we would start by editing line 110 to define the new function by typing

```
110 DEF FNF(X)=6*X+17
```

then we would type RUN to execute the program. A more "elegant", but longer way to alter the DEF statement is presented in the following program which is a modification of Example 3.

Example 4:

```
100 REM ** CHANGING THE DEF STATEMENT **
110 PRINT " MENU"
120 PRINT "1.   CHANGE FUNCTION."
130 PRINT "2.   RUN PORGRAM."
140 PRINT "WHAT DO YOU WANT TO DO?(1 OR 2)"
150 INPUT N
160 IF N=2 THEN 220
170 PRINT
180 PRINT "TYPE 210 DEF FNF(X)=NEW FUNCTION"
190 PRINT "PRESS RETURN AND TYPE RUN."
200 GO TO 300
210 DEF FNF(X)=X*X-9
220 LET X=5
230 PRINT
240 PRINT "X+H","DIFFERENCE QUOTIENT"
250 FOR I=1 TO 5
260 LET H=.6^I
270 LET D=(FNF(X+H)-FNF(X))/H
280 PRINT H,D
290 NEXT I
300 END
```

```
RUN
          MENU
1. CHANGE FUNCTION.
2. RUN PROGRAM.
WHAT DO YOU WANT TO DO?(1 OR 2)
?1

TYPE 210 DEF FNF(X)=NEW FUNCTION
PRESS RETURN AND TYPE RUN.
210 DEF FNF(X)=X*X*X+X*X+2
          MENU
1. CHANGE FUNCTION.
2. RUN PROGRAM.
WHAT DO YOU WANT TO DO?(1 OR 2)
?2
X+H                 DIFFERENCE QUOTIENT
5.5                 93.25
5.25                89.0625
5.125               87.015625
5.0625              86.0039063
5.03125             85.5009766
```

Ch. 3 PROGRAMMING IN BASIC PART III

Exercise Set 3.2

1. What is a user-defined function? Give one advantage of a user-defined function.

2. Where is the dummy argument coded in a program? What is its purpose?

3. What are the rules for naming a user-defined function?

4. Which of the symbols X, Y, or Z are dummy arguments in the DEF statement DEF FNF(X)=X*Y*(X+Z) ?

5. What does the function FNZ compute?

        FNX(X)=X*X
        FNZ(X)=FNX(FNX(X))

*
6. Give a user-defined function which computes the smallest integer greater than X. That is, find a function FNF(X) such that FNF(7.8)=8 and FNF(-2.8)=-2.

7. Give a user-defined function which removes the integral part of its argument, that is find a function FNF(X) such that FNF(48.79)=.79 and FNF(-12.45)=-.45.

8. Give a user-defined function which removes the fractional part of its argument, that is find a function FNF(X) such that FNF(78.9)=78 and FNF(-12.34)=-12.

**
9. Write and run a program which produces a three column table showing the values of T, X, and Y for T=0,1,2....,10, where $X=(3T)/(1+T^3)$ and $Y=(3T^2)/(1+t^3)$. Use user-defined functions for X and Y.

***
10.   a. Give a user defined function which computes the area of a circle if the circumference is known.

   b. Give a user-defined function which computes the area of a square if the perimeter is known.

   c. Write and run a program which uses the functions found in parts a and b to produce a three column table giving the length l, the area of a circle of circumference L, and the area of the square of perimeter L.

   d. Use the output from part c to make a conjecture about how a wire of length L should be cut into two pieces, L1 and L2, so that the sum of the areas of a circle of circumference L1 and a square of perimeter L2 is as large as possible.

CHALLENGE ACTIVITY

Write and run a program which can compute the value of $f(x)+g(x)$, $f(x)-g(x)$, $f(x)g(x)$, $f(x)/g(x)$, or $f(g(x))$ for functions $f(x)$ and $g(x)$ at a specified value for x entered by the user. Your program should use a "menu" to allow the user to select the desired calculation. In addition, include in the menu an option to enter the function as illustrated in Example 4 of Section 3.2. Be sure to check for $g(x)=0$ when performing $f(x)/g(x)$. Test your program with $f(x)=3x+4$ and $g(x)=x^2+1$ at $x=1$ and $x=2$.

x=1	x=2
f(x)+g(x)=7	f(x)+g(x)=13
f(x)-g(x)=7	f(x)-g(x)=7
f(x)g(x)=0	f(x)g(x)=30
f(x)/g(x) undefined	f(x)/g(x)=3.3333333
f(g(x))=4	f(g(x))=13

CHAPTER 4
Computations And The Computer

One of the major uses of the computer is numeric calculations producing a result which is a number. The degree of precision expected or desired depends upon the context of the problem. Although the layman would see little difference between 17.00000 and 17.00001, to the scientist and the engineer the difference may be significant indeed. In fact, small differences occurring during the solution of a problem can lead to quite unexpected and even false results. Unfortunately, numeric calculations with a computer are always subject to error. Although an in-depth study of these considerations is beyond the scope to this book, it is important to understand how they can affect the results obtained by the computer. In this chapter we will investigate briefly several of the basic ideas which affect numerical computing.

4.1 Computer Error

Consider the following program.

Example 1:

```
100 REM ** FINDING A ROOT **
110 DEF FNF(X)=X^2+2*X-35
120 PRINT "ENTER A VALUE FOR X."
130 INPUT X
140 IF FNF(X)=0 THEN 170
150 PRINT X;" IS NOT A ROOT"
160 GO TO 180
170 PRINT X;" IS A ROOT"
180 END
```


It looks rather simple--a program to determine if a number is a root of $x^2+2x-35$. However, running this program for several values of x gives the following results.

```
RUN
ENTER A VALUE FOR X.
?0
0 IS NOT A ROOT


RUN
ENTER A VALUE FOR X.
?-7
-7 IS NOT A ROOT


RUN
ENTER A VALUE FOR X.
?4
4 IN NOT A ROOT


RUN
ENTER A VALUE FOR X.
?5
5 IS NOT A ROOT
```

Have you noticed anything wrong? What about the value of -7 for x?

$$(-7)^2+2(-7)-35$$
$$49-14-35$$
$$0$$

-7 is a root! Is something wrong with the computer? The answer is yes and no. No, nothing is mechanically amiss, but, yes, something is definitely not right.

When faced with results which don't make sense the best procedure is to analyze the program and try to determine the source of trouble. But where do we start? One good method is to trace the program through by hand until a line is found which is not doing what we expect. Since an x-value of -7 caused a problem we will trace the program for this value.

LINE	WHAT WE EXPECT TO HAPPEN
110	FNF(X)=X^2+2*X-35
130	X=-7
140	FNF(-7)=0, therefore, the next line executed should be 170

However, the computer didn't go to line 170, instead it went to line 150. Yet this would only happen if FNF(-7) did not equal zero. Is it possible that the calculated value of FNF(-7) is not zero? To find out, we add a new line to our program.

Example 2:

```
135 PRINT "FNF(X)= ";FNF(X)

LIST

100 REM ** FINDING A ROOT **
110 DEF FNF(X)=X^2+2*X-35
120 PRINT "ENTER A VALUE FOR X."
130 INPUT X
135 PRINT "FNF(X)= ";FNF(X)          adding new line
140 IF FNF(X)=0 THEN 170
150 PRINT X;" IS NOT A ROOT"
160 GO TO 180
170 PRINT X;" IS A ROOT"
180 END
```

Running the program yields:

```
RUN
ENTER A VALUE FOR X.
?-7
FNF(X)= 2.98023224E-08
-7 IS NOT A ROOT
```

As can be seen, the computer calculates FNF(-7) to be 2.98023224E-08, a number in exponential form meaning 0.000000298023224. This is unquestionably a small number, but still not zero. This at least indicates that the logic of the program is correct; the error must be in the way the computer calculates. Expanding on this idea, include the following lines in the program.

Example 3:

```
131 PRINT "X^2= ";X^2
132 PRINT "2*X= ";2*X
133 PRINT "X^2+2*X= ";X^2+2*X
134 PRINT "X^2+2*X-35= ";X^2+2*X-35

LIST

100 REM ** FINDING A ROOT **
110 DEF FNF(X)=X^2+2*X-35
120 PRINT "ENTER A VALUE FOR X."
130 INPUT X
131 PRINT "X^2= ";X^2
132 PRINT "2*X= ";2*X
133 PRINT "X^2+2*X= ";X^2+2*X
134 PRINT "X^2+2*X-35= ";X^2+2*X-35
135 PRINT "FNF(X)= ";FNF(X)
140 IF FNF(X)=0 THEN 170
150 PRINT X;" IS NOT A ROOT"
160 GO TO 180
170 PRINT X;" IS A ROOT"
180 END
```

Running the program we have the output below.

```
RUN
ENTER A VALUE FOR X.
?-7
X^2= 49.0000001
2*X= -14
X^2+2*X= 35
X^2+2*X-35= 2.98023224E-08
FNF(X)= 2.98023224E-08
-7 IS NOT A ROOT
```

These are indeed unexpected results!

We are now in a position to answer our original question. What went wrong is the apparent fact that a computer does not do very accurate arithmetic calculations. We should always be aware of this limitation of the computer when dealing with numerical calculations, always questioning whether it has affected our results in a significant manner. Additionally, we should strive, whenever possible, to write programs which attempt to minimize such errors. Note that we say minimize, for elimination of these errors, called "round-off errors", is impossible.

At this point the most natural question to ask is, "In what ways can we minimize round-off errors?". This is a rather difficult question to answer. In fact, a significant portion of the subject of numerical analysis is devoted to resolving this question. For our part, we will limit ourselves to the use of a few techniques to reduce round-off error. One of the most basic is suggested by the program below.

Example 4:

```
100 REM ** SQUARES **
110 PRINT "X";TAB(5);"X^2";TAB(22);"X*X"
120 FOR X=1 TO 30
130 PRINT X;TAB(5);X^2;TAB(22);X*X
140 NEXT X
150 END
```

```
RUN
X    X^2              X*X
1    1                1
2    4                4
3    9                9
4    16               16
5    25               25
6    36               36
7    49.0000001       49
8    64               64
9    81.0000001       81
10   100              100
11   121              121
12   144              144
13   169              169
14   196              196
15   225              225
16   256              256
17   289              289
18   324              324
19   361              361
20   400              400
21   441.000001       441
22   484.000001       484
23   529              529
24   576.000001       576
25   625.000001       625
26   676.000001       676
27   729.000001       729
28   784.000001       784
29   841              841
30   900              900
```

RULE: Using $\underbrace{X*X*...*X}_{N \text{ times}}$ for X^N

will generally reduce round-off error.

Notice the improvement when we rerun our original program using this rule.

70

Example 5:

```
100 REM ** FINDING A ROOT **
110 DEF FNF(X)=X*X+2*X-35
120 PRINT "ENTER A VALUE FOR X."
130 INPUT X
140 IF FNF(X)=0 THEN 170
150 PRINT X;" IS NOT A ROOT"
160 GO TO 180
170 PRINT X;" IS A ROOT"
180 END
```

```
RUN
ENTER A VALUE FOR X.
?0
0 IS NOT A ROOT
```

```
RUN
ENTER A VALUE FOR X.
?-7
-7 IS A ROOT
```

```
RUN
ENTER A VALUE FOR X.
?4
4 IS NOT A ROOT
```

```
RUN
ENTER A VALUE FOR X.
?5
5 IS A ROOT
```

Although we have managed to "clean-up" this program, it will not always be possible to do so with ease. Therefore, we should view our output in terms of accuracy desired and required by the problem.

Ch. 4   COMPUTATIONS AND THE COMPUTER

Exercise Set 4.1

1. Do you expect that your computer will calculate 2*(1/2) accurately? Same for 3*(1/3) and 20*(1/20). Explain.

2. Round-off errors are machine related with different machines not necessarily giving the same error. Run the program in Example 1 on your computer and see if you obtain an error in the calculations. If you do, run the programs in Examples 2 through 5 to determine if the errors caused by your computer are the same as in the text or different, and if using X*X in place of x^2 corrects the error.

3. Consider the following program:

```
100 LET S=0
110 FOR I=1 TO 1000
120 LET S=S+.001
130 NEXT I
140 PRINT S
150 END
```

```
RUN
0.999999976
```

What should the result be? Run this program on your computer and see if you get the same result.

*
4. Write and run a computer program to compare using SQR(X) versus X^.5 for computing square roots. Which expression appears to be more accurate?

**
5. A procedure which is often used in calculus is that of finding a difference between functional values when the values at which the function is evaluated are close together. Symbolically this becomes f(x+h)-f(x) where h is small. The program below illustrates this procedure. However, something is wrong with the output. What is the error? If possible explain why you think the error occurred.

```
100 DEF FNF(X)=X*X+2*X-35
110 FOR I=3 TO 15
120 LET H=(.2)^I
130 LET D=FNF(2+H)-FNF(2)
140 PRINT H,D
150 NEXT I
160 END
```

```
RUN
8.00000001E-03    .0480640009
1.6E-03           9.60256159E-03
3.20000001E-04    1.92009658E-03
6.40000002E-05    3.84002924E-04
1.28E-05          7.68005848E-05
2.56000001E-06    1.53630972E-05
5.12000002E-07    3.06963921E-06
1.02400001E-07    4.09781933E-07
2.04800001E-08    8.19563866E-08
4.09600002E-09    1.49011612E-08
8.19200006E-10    0
1.63840001E-10    0
3.27680002E-11    0
```

6.   Verify that $a^b = e^{b*\ln a}$.   On your computer   try   and   find values for a and b so that the expressions are not equal.

7.   Same as exercise 6 except use the expression

$$\sin(x-y)=(\sin x)(\cos y)-(\cos x)(\sin y).$$

***
8.   Experiment   with   your   computer to determine how many of the sums below add to exactly 1.

```
1/2 +1/2
1/3 +1/3 +1/3
         .
         .
         .
1/N +  ...   +1/N
```

9.   Consider the programs below.

PROGRAM 9A.

```
100 LET S=0
110 LET F=1/(11*11*11*11*11*11*11)
120 FOR I=1 TO 8
130 LET S=S+F
140 LET F=F*11
150 NEXT I
160 PRINT S
170 END
```

RUN
1.1

PROGRAM 9B.

```
100 LET S=0
110 LET F=1
120 FOR I=1 TO 8
130 LET S=S+F
140 LET F=F/11
150 NEXT I
160 PRINT S
170 END
```

RUN
1.09999999

a) What is the purpose of program 9A?

b) What is the purpose of program 9B?

c) Are the programs doing the same thing?

d) Run both of these programs on your computer.   What are your results?

e)   Which program do you think is more accurate? If possible, explain why.   <u>Hint</u>: Each arithmetic operation has the potential for round-off error.   Which program has more arithmetic operations?

4.2 Causes Of Round-Off Error

As illustrated in the previous section numerical calculations by a computer are subject to what is referred to as round-off errors. These errors are not due to the manner in which a computer performs its arithmetic but limitations in the way in which numbers must be represented. Every irrational number and many rational numbers have infinite decimal expansions. However, the memory of a computer is finite in size and thus every number stored must consist of a finite number of digits. This inability of the computer to store the complete decimal expansion of irrational and rational numbers is the cause of round-off error.

Before illustrating the more common situations in which round-off error plays a significant role it is necessary to understand how decimal numbers are stored by a computer. The usual way to represent a real number in decimal form is by an integer part, a decimal point, and a fractional part, as in 105.67 and .0219 . However, a computer stores this type of number in a different format, known as normalized scientific notation or normalized floating-point representation. In this form the numbers 105.67 and .0219 would be written as:

$$.10567 \times 10^3 \qquad \text{and} \qquad .219 \times 10^{-1}$$

Numbers in this form are represented by a fractional part, where the lead digit is non-zero(except when the number is zero), multiplied by a power of 10. The fractional part is referred to as the mantissa and the power the characteristic. Now suppose our computer can store only a four digit mantissa and a one digit characteristic. Then depending on whether our computer actually rounds or truncates the number 105.67 would be stored as were used in an arithmetic computation a loss of accuracy would result.

In the remaining part of this section we will examine several of the more common situations in which round-off error plays a significant part in the calculations.

Testing For Equality

Using a computer with a four digit mantissa the numbers .987612 and .987642 would be stored as the same number, $.9876 \times 10^0$, and the logical expression

$$.987612=.987642$$

would be erroneously evaluated as true. Thus testing for equality between two decimal numbers can be pointless.

Although all computers store more than four digits for the mantissa, the fact that the calculus often deals with limiting processes and comparisons of values which are extremely close to each other means that round-off error when testing for equality may affect our programs. The program in the following example illustrates an occurrence of this form of round-off error.

Example 1:

```
100 REM ** ROUND-OFF ERROR IN **
110 REM ** TESTING FOR EQUALITY
120 DEF FNF(X)=X*X*X-7*X*X+5*X-8
130 LET H=.2^13
140 IF FNF(2+H)=FNF(2) THEN 170
150 PRINT "FNF(2+H) AND FNF(2) ARE NOT EQUAL."
160 GO TO 180
170 PRINT "FNF(2+H) AND FNF(2) ARE EQUAL."
180 END


RUN
```
FNF(2+H) AND FNF(2) ARE EQUAL.

By graphing or using derivatives it can be shown that x^3-7x^2+5x-8 is decreasing on the interval $(1,3)$ and thus no two different values of x yield the same functional value. Thus the results of the program are incorrect.

To compensate for round-off error when testing for equality, we often consider two quantities equal if their difference is less than some specific value called the tolerance or precision. For example, the statement

IF ABS(FNF(X)-FNG(Y)) < .00001 THEN .....

could be used to determine if FNF(X) and FNG(Y) are equal to each other within a tolerance of .00001

Addition Or Subtraction Of A Large And Small Number

In certain situations, the loss of accuracy in addition and subtraction can be dramatic. Again assuming a computer with a four digit mantissa and one digit characteristic consider adding 1000 and .5 Stored in normalized form these values are

$$.8 \times 10^4 \qquad \text{and} \qquad .5 \times 10^0$$

However, most computers can not perform addition or subtraction on two numbers unless the characteristic of both are the same. Thus, the number $.5 \times 10^0$ would be changed to $.00005 \times 10^4$ before the addition. Adding we obtain:

$$
\begin{array}{r}
.80000 \times 10^4 \\
+ \underline{.00005 \times 10^4} \\
.80005 \times 10^4
\end{array}
$$

However, since the computer can only store four digits in the mantissa the result would be $.8000 \times 10^4$, or 8000, one of the original numbers. The following program illustrates the effects of round-off error when adding large values to small values.

Example 2: The sum of the first N terms of a geometric sequence is given by the formula

$$S = 1 + r + r^2 + r^3 + \ldots + r^{N-1}$$

where r is the common ratio between successive terms and 1 is the first term. The sum of the infinite geometric sequence with the the same ratio and first term is given by

$$SUM = 1/(1-r).$$

Now for r=1/3 the sum of the infinite geometric sequence would be

$$SUM = 1/(1-(1/3)) = 1/(2/3) = 3/2 = 1.5$$

To approximate the sum of the infinite sequence the program below adds the first fifteen terms of the sum.

```
100 REM ** SUM OF FIRST FIFTEEN TERMS **
110 REM ** OF THE GEOMETRIC SEQUENCE **
120 REM ** WITH COMMON RATIO 1/3 **
130 LET SUM=1
140 LET A=1
150 REM ** LOOP FOR FINDING SUM **
160 FOR I=1 TO 14
170 LET A=A/3
180 LET SUM=SUM+A
190 NEXT I
200 REM ** OUTPUT **
210 PRINT "THE SUM OF THE FIRST 15 TERMS IS ";SUM
220 END


RUN
THE SUM OF THE FIRST 15 TERMS IS 1.49999989
```

Now the sixteenth term of the sequence is $(1/3)^{15}$ which is approximately 6.969172×10^{-8}. Thus the sum of the first 15 terms should be 1.49999993030828 and as we see a small error has resulted. Although the error in terms of magnitude is quite small, as we will see later in this section, even small errors in one portion of series of operations can result in much more significant errors for the entire computation.

This occurrence of round-off error is very common, and can only be avoided by not adding or subtracting numbers of vastly different magnitudes. Sometimes this can be accomplished by rearranging the order of the additions or subtractions. In the example below we give a program which again adds the first fifteen terms of the geometric sequence with common ratio 1/3 and first term 1, but we do the addition starting with the 15th term and adding backwards instead of starting with the first term and adding forwards as in the previous example.

Example 3:

```
100 REM ** SUM OF THE FIRST FIFTEEN **
110 REM ** TERMS OF THE GEOMETRIC **
120 REM ** SEQUENCE WITH COMMON **
130 REM ** RATIO 1/3 AND FIRST TERM **
140 REM ** 1 ADDING BACKWARDS **
150 LET SUM=0
160 LET A=1/3/3/3/3/3/3/3/3/3/3/3/3/3/3
170 REM ** LOOP FOR FINDING SUM **
180 FOR I=1 TO 15
190 LET A=A*3
200 LET SUM=SUM+A
210 NEXT I
220 REM ** OUTPUT **
230 PRINT "THE SUM OF THE FIRST 15 TERMS IS ";SUM
240 END
```

```
RUN
```
THE SUM OF THE FIRST 15 TERMS IS 1.4999999

As can be seen this result is a slight improvement over the previous program.

Subtraction Of Almost Equal Numbers (Subtractive Cancellation)

Subtracting almost equal numbers in complicated calculations can introduce errors which are potentially quite serious. The more nearly equal the numbers the more pronounced the effects. The problem we encounter is not in the actual subtraction but in its effect on later calculations. The source of the problem is that when two nearly equal numbers are subtracted if either number is in error then that error could represent a large proportion of the result. To illustrate suppose that through an earlier error instead of storing the correct value of .9877 for a variable, a value of .9878 was stored. Now the absolute error at this point is .0001 which represents a percentage of error of approximately

$$(.9878-.9877)/.9877 \ X \ 100=.01\%$$

Now suppose we want to subtract .9876 from this value. The true value is .0001, (.9877-.9876). However, the value which will be computed is .0002, (.9878-.9876), which gives a percentage of error of

$$(.0002-.0001)/.0001 \ X \ 100=100\%$$

Thus, a rather modest error of only .01% has become and error of 100%. The formula for computing percent of error is:

%-error=(computed value-real value)/(real value) X 100

In the program of example 4 we illustrate the problems that can result when two nearly equal numbers are subtracted by evaluating the function

$$f(x)=\frac{\sqrt{x^2+4} - 2}{x^3}$$

for values close to 0.

Example 4:

```
100 REM ** DIFFICULTIES IN SUBTRACTING **
110 REM ** TWO NEARLY EQUAL NUMBERS **
120 DEF FNF(X)=(SQR(X*X+4)-2)/(X*X*X)
130 PRINT "ENTER A VALUE FOR X."
140 INPUT X
150 PRINT "FOR X=";X;" F(X)=";FNF(X)
160 END


RUN
ENTER A VALUE FOR X.
?.01
FOR X=.01 F(X)=25.0004232

RUN
ENTER A VALUE FOR X.
?.0001
FOR X=1E-04 F(X)=1862.64515

RUN
ENTER A VALUE FOR X.
?.00001
FOR X=1E-05 F(X)=0
```


As can be seen for x values of .01 and .0001 the functional value is increasing, but for x=.00001 the functional value is 0.

When this type of error occurs it is sometimes possible to compensate for it by locating the "sensitive" subtraction and eliminating it by algebraic manipulations. In the previous example the difficulty arose in the numerator where the quantities $\sqrt{x^2+4}$ and 2 are nearly equal for values of x near 0. To eliminate this subtraction we could rationalize the numerator by multiplying both numerator and denominator by the quantity

$$\sqrt{x^2+4} \ +2$$

obtaining

$$\frac{\sqrt{x^2+4} \ -2}{x^3} \ \frac{\sqrt{x^2+4} \ +2}{\sqrt{x^2+4} \ +2} \ = \ \frac{1}{x(\sqrt{x^2+4}+2)}$$

In Example 5 below we use this equivalent expression for f(x) to compute functional values for values of x close to 0.

Example 5:

```
100 REM ** REDUCING THE ERROR **
110 REM ** DUE TO SUBTRACTING **
120 REM ** TWO NEARLY EQUAL **
130 REM ** NUMBERS          **
140 DEF FNF(X)=1/(X*(SQR(X*X+4)+2))
150 PRINT "ENTER A VALUE FOR X."
160 INPUT X
170 PRINT "FOR X=";X;" F(X)=";FNF(X)
180 END


RUN
ENTER A VALUE FOR X.
?.01
FOR X=.01 F(X)=24.9998438

RUN
ENTER A VALUE FOR X.
?.0001
FOR X=1E-04 F(X)=2500

RUN
ENTER A VALUE FOR X.
?.00001
FOR X=1E-05 F(X)=25000
```

Note the difference between the two values of the function at an x value of .00001 in example 4 and example 5.

Overflow And Underflow

At times calculations will be performed which involve valid numbers but will result in values which are outside the range of values which can be stored by the computer. When this occurs we have what is referred to as overflow or underflow. Overflow occurs when the magnitude of the number is too large and underflow when the magnitude is too small.

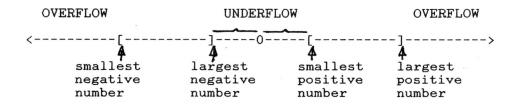

To illustrate, consider multiplying 4×10^6 by 3×10^5 on our hypothetical computer which stores a four digit mantissa and a one digit characteristic.

$$
\begin{array}{r}
.4000 \times 10^7 \\
x \ \underline{.3000 \times 10^6} \\
.1200 \times 10^{13}
\end{array}
$$

Even though both numbers being multiplied were within the range of our computer, the product exceeded the range of the computer and an overflow has occurred .

On the otherhand, underflow occurs when a computation produces a number whose magnitude is less than the smallest positive number which can be stored by the computer. On our imaginary machine this value would be $.1 \times 10^{-9}$, assuming the lead digit must be non-zero. In general an overflow will result in a fatal error and the execution of the program will halt. Unfortunately, an underflow, on most computers, is handled automatically by setting a value of zero to the variable without any interruption of the program or any message to the operator.

Although nothing can be done to compensate for overflow or underflow of the final result in a computation, sometimes it is possible to avoid these types of round-off error in intermediate calculations by rearranging the order in which the computations are performed. For example, suppose $X = .5 \times 10^{-7}$, $Y = .6 \times 10^{-6}$, and $Z = .4 \times 10^7$ and that the expression X*Y*Z is to be evaluated. Then by the order of operations the computation would be performed in the following manner.

$$(.5 \times 10^{-7} \times .6 \times 10^{-6}) \times .4 \times 10^7$$
$$.3 \times 10^{-13} \times .4 \times 10^7$$

However, the number $.3 \times 10^{-13}$ is smaller in magnitude than the smallest number which can be represented by the computer($.1 \times 10^{-9}$) and an underflow has occurred . However, if we simply rearrange the product and evaluate the expression in the order X*Z*Y we obtain

$$(.5 \times 10^{-7} \times .4 \times 10^7) \times .6 \times 10^{-6}$$
$$.2 \times 10^0 \times .6 \times 10^{-6}$$
$$.12 \times 10^{-6}$$

which is the correct value without an underflow error.

In Example 6, we illustrate a common occurrence of overflow when using the computer for mathematical computations. This occurs when we are trying to compute what are referred to as factorials, which are very important in the study of power series and in probability theory. By definition N-factorial, denoted by N!, is the product

$$N! = N*(N-1)*(N-2)*...*(2)*(1)$$

where N is any non-negative integer(if N=0 then 0!=1).

Example 6:

```
100 REM ** FACTORIALS **
110 PRINT "ENTER A VALUE FOR N."
120 INPUT N
130 LET NF=1
140 REM ** LOOP TO COMPUTE THE PRODUCT **
150 FOR I=1 TO N
160 LET NF=NF*I
170 NEXT I
180 PRINT N;"!=";NF
190 PRINT
200 PRINT "ANOTHER VALUE FOR N?"
210 INPUT A$
220 IF A$="Y" THEN 110
230 END
```

```
RUN
ENTER A VALUE FOR N.
?4
4!=24

ANOTHER VALUE FOR N?
?Y
ENTER A VALUE FOR N.
?30
30!=2.6525286E+32

ANOTHER VALUE FOR N?
?Y
ENTER A VALUE FOR N.
?35

OVERFLOW ERROR IN 160
```

Error Propagation

As we have seen there are many different instances which give rise to round-off error. Fortunately, the majority of these errors are rather small. However, we must be aware that in complicated computations or when a computation is being performed over and over as in a loop these small errors can be compounded to produce large inaccuracies as illustrated in the discussion of subtraction two nearly equal numbers. This compounding of small errors into more significant errors is referred to as error propagation.

Instances of error propagation as well as the other types of round-off error illustrated in this section will affect the results of almost every calculation performed by a computer. Since there is no way in which these errors can be avoided it is important to analysis computer results carefully always questioning their accuracy.

Exercise Set 4.2

1. Give an example to illustrate that on a computer which stores a four digit mantissa the expressions x+(y+z) and (x+y)+z may have different values.

2. For large values of x the expression $\sqrt{x} - \sqrt{x+5}$ will encounter round-off error because we are subtracting two nearly equal numbers. How can we re-write this expression to avoid this form of round-off error?

3. For your computer, refer to the operating manual or experiment to determine the magnitude of the largest and smallest values which can be stored.

**
4. The exact value of $\pi/4$ is given by the infinite sum

$$1-(1/3)+(1/5)-(1/7)+.....$$

   a. Write and run a program to add the first one hundred terms of the sum starting with the first term. <u>Hint</u>: Successive terms in the sum can be found using a loop and the expression $T=(-1)^{(I+1)}/(2*I-1)$ where I is the index of the loop.

   b. Same as part a but sum the terms backwards.

   c. Are the results of parts a and b the same or different? If the same explain why both sums gave the same results whereas in the section summing forwards and backwards gave different results. If different, which sum is more accurate. The value of PI to 15 decimals places is 3.141592653589793.

   d. Same as part a except rearrange the sum in the following manner

$$(1-(1/3))+((1/5)-(1/7))+((1/9)-(1/11))+....$$

Do each subtraction as a seperate step and then add the result to the sum. Compare the result to the results in parts a and b Comment on which of the three methods (if any) are better in terms of producing a more accurate result.

e. In parts a through c we summed a specific number of terms to approximate $\pi/4$. However, this method gives us no ideal of how many decimal places the result is accurate. To compensate for this write and run a program which sums the terms until the next term is less than .001 in magnitude. Using this method how accurate is the result. What would be the drawback to this method if we wanted accuracy to .000001?

***
5. The roots of a quadratic function, $f(x)=ax^2+bx+c$, can be found using the quadratic formula:

$$x=(-b\pm\sqrt{b^2-4ac})/(2a).$$

a. Write and run a program which finds the roots of a quadratic function using the quadratic formula. Be sure to test for double roots and complex roots. Test your program using $f(x)=x^2-3x+2$, which has roots 1 and 2.

b. Rearrange the quadratic formula so as to eliminate having to compute $-b+\sqrt{b^2-4ac}$, when b is positive, which could result in the subtraction of almost equal numbers.

c. Modify the program of part a to incorporate the changes of part b.

d. If b is negative what changes have to be made in the quadratic formula to avoid the subtraction of almost equal numbers?

6. The number of combinations of n distinct objects taken r at a time is given by the formula:

$$\binom{n}{r}=\frac{n!}{r!(n-r)!}$$

a. Write and run a program which will compute the number of combinations of n things taken r at a time. Use n=6 and r=3 to test your program. The answer is 20.

b. Experiment to determine the largest n and r for which your program can compute the number of combinations.

c. If possible rearrange the formula for combinations to allow for the computation of combinations for larger n and r than found in part b.

CHALLENGE ACTIVITY

It is not always practical to use X*X*....*X in place of X^N when writing a computer program. For instance, we may need a general program for evaluating polynomials which we want to run for many different polynomials by only having to enter the degree and the coefficients. One method which avoids using X^N and the associated round-off error is that of evaluation by nested intervals or Horner's method. For example,

$$x^3 + 5x^2 + 7x - 8 = x(x(x+5)+7) - 8.$$

a. Determine the number of arithmetic operations involved in calculating the above polynomial by nested intervals and by direct evaluation. (x^3 requires two operations)

b. Generalize part a for a polynomial of nth degree.

c. Use the results of part b to argue that using nested intervals should result in less round-off error.

d. Write and run a program to evaluate a polynomial by nested intervals. Your program should be general, with the degree and coefficients of the polynomial entered by an INPUT statement. Test your program using the polynomial above for an x-value of 2, which evaluates to 34. <u>Hint</u>: The use of a one dimensional array will be helpful. See APPENDIX D.

CHAPTER 5
Writing Computer Programs

Designing a computer program is a problem solving activity. It is a creative process with no complete set of rules telling us how to write the program. However, as we shall see in this chapter, there are some guidelines and useful hints which can help this process along.

5.1 Writing A Computer Program

Knowing the correct syntax of BASIC is necessary, but designing and writing correct programs requires careful planning and an understanding of what a computer can and cannot do. To illustrate, consider the following problem.

Example 1:

A long time ago, in a faraway land, a king and his trusted counselor were playing chess. As a reward for faithful service, the king announced that he would fulfill any reasonable wish his counselor might have. The counselor thought for a moment and then made the following request. "Dear Sire, all that I wish for is the amount of grain required to fill the chessboard by placing one grain of wheat on the first square, two grains on the second square, four grains on the third, and so on, each square having twice the number of grains as the square before." The king quickly agreed to such a modest-sounding request and sent a servant for a sack of wheat.

Unfortunately, the king had a terrible temper, and the counselor soon lost his head.

Why did the king have the counselor beheaded?

When we approach the computer with our problem, we find our machine limited indeed! It certainly cannot answer our question. However, we could utilize the computer if we could first phrase a numeric question the computer is able to answer. Suppose we ask, "How many grains of wheat did the counselor receive?" Here we have a programming problem. But how do we proceed with the solution? Our machine is still of no assistance unless we supply the "program". A computer program is composed essentially of three types of instructions: entering given information, processing the information, and outputting the results of the processing. In our problem we know how the number of grains on each square is computed, and we want to determine the total number of grains received. Yet we still may not be sure how to write the program.

In general, computer programs are often developed by investigating the instructions in a different order from that actually executed by the machine. In this book we will use the following scheme to analyze computer programming problems.

Programming Problem: a statement of the problem in terms of something a computer can do.

Output: the results we want to obtain or have printed out. This is the goal of the program and should be used to guide us in the designing of the program.

Input: the given information.

Strategy: a development of the steps--the algorithm--which will describe the sequence of instructions to be performed in carrying out the problem solution.

Program: a language(BASIC) implementation of the algorithm.

Test Data: a simple set of data to verify that the program is indeed producing the desired results.

Ch. 5 WRITING COMPUTER PROGRAMS

Applying the above analysis to our problem we have:

Programming Problem: Determine the total number of grains of wheat the counselor would receive.

Output: The total number of grains of wheat.

Input: Starting with one grain on the first square, double the amount on each successive square for a total of 64 squares.

Strategy: The most direct method for finding the total number of grains of wheat received is to add together the number of grains on each square. From the input, we know that there is one grain on the first square with the number doubling on each successive square. Thus, if a square has G grains of wheat, the next square would have 2G grains. In BASIC, the statement:

```
        LET G=2*G
```

will generate the number of grains on successive squares.

Adding the number of grains on successive squares can be accomplished by a loop:

```
        FOR I=1 TO 64          since there are 64 squares
        LET S=S+G             storing the sum in S
             .
             .
             .
        NEXT I
```

Program:

```
    100 REM ** COUNSELOR'S DILEMMA **
    110 REM ** INITIALIZING VARIABLES **
    120 LET G=1
    130 LET S=0
    140 REM ** LOOP TO COMPUTE THE SUM **
    150 FOR I=1 TO 64
    160 LET S=S+G
    170 LET G=2*G
    180 NEXT I
    190 REM ** PRINTING THE OUTPUT **
    200 PRINT "THE TOTAL NUMBER OF GRAINS IS ";S
    210 END
```

Test Data: If the board had only four squares, the counselor would receive 1+2+4+8=15 grains of wheat. Using this data we will test our program by changing line 150 as follows.

```
150 FOR I=1 TO 4
```

```
LIST
```

```
100 REM ** COUNSELOR'S DILEMMA **
110 REM ** INITIALIZING VARIABLES **
120 LET G=1
130 LET S=0
140 REM ** LOOP TO COMPUTE THE SUM **
150 FOR I=1 TO 4
160 LET S=S+G
170 LET G=2*G
180 NEXT I
190 REM ** PRINTING THE OUTPUT **
200 PRINT "THE TOTAL NUMBER OF GRAINS IS ";S
210 END
```

```
RUN
```
THE TOTAL NUMBER OF GRAINS IS 15

Since our test data is accurate, we can feel fairly confident that our program will "work". Now running the program for sixty-four squares we obtain the following result.

```
150 FOR I=1 TO 64
```

```
LIST
```

```
100 REM ** COUNSELOR'S DILEMMA **
110 REM **INITIALIZING VARIABLES **
120 LET G=1
130 LET S=0
140 REM ** LOOP TO COMPUTE THE SUM **
150 FOR I=1 TO 64
160 LET S=S+G
170 LET G=2*G
180 NEXT I
190 REM ** PRINTING THE OUTPUT **
200 PRINT "THE TOTAL NUMBER OF GRAINS IS ";S
210 END
```

```
RUN
```
THE TOTAL NUMBER OF GRAINS IS 1.84467441E+19

As we can see, the result is 1.84467441E+19, a number in
exponential notation meaning 18,446,744,100,000,000,000 grains of
wheat! Now did the counselor loose his head because he bankrupted
the kingdom---or was it because he beat the king at chess?

Exercise Set 5.1

*
1. Modify and run the program COUNSELOR'S DILEMMA to output not
only the number of grains of wheat required but also the cost of
wheat. Assume that a bushel of wheat holds approximately 550,000
grains and sells for $2.50 a bushel.

2. An idea used quite often in writing computer programs
involving calculus topics is that of increments or decrements. An
increment is a fixed value which is repeatedly added to a
quantity; decrements are subtracted instead of added.   Write and
run a computer program which will print a table of functional
values for a given function.   Your program should enter the
starting and ending value of x and the increment. The function
should be in a DEF statement. Use your program to generate a
table for the following:

    a. f(x)=sin 3x, x starting at 0 and ending at 2.0, increment
.2

    b. f(x)=(x²-3x+1)/(x-2), x starting at 4 and ending at 8,
increment .5

**
3. Ulam's conjecture hypothesizes that, starting with any
positive integer, the sequence of numbers generated by dividing an
even number in half and multiplying an odd number by three and
adding one would converge to one. For example, starting with 17,
the following sequence is produced:

    17   52   26   13   40   20   10   5   16   8   4   2   1

Write and run a computer program which will produce the sequence
for Ulam's conjecture. Your input should be a positive   integer.
As test data use 17.

***
4. Shipwrecked   on   a   deserted   island,   three   sailors   have
collected a pile of coconuts which are to be divided the next day.
However, one of the sailors does not trust the other two, so he
awakens in the middle of the night and divides the pile into three
equal parts with one coconut left over, which he throws in the
bushes.   He then hides his share and goes back to sleep.   Later in
the night each of the other two sailors repeats the performance of
the first sailor exactly.   In the morning the three sailors divide
the pile into three equal parts and find that they have one left
over which they eat for breakfast.

Write and run a computer program which accepts as input a
potential value for the original number of coconuts and outputs a
message as to whether the number satisfies the conditions of the
story.   Also, by trial and error determine the smallest positive
integer which will work.

5.2 DEVISING AN ALGORITHM

Of the five steps listed in our scheme for writing a computer program by far the most difficult is the strategy or formulation of the algorithm. One effective method for designing algorithms is to subdivide the task to be accomplished into several subtasks, then subdivide each subtask into even smaller subtasks, which are subdivided again, and so on. At some point the subtasks become so simple that they are easily implemented in BASIC. Such a method is usually referred to as top-down program design. As an example we will apply the top-down program design philosophy to the design and writing of a program to determine the greatest common divisor of two positive integers using the Euclidean Algorithm.

The Euclidean Algorithm: To find the greatest common divisor of two positive integers, start by dividing the smaller of the two numbers into the larger. If the remainder is non-zero, replace the numbers by the remainder and the smaller of the two numbers and repeat the process. If the remainder is zero, then the greatest common divisor is the last non-zero remainder.

Example: Find the greatest common divisor of 720 and 525 using the Euclidean Algorithm.

Step 1. Divide the smaller number into the larger.

720/525 = 1 rem 195

2. Is the remainder 0? No.

3. Replace the numbers with the remainder and the smaller of the two numbers and repeat the process.
525 and 195

Repeating the process yields the following results.

Step 1. 525/195 = 2 rem 135
2. Non-zero remainder.
3. Replace the numbers with 195 and 135.

    Step 1.   195/135 = 1 rem 60
         2.   Non-zero remainder.
         3.   Replace the numbers with 135 and 60.

    Step 1.   135/60 = 2 rem 15
         2.   Non-zero remainder.
         3.   Replace the numbers with 60 and 15.

    Step 1.   60/15 = 4 rem 0
         2.   Zero remainder.
         3.   Greatest common divisor is the last non-zero
    remainder, that is, 15.

Programming Problem: Determine the greatest common divisor of two
positive integers using the Euclidean Algorithm.

Output: The greatest common divisor of the two numbers.

Input: Two numbers, A and B.

Strategy: The overall task is to determine the greatest common
divisor of A and B.   However, the Euclidean Algorithm already
provides a subdivision of the task into the following subtasks.

        Subtask 1. Divide the smaller number into
        the larger and determine the remainder.

        Subtask 2.   If the remainder is non-zero,
        replace the numbers by the remainder and the
        smaller of the two numbers and return to step
        1.

        Subtask 3. If the remainder is zero, then
        the greatest common divisor is the last
        non-zero remainder.

Now the first subtask can be further subdivided into   the
following.

        Subtask 1a.   Interchange the numbers, if
        necessary, so that B is the largest.

        Subtask 1b. Determine the remainder when B
        is divided by A. Let R denote the remainder.

Implementing in BASIC the code for the first subtask is:

```
IF B>=A THEN (line number for finding remainder)
LET C=B
LET B=A                        interchanging so B is largest
LET A=C
LET R=B-A*INT(B/A)             finding remainder
```

To see that the last statement does give the remainder let B=720 and A=525 which has a remainder of 195 when divided.

```
R=B-A*INT(B/A)
R=720-525*INT(720/525)
R=720-525*INT(1.3714286)
R=195
```

Likewise, we can further subdivide the second subtask.

2a.   Test to see if the remainder is zero.
2b.   If yes, go to subtask 3.
2c.   If no, assign the value of A to B and R to A.   Remember, A is the smaller of the two numbers.
2d.   Return to subtask 1.

Now, the BASIC code for the second subtask becomes:

```
IF R=0 THEN (line no.  for subtask 3) testing rem.  for 0
LET B=A
LET A=R                              replacing the nos.
GOTO (line no.  for subtask 1)
```

The third subtask does not need to be broken down into simpler subtasks since the value of the remainder was tested for zero in subtask 2.   However, only the value of the current remainder is known, but it is necessary to print the last non-zero remainder. This value can be saved in subtask 2 using the BASIC statement

```
LET LR=R
```

and the result printed using the statement

```
PRINT "THE GREATEST COMMON DIVISOR IS";LR
```

Now, the writing of the final program is simply a matter of putting the statements for the various subtasks together with the necessary input statements.

Program:

```
100 REM ** EUCLIDEAN ALGORITHM **
110 REM ** INPUT **
120 PRINT "ENTER TWO POSITIVE INTEGERS" prompting input
130 INPUT A,B
140 PRINT "FOR THE VALUES";A;"AND";B echoing data
150 REM ** MAIN PROGRAM **
160 REM ** SUBTASK 1 **
170 IF B>=A THEN 210
180 LET C=B
190 LET B=A
200 LET A=C
210 LET R=B-A*INT(B/A)
220 REM ** SUBTASK 2 **
230 IF R=0 THEN 280
240 LET B=A
250 LET A=R
260 LET LR=R                    saving previous remainder
270 GO TO 210                   returning to compute next remainder
280 REM ** SUBTASK 3 **
290 PRINT "THE GREATEST COMMON DIVISOR IS";A
300 END
```

Test Data: From the example we know that the greatest common divisor of 720 and 525 is 15.

```
RUN
ENTER TWO POSITIVE INTEGERS.
?720,525
FOR THE VALUES 720 AND 525
THE GREATEST COMMON DIVISOR IS 15


RUN
ENTER TWO POSITIVE INTEGERS.
?525,720
FOR THE VALUES 525 AND 720
THE GREATEST COMMON DIVISOR IS 15
```

Notice that the order in which the values are entered do not affect the result.

Top-down program design has many features of outlines in writing. In fact, programming is a form of writing. However, since programs are read by machines, a high degree of precision is needed. Using an outline, as in top-down program design, can be of immense aid in producing this precision. The outline provides a structure of increasing detail which not only tends to force us to "stick" to the topic at hand, but also allows us to delve into the details of one part of the program without getting "lost" in the details of another part.

Exercise Set 5.2

**

1. Write and run a program which accepts as input two fractions and outputs their sum reduced to lowest terms. The fractions are not to be entered in decimal form, but they should be entered by giving the numerator and denominator separately. For example, 4/5 would be entered as 4,5. Test your program with 2/15 +2/3 = 4/5. Hint: a/b + c/d = (ad + cb)/bd.

A Pythagorean Triple is a set of three positive integers a,b and c which satisfy the Pythagorean Theorem $a^2+b^2=c^2$. For example, (5,12,13) is a Pythagorean Triple since $5^2+12^2=13^2$. In exercises 2-5, write and run a program to output the desired information or verify the indicated result.

2. Determine all Pythagorean Triples with all components less than 50. Assume A<B<C.

3. Modify the program of exercise 2 to verify that for the Pythagorean Triples found their product is divisible by 60. Do you think this result holds for any Pythagorean Triple? Prove your answer if possible.

***

4. Modify the program of exercise 2 to determine only primitive Pythagorean Triples whose components are all less than 50. For a primitive Pythagorean Triple the components are relatively prime; that is, all three have no common factors except 1. Hint: The greatest common factor of a,b and c can be found by first finding the greatest common factor of a and b, call it d. Then find the greatest common factor of d and c.

5. The program EUCLIDEAN ALGORITHM does not yield the correct result for a pair of number such as 2 and 4. Explain why not and modify the program to correct this problem.

6. For certain fractions illegal cancellation can produce correct results. For example, 16/64=16/64=1/4. In addition to 16/64 there are three other proper fractions with two digit numerators and denominators where this type of illegal cancellation gives the correct result. Write and run a program to determine these four fractions.

CHALLENGE ACTIVITY

Synthetic division simplifies the division of a polynomial by a divisor of the form x-a. For example, to determine the quotient and remainder when $x^5 - 3x^4 + 2x^3 + 2x^2 - x + 1$ is divided by x-2 using synthetic division we have:

```
 2⌋    1    -3    2    2   -1    1

            2    -2    0    4    6
       _____
       1    -1    0    2    3    7
```

Quotient: $x^4 - x^3 + 0x^2 + 2x + 3$ Remainder: 7

Write and run a program to perform synthetic division. Your program should be general, accepting as input the degree and coefficients of the polynomial to be divided and the value of a in the divisor x-a. As test data use the polynomial and divisor in the example above. <u>Hint</u>: You will need to use an array. See Appendix D.

CHAPTER 6
Limits

Central to the calculus is the notion of the limit of a function. The problem of finding the limit of a function as x approaches "a" reduces to two distinct cases. First, some functions are continuous at the limit point, and, in this case, lim x->a f(x)=f(a). That is, the evaluation of the limit is obtained by direct substitution. Secondly, some functions are not continuous at the point. Evaluation of this type of limit often involves algebraic manipulation or the application of special theorems about limits. Unfortunately, we cannot always determine the auxiliary functions necessary to apply many of these theorems. Faced with limits such as

$$\lim_{x\to 0} \frac{4+7^{1/x}}{7^{1/x}-3} \qquad \text{or} \qquad \lim_{x\to\infty} \frac{x^x}{(x-1)^x}$$

it is not readily apparent how to proceed with their evaluation. However, the computer can be used to approximate the limit, if one exists.

6.1 APPROXIMATION OF LIMITS

Programming Problem: Approximate
$$\lim_{x\to a} f(x)$$

Output: An approximation of
$$\lim_{x\to a} f(x)$$

Input: A function f(x) and a number "a".

Remembering to use a prompt for the INPUT statement, we can accomplish the input part of the program in the following way.

```
DEF FNF(X)=.....
PRINT "THE LIMIT IS TO BE TAKEN AS X->?"
INPUT A
```

The symbol -> is formed by a minus sign followed by a greater than symbol.

Strategy: If the limit of f(x) as x approaches "a" is some number L, then for values of x "close to a", the values of f(x) will be "close to L". Thus, an approximation of the limit can be obtained by creating a table containing values of x "close to a" and the evaluation of f(x) at these values.

Basically, this strategy entails two tasks. First, we must generate values of x near "a", and, secondly, we must print a table of values for x and f(x). We can produce values of x "close to a" by utilizing the fact that for a number between 0 and 1 successive powers of the number become smaller. For example, $.5^2=.25$, $.5^3=.125$, $.5^4=.0625$, etc. Thus, we can produce values of x approaching "a" by increasing the value of I in the statement

$$\text{LET } X=A+(-.5)^I$$

The negative sign on the .5 is needed so that we will have values both to the left and right of "a".

Program:

```
100 REM ** LIMIT-1 **
110 REM ** FUNCTION **
120 DEF FNF(X)=.....
130 REM ** INPUT OF "A" **
140 PRINT "THE LIMIT IS TO BE TAKEN AS X->?"
150 INPUT A
160 PRINT
170 REM ** HEADINGS FOR THE TABLE **
180 PRINT "VALUE OF X","VALUE OF F(X)"
190 REM ** MAIN PROGRAM **
200 REM ** LOOP FOR SUCCESSIVE VALUES **
210 REM ** OF X CLOSE TO A **
220 FOR I=1 TO 20
230 LET X=A+(-.5)^I
240 REM ** OUTPUT **
250 PRINT X,FNF(X)
260 NEXT I
270 END
```

Test Data:

$$\lim_{x \to 5} \ 2x^2 + 7 = 57.$$

120 DEF FNF(X)=2*X*X+7

RUN
THE LIMIT IS TO BE TAKEN AS X->?
?5

X	F(X)
4.5	47.5
5.25	62.125
4.875	54.53125
5.0625	58.2578125
4.96875	56.3769531
5.015625	57.3129883
4.9921875	56.8438721
5.00390625	57.0781555
4.99804688	56.9609451
5.00097656	57.0195332
4.99951172	56.9902348
5.00024414	57.004883
4.99987793	56.9975587
5.00006104	57.0012207
4.99996948	56.9993897
5.00001526	57.0003052
4.99999237	56.9998474
5.00000381	57.0000763
4.99999809	56.9999619
5.00000095	57.0000191

Considering round-off error, it appears that the output of the program is yielding a reasonable approximation of the actual limit. However, the output is difficult to follow with the value of x alternating above and below 5, and the values of f(x) alternating above and below 57. A more readable output could be obtained if we separated the values of x below 5 from those above 5 and print two tables. This can be accomplished by a slight modification to our present program.

```
PRINT "VALUE OF X","VALUE OF F(X)",
PRINT "VALUE OF X","VALUE OF F(X)"
LET H=.5^I
LET X1=A-H                          finding values < A
LET X2=A+H                          finding values > A
PRINT X1,FNF(X1),X2,FNF(X2)
```

Incorporating these changes into our program yields the following program which we shall call LIMIT-2. <u>Note</u>: If the computer system being used has only a 40 column display do not use the program LIMIT-2 since the output will not fit on the display. Instead use program LIMIT-1.

```
100 REM ** LIMIT-2 **
110 REM ** FUNCTION **
120 DEF FNF(X)=.....
130 REM ** INPUT OF "A" **
140 PRINT "THE LIMIT IS TO BE TAKEN AS X->?"
150 INPUT A
160 PRINT
170 REM ** HEADINGS FOR THE TABLE **
180 PRINT "VALUE OF X","VALUE OF F(X)",
190 PRINT "VALUE OF X","VALUE OF F(X)"
200 REM ** MAIN PROGRAM **
210 REM ** LOOP FOR SUCCESSIVE VALUES **
220 REM ** OF X CLOSE TO A **
230 FOR I=1 TO 20
240 LET H=.5^I
250 LET X1=A-H
260 LET X2=A+H
270 REM ** OUTPUT **
280 PRINT X1,FNF(X1),X2,FNF(X2)
290 NEXT I
300 END
```

Running LIMIT-2 with the test data results in a more readable printout as illustrated below.

```
120 DEF FNF(X)=2*X*X+7

RUN
```
THE LIMIT IS TO BE TAKEN AS X->?
?5

VALUE OF X	VALUE OF F(X)	VALUE OF X	VALUE OF F(X)
4.5	47.5	5.5	67.5
4.75	52.125	5.25	62.125
4.875	54.53125	5.125	59.53125
4.9375	55.7578125	5.0625	58.2578125
4.96875	56.3769531	5.03125	57.6269531
4.984375	56.6879883	5.015625	57.3129883
4.9921875	56.8438721	5.0078125	57.1563721
4.99609375	56.9219055	5.00390625	57.0781555
4.99804688	56.9609451	5.00195313	57.0390701
4.99902344	56.9804707	5.00097656	57.0195332
4.99951172	56.9902348	5.00048828	57.0097661
4.99975586	56.9951173	5.00024414	57.004883
4.99987793	56.9975587	5.00012207	57.0024415
4.99993897	56.9987793	5.00006104	57.0012207
4.99996948	56.9993897	5.00003052	57.0006104
4.99998474	56.9996948	5.00001526	57.0003052
4.99999237	56.9998474	5.00000763	57.0001526
4.99999619	56.9999237	5.00000381	57.0000763
4.99999809	56.9999619	5.00000191	57.0000382
4.99999905	56.9999809	5.00000095	57.0000191

Example 1: Approximate

$$\lim_{x \to 0} \frac{\cos(x) - 1}{x^2 + x} \text{ using LIMIT-2.}$$

```
120 DEF FNF(X)=(COS(X)-1)/(X*X+X)
```

RUN
THE LIMIT IS TO BE TAKEN AS X->?
?0

VALUE OF X	VALUE OF F(X)	VALUE OF X	VALUE OF F(X)
-.5	.489669752	.5	-.1632232508
-.25	.165800419	.25	-.0994802505
-.125	.0713356158	.125	-.0554832553
-.0625	.0333224853	.0625	-.0294021918
-.03125	.0161277248	.03125	-.0151502822
-.015625	7.93635656E-03	.015625	-7.69216097E-03
-7.8125E-03	3.93700787E-03	7.8125E-03	-3.87596899E-03
-3.90625E-03	1.96084415E-03	3.90625E-03	-1.94558466E-03
-1.953125E-03	9.78593024E-04	1.953125E-03	-9.74777846E-04
-9.765625E-04	4.88997205E-04	9.765625E-04	-4.88043064E-04
-4.8828125E-04	2.44736963E-04	4.8828125E-04	-2.44498078E-04
-2.44140625E-04	1.23054029E-04	2.44140625E-04	-1.22993959E-04
-1.22070313E-04	6.29501892E-05	1.22070313E-04	-6.29348224E-05
-6.10351563E-05	3.05194409E-05	6.10351563E-05	-3.05157156E-05
-3.05175781E-05	2.28888821E-05	3.05175781E-05	-2.28874851E-05
-1.52587891E-05	0	1.52587891E-05	0
-7.62939453E-06	0	7.62939453E-06	0
-3.81469727E-06	0	3.81469727E-06	0
-1.90734863E-06	0	1.90734863E-06	0
-9.53674317E-07	0	9.53674317E-07	0

The output suggests that the limit is 0. However, this does not constitute a proof; only an analytical argument will suffice for a proof.

Example 2: Approximate

$$\lim_{x \to 1} \frac{(x-1)^2}{\sqrt{x} - 1/2 - x/2}$$

```
120 DEF FNF(X)=((X-1)*(X-1))/(SQR(X)-.5-(X/2))
```

```
RUN
THE LIMIT IS TO BE TAKEN AS X->?
?1
```

VALUE OF X	VALUE OF F(X)	VALUE OF X	VALUE OF F(X)
0.5	-5.82842713	1.5	-9.89897962
0.75	-6.96410166	1.25	-8.97213611
0.875	-7.49165801	1.125	-8.49264073
0.9375	-7.74798521	1.0625	-8.2481097
0.96875	-7.87451258	1.03125	-8.12452833
0.984375	-7.9374437	1.015625	-8.06237218
0.9921875	-7.96887159	1.0078125	-8.03137255
0.99609375	-7.98440546	1.00390625	-8.01565558
0.998046875	-7.99609566	1.00195313	-8.00782014
0.999023438	-8	1.00097656	-8
0.999511719	-8.06299213	1.00048828	-8
0.999755859	-8	1.00024414	-8
0.99987793	-8	1.00012207	-8
0.999938965	-8	1.00006104	-8
		1.00003052	

DIVISION BY ZERO ERROR IN 210

Observing this output we would most likely conclude that the limit is -8. Yet, how much confidence do we have considering the error message? The actual value of the limit is -8; the error message is the result of round-off error caused by subtracting nearly equal numbers in the denominator.

We next consider the case of limits as x approaches infinity.

Programming Problem: Approximate
$$\lim_{x\to\infty} f(x)$$

Output: An approximation of
$$\lim_{x\to\infty} f(x)$$

Input: A function f(x).

Strategy: As in the programs LIMIT-1 and LIMIT-2 the output will not be a single value but a table of values for f(x) as x increases. However, we are now interested in approaching the limit from only one direction, namely, increasing values of x. Since LIMIT-1 yielded a single table of values, we can modify this program to satisfy our programming problem.

Analysis of LIMIT-1 indicates that two changes will be required. Since the limit is taken as x->∞ it is not necessary to input a value of "a". Thus, lines 130-150 should be deleted. Also, line 230

$$LET\ X=A+(-.5)^I$$

must be changed to create values of x which are increasing. One such expression is

$$LET\ X=10^I$$

Program:

```
100 REM ** LIMIT-3 **
110 REM ** FUNCTION **
120 DEF FNF(X)=.....
130 PRINT
140 REM ** HEADINGS FOR THE TABLE **
150 PRINT "VALUE OF X","VALUE OF F(X)"
160 REM ** MAIN PROGRAM **
170 REM ** LOOP FOR SUCCESSIVE VALUES OF X **
180 FOR I=1 TO 15
190 LET X=10^I
200 REM ** OUTPUT **
210 PRINT X,FNF(X)
220 NEXT I
230 END
```

Test Data: Find
$$\lim_{x->\infty}\ (3x^2-4x+1)/(x^2+2x-9)$$

The limit can be found analytically to be 3.

```
120 DEF FNF(X)=(3*X*X-4*X+1)/(X*X+2*X-9)
```

```
RUN
VALUE OF X        VALUE OF F(X)
10                2.35135135
100               2.90462173
1000              2.99004782
10000             2.99900048
100000            2.9999
1000000           2.99999
10000000          2.999999
100000000         2.9999999
1E+09             2.99999999
1E+10             3
1E+11             3
1E+12             3
1E+13             3
1E+14             3
1E+15             3
```

In the last six lines of the output we have an occurrence of round-off error. The actual functional value lies somewhere between 2.99999999 and 3.

Example 3: Approximate

$$\lim_{x \to \infty} x^{1/x}$$

using LIMIT-3. Before running the program try to determine the value of the limit.

As x increases, $1/x$ approaches 0. Thus, $x^{1/x}$ will take on the form ∞^0. An expression of this type is known as an indeterminant form since it can yield different results. We might guess that the limit is 1 or 0.

```
120 DEF FNF(X)=X^(1/X)
```

```
RUN
VALUE OF X          VALUE OF F(X)
10                  1.25892541
100                 1.04712855
1000                1.00693167
10000               1.00092146
100000              1.00011514
1000000             1.00001382
10000000            1.00000161
100000000           1.00000018
1E+09               1.00000002
1E+10               1
1E+11               1
1E+12               1
1E+13               1
1E+14               1
1E+15               1
```

In this instance the limit is apparently 1. For other indeterminant forms see problem 14 of Exercise Set 6.1

Ch. 6   LIMITS

Exercise Set 6.1

In problems 1-5 use LIMIT-1 or LIMIT-2 to approximate the indicated limit. If possible, verify the results analytically without the aid of a computer.

1. $\lim_{x \to -2}$ $4x^2 - x + 4$

2. $\lim_{x \to 0}$ $(\sin 3x)/(\sin 4x)$

3. $\lim_{x \to 4}$ $(x^2 - 2x - 8)/(x^2 - 3x - 4)$

4. $\lim_{x \to 1}$ $(x^5 - x)/(x^4 - 1)$

5. $\lim_{x \to 0}$ $x\,[1/x]$
     Hint: [ ] is the greatest integer function.
     In BASIC use LET F(X)=X*INT(1/X)

In problems 6-10 use LIMIT-3 to approximate the indicated limit. Explain your result and, if possible, verify the result analytically without the aid of a computer.

6. $\lim_{x \to \infty}$ $\sqrt{9x+1}/\sqrt{4x-1}$

7. $\lim_{x \to \infty}$ $10\sqrt{x}\,/4\sqrt{x}$

8. $\lim_{x \to \infty}$ $(\sqrt{2x}+3)/(4x-5)$

9. $\lim_{x \to \infty}$ $(5x^3 - 4x^2 + 2x - 1)/(3x^3 + 10x^2 - 10)$

10. $\lim_{x \to \infty}$ $(x^7 + 1)/(x^8 - 1)$

*
11. Make the adjustments necessary to LIMIT-1 to investigate

$$\lim_{x \to a^+} f(x)$$

Use as test data

$$\lim_{x \to -3^+} (2x|x+3|)/(x+3) = -6$$

to verify your program.
<u>Hint</u>: In BASIC use LET F(X)=(2*X*ABS(X+3))/(X+3).

12. Make the necessary adjustments to LIMIT-1 to investigate

$$\lim_{x \to a^-} f(x)$$

Use as test data

$$\lim_{x \to -3^-} (2x|x+3|)/(x+3) = 6$$

to verify your program. Compare the results with that of exercise
number 11. What can you conclude about the value of

$$\lim_{x \to /3} (2x|x+3|)/(x+3) ?$$

13. Make the adjustments necessary to LIMIT-3 to investigate

$$\lim_{x \to -\infty} f(x)$$

Use as test data

$$\lim_{x \to -\infty} (3x^3-x)/(x^3+1) = 3$$

to verify your program. Rerun the program for

$$\lim_{x \to -\infty} (8x^{1/7}-1)/(4x^{1/7} - x^{1/9})$$

What can you conclude about the value of the limit?

**
14. Besides ∞^0 other indeterminant forms are 0/0, ∞/∞, 0^0,
$0*\infty$, $\infty-\infty$, and 1^∞. Both of the limits below are of the form 1^∞.
Use LIMIT-3 to approximate the following limits. What conclusion
can you make about indeterminant forms?

$$\lim_{x \to \infty} (1 + 1/x)^x \qquad \text{and} \qquad \lim_{x \to \infty} (1 - 1/x^2)^x$$

***
15.   Approximate the value of

$$\lim_{x \to 1} a(1-x^n)/(1-x)$$

for a>0 and n any integer greater than 1.
<u>Hint</u>: Let a=1 and use LIMIT-1 or LIMIT-2 with different values of
n.   Then fix n and use LIMIT-1 or LIMIT-2 with different values of
a.

16.   Run program LIMIT-2 to evaluate

$$\lim_{x \to 0} \sin(\pi/x)$$

Use π=ATN(1)*4.   Are you able to conclude from your output that
$\lim_{x \to 0} \sin(\pi/x)$ does   not exist? If not, explain what is wrong
with the program, make the necessary adjustments,   and   run   your
modified program for $\lim_{x \to 0} \sin(\pi/x)$.

6.2 DISCONTINUITIES

The functions $f(x)=x^2/x$ and $f(x)=1/x$ are both discontinuous at $x=0$.

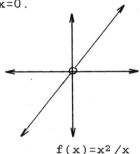

f(x)=x²/x

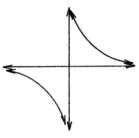

f(x)=1/x

However, the discontinuities are of a different nature. The function $f(x)=x^2/x$ is discontinuous at $x=0$, but the one-sided limits are the same. Thus, the graph as a "hole" in it. The function could be made continuous at $x=0$ by redefining it so that it has the same value as its limit at 0, namely 0. The new function would then be continuous at $x=0$, removing the discontinuity of $f(x)=x^2/x$. Discontinuities of this type are called <u>removable</u>.

The function $f(x)=1/x$, on the other hand, makes a tremendous jump across 0. The one-sided limits are $-\infty$ and ∞. There is no way to define a function value at $x=0$ to make the function become continuous at $x=0$. Discontinuities of this type are called <u>essential</u>.

Our problem in this section is to determine the type of discontinuity given a function and a point of discontinuity. If the discontinuity is removable, we will find the value which should be assigned to the function at the point to remove the discontinuity.

Programming Problem: Given a function and a point of discontinuity determine the type ;of discontinuity, removable or nonremovable. If the discontinuity is removable, determine the value which should be assigned to the function to remove the discontinuity.

Output: The output will consist of one of two messages.

For a removable discontinuity:

```
AT ____ THE FUNCTION HAS A
REMOVABLE DISCONTINUITY.  IT IS REMOVED
BY ASSIGNING THE VALUE ____
TO THE FUNCTION.
```

For a nonremovable discontinuity:

```
AT ____ THE FUNCTION HAS A
NONREMOVABLE DISCONTINUITY.
```

Input: A function f(x) and a point of discontinuity "a". We will enter the function and prompt the user to enter the point of discontinuity in the following manner.

```
DEF FNF(X)=.....
PRINT "WHAT IS THE POINT OF DISCONTINUITY?"
INPUT A
```

Strategy: By definition we know that a function is continuous at "a" if and only if all three of the following conditions are satisfied.

1. f(a) exists
2. $\lim_{x \to a} f(x)$ exists
3. $\lim_{x \to a} f(x) = f(a)$

Thus, a function will have a point of discontinuity if any one of the three conditions fail. However, if a function, f(x), is discontinuous at "a" but the $\lim_{x \to a} f(x)$ exists, then it is possible to define a new function, g(x), which will be continuous at "a" and behave like f(x) elsewhere by defining g(x) as:

$$g(x) = \begin{cases} \lim_{x \to a} f(x), & x=a \\ f(x), & x \neq a \end{cases}$$

This type of discontinuity is referred to as a removable discontinuity. On the other hand, if f(x) is discontinuous at "a" and $\lim_{x \to a} f(x)$ does not exist, then we say that the function has a nonremovable discontinuity.

Summarizing, if $\lim_{x \to a} f(x)$ exists, then the discontinuity is removable. Otherwise, it is nonremovable. Further, the value of the limit is the value which should be assigned to the function to remove the discontinuity. Thus, we need a program which will determine if $\lim_{x \to a} f(x)$ exists, and, if so, determine its value.

In the previous section we used the computer to print a table of functional values as x approached "a". It was up to us to determine from the table if the limit existed. However, is there some way to have the computer make this decision for us? The answer is yes if we can give a numerical condition which the computer can test. In examining a table we felt confident that the limit existed if successive values were getting "close". The following statements illustrate one method of doing this in BASIC.

```
LET X1=A+(-.6)^(I-1)          producing two successive
LET X2=A+(-.6)^I              values of x
IF ABS(FNF(X1)-FNF(X2))<.0001 THEN .....
```

The number .0001 is called the tolerance or precision. For a discussion of tolerance to test for equality see section 4.2 We will consider two values "close" when their difference falls within the desired tolerance. The choice of .0001 for the desired tolerance is completely arbitrary and any value could be used. Also, the values -.6 and .6 are also arbitrary. These values were selected to avoid the difficulties which can occur when using .5 For an example of the difficulty using .5 see problem 16 of exercise set 6.1

Program:

```
100 REM ** DISCONTINUITY **
110 REM ** FUNCTION **
120 DEF FNF(X)=.....
130 REM ** INPUT **
140 PRINT "WHAT IS THE POINT OF DISCONTINUITY?"
150 INPUT A
160 REM ** MAIN PROGRAM **
170 REM ** FINDING LIMIT **
180 FOR I=1 TO 35
190 LET X1=A+(.6)^(I-1)
200 LET X2=A+(-.6)^I
210 IF ABS(FNF(X2)-FNF(X1))<.0001 THEN 280
220 NEXT I
230 REM ** MESSAGE NONREMOVABLE **
240 PRINT "AT ";A;" THE FUNCTION HAS A"
250 PRINT "NONREMOVABLE DISCONTINUITY."
260 GOTO 320
270 REM ** MESSAGE REMOVABLE **
280 PRINT "AT ";A;" THE FUNCTION HAS A"
290 PRINT "REMOVABLE DISCONTINUITY.  IT IS REMOVED"
300 PRINT "BY ASSIGNING THE VALUE ";FNF(X2)
310 PRINT "TO THE FUNCTION."
320 END
```

Test Data: We will use two different sets of test data, one where the function has a removable discontinuity and one where the discontinuity is nonremovable. Since

$$\lim_{x \to 3} (x^2-9)/(x-3) = \lim_{x \to 3} x+3 = 6$$

then $f(x)=(x^2-9)/(x-3)$ has a removable discontinuity at 3. On the other hand, at 3 the function $f(x)=1/(x-3)$ has a nonremovable discontinuity. We will test both functions in our program.

```
120 DEF FNF(X)=(X*X-9)/(X-3)

RUN
WHAT IS THE POINT OF DISCONTINUITY?
?3
AT 3 THE FUNCTION HAS A
REMOVABLE DISCONTINUITY. IT IS REMOVED
BY ASSIGNING THE VALUE 6
TO THE FUNCTION.
```

```
120 DEF FNF(X)=1/(X-3)

RUN
```
WHAT IS THE POINT OF DISCONTINUITY?
?3
AT 3 THE FUNCTION HAS A
NONREMOVABLE DISCONTINUITY.

Example 1: The function $f(x)=(1-\cos x)/(\sin x)$ is undefined at $x=0$, and, thus, has a discontinuity at that point. Use the program DISCONTINUITY to determine the type of discontinuity and, if removable, the value to assign to remove the discontinuity.

```
120 DEF FNF(X)=(1-COS(X))/SIN(X)

RUN
```
WHAT IS THE POINT OF DISCONTINUITY?
?0
AT 0 THE FUNCTION HAS A
REMOVABLE DISCONTINUITY. IT IS REMOVED
BY ASSIGNING THE VALUE -3.21791986E-05
TO THE FUNCTION.

Again, care must be taken in interpreting the output. This function does have a removable discontinuity at 0, but the actual value which should be assigned to the function is 0, not -3.21791986E-05. You should always remember that answers produced by a computer are subject to round-off error and in many instances are only approximations of the actual values.

Example 2: Investigate the discontinuities of

$$f(x)=(x^2+x-2)/(x^2-4x+3)$$

Since $f(x)$ is a rational function, we know that the only points of discontinuity will be roots of the denominator. Thus,

$$x^2-4x+3=0$$
$$(x-3)(x-1)=0$$
$$x=3 \quad \text{or} \quad x=1$$

The points of discontinuity are 3 and 1.

```
120 DEF FNF(X)=(X*X+X-2)/(X*X-4*X+3)
```

RUN
WHAT IS THE POINT OF DISCONTINUITY?
?1
AT 1 THE FUNCTION HAS A
REMOVABLE DISCONTINUITY. IT IS REMOVED
BY ASSIGNING THE VALUE -1.49998939
TO THE FUNCTION.

RUN
WHAT IS THE POINT OF DISCONTINUITY?
?3
AT 1 THE FUNCTION HAS A
NONREMOVABLE DISCONTINUITY.

For the removable discontinuity at a=1 what value should actually be assigned, -1.49998939 or -1.5?

Exercise Set 6.2

In exercises 1-5 use program DISCONTINUITY to determine the type of discontinuity , removable or nonremovable, the function has at the given point a. If the discontinuity is removable, determine the value to assign to remove the discontinuity.

1. $f(x)=x^4/\sin x$; $a=0$

2. $f(x)=(\sin 2x)/(\sin 5x)$; $a=0$

3. $f(x)=(\tan x)/(\tan 2x)$; $a=0$

4. $f(x)=(\tan x)/x$; $a=0$

5. $f(x)=x\ 3^x/(1-3^x)$; $a=0$

In exercises 6-10 determine all points of discontinuity. Then use program DISCONTINUITY to decide if the discontinuity is removable or nonremovable. If the discontinuity is removable, determine the value to assign to remove the discontinuity.

6. $f(x)=(x-4)/(x^2-26)$

7. $f(x)=(x-3)/(x^2-2x-3)$

8. $f(x)=(3x-5)/(2x^2-x-3)$

9. $f(x)=\ x+3\ /(x+3)$

10. $f(x)=(x+10)/\sqrt{(x+10)^2}$

CHALLENGE ACTIVITY

The formal definition of the limit of a function is

limx->a f(x)=L if and only if, for every positive number ε, there exists a positive number d such that, if 0< x-a <d, then f(x)-L <ε.

In general, a computer cannot be used to determine the value of d for a given ε since there would be an infinite numbers of values for x satisfying 0< x-a <d which would have to be tested in the inequality f(x)-L <ε. However, under certain assumptions the computer can be used to determine d for a given ε and function.

Suppose we make the following assumption. When d<.1, .1 being arbitrarily chosen, if the endpoints of the open interval determined by x-a <d satisfy the inequality f(x)-L <ε, then all values of x in 0< x-a <d satisfy f(x)-L <ε.

Using this assumption write a computer program which will determine a value of d so that the definition of the limit is satisfied for a given function f(x), a, L, and ε. Use the program to determine d for the given information below.

1. f(x)=3x+2, a=3, L=1, ε=.06

2. f(x)=3x²+2x-1, a=2, L=15, ε=.5

3. f(x)=(six x)/x, a=0, L=1, ε=.02

CHAPTER 7
Derivatives

The slope of a line tangent to a curve at a point has many useful applications. The Greeks recognized the importance of this line and knew, for example, how to find the line tangent to a circle at a point by using the idea that it is perpendicular to the radius. In fact, the Greek scientist Archimedes (287-212 BC) devoted a large portion of one of his books, On Spirals, to finding a line tangent to the spiral of Archimedes. As the number of recognized curves increased, a general method was sought, all doomed to failure until the work of Issac Newton and Gottfried Liebnitz resolved the problem. These two eminent mathematicians developed a definition of derivative based on limits which, fortunately, can be implemented on the computer. However, the definition of the derivative as a limit poses unique problems for computer application as you shall see in this chapter.

7.1 DERIVATIVE AT A POINT

Programming Problem: Approximate the derivative of a function at a point.

Output: Since a computer cannot find a limit, but only approximate it, our output will consist of a table of values similar to those used in Chapter 6 for limits. We will need the following headings:

        H   LEFT-HAND LIM.   RIGHT-HAND LIM.

The column headings can be accomplished by the statement

PRINT "H";TAB(9);"LEFT-HAND LIM.";TAB(25);"RIGHT-HAND LIM."

Input: A function f(x) and a point at which the derivative will be taken. We will prompt the input as follows.

```
DEF FNF(X)=.....
PRINT "AT WHAT VALUE DO YOU WANT"
PRINT "TO TAKE THE DERIVATIVE?"
INPUT A
```

Strategy: The derivative is defined by

$$f'(a) = \lim_{x \to 0} \frac{f(a+h)-f(a)}{h}$$

Thus, we need to write a program which will approximate a limit as h approaches 0. Using the concepts developed in Chapter 6 this can be accomplished by the following statements.

```
FOR I=1 TO 20
LET H=(.3)^I
LET D1=(FNF(A-H)-FNF(A))/(-H)
LET D2=(FNF(A+H)-FNF(A))/H
PRINT H;TAB(9);D1:TAB(25);D2
NEXT I
```

The choice of .3 used in calculating H is again arbitrary and could be replaced by any value between 0 and 1.

Program:

```
100 REM ** FIRST DERIVATIVE-1 **
110 REM ** FUNCTION **
120 DEF FNF(X)=.....
130 REM ** INPUT **
140 PRINT "AT WHAT VALUE DO YOU WANT"
150 PRINT "TO TAKE THE DERIVATIVE?"
160 INPUT A
170 PRINT
180 REM ** HEADINGS **
190 PRINT "H";TAB(9);"LEFT-HAND LIM.";
    TAB(25);"RIGHT-HAND LIM."
200 REM ** MAIN PROGRAM **
210 REM ** LOOP FOR COMPUTING LIMIT **
220 FOR I=1 TO 20
230 LET H=(.3)^I
240 LET D1=(FNF(A-H)-FNF(A))/(-H)
250 LET D2=(FNF(A+H)-FNF(A))/H
260 REM ** OUTPUT **
270 PRINT "(.3^)";I;TAB(9);D1;TAB(25);D2
280 NEXT I
290 END
```

Test Data:

If $f(x)=x^2-9$, then $f'(4)=8$.

```
120 DEF FNF(X)=X*X-9

RUN
AT WHAT VALUE DO YOU WANT
TO TAKE THE DERIVATIVE?
?4
```

H	LEFT-HAND LIM.	RIGHT-HAND LIM.
$(.3)^1$	7.7	8.30000003
$(.3)^2$	7.91	8.09000002
$(.3)^3$	7.97299999	8.02700028
$(.3)^4$	7.99190023	8.00809926
$(.3)^5$	7.99757126	8.00242946
$(.3)^6$	7.99927139	8.00072778
$(.3)^7$	7.99976708	8.00022699
$(.3)^8$	7.99992038	8.00009072
$(.3)^9$	7.99982574	7.99982574
$(.3)^{10}$	7.99957339	7.99957339
$(.3)^{11}$	7.99957339	7.99957339
$(.3)^{12}$	8.00518123	7.99116164
$(.3)^{13}$	7.99116164	6.02842018
$(.3)^{14}$	7.9444297	6.07515212
$(.3)^{15}$	7.78865656	6.23092525
$(.3)^{16}$	8.65406285	5.19243771
$(.3)^{17}$	5.76937523	11.5387505
$(.3)^{18}$	0	0
$(.3)^{19}$	0	0
$(.3)^{20}$	0	0

This output, to say the least, is not quite what we expected. It begins as we expect but at line 9 of the output the right-hand limit jumps past the limit value, 8, and continues to behave erratically. Similar behavior is also exhibited by the left-hand limit.

This is yet another example of round-off error. The difficulty results from the subtraction of almost equal numbers, $f(a+h)-f(a)$ or $f(a-h)-f(a)$, in the numerator when h is small. See section 4.2 for a more detailed discussion of this type of round-off error, also known as subtractive cancellation. Unfortunately, the limit definition of the derivative necessitates small values since the numerator and denominator of the difference quotient are approaching zero.

Ch. 7 DERIVATIVES

What we will do, then, is modify our program, FIRST DERIVATIVE-1, so that the useful information is displayed, but the program stops when the round-off error begins to have a drastic effect on the results.

Notice that in the output for the derivative of x^2-9 at $x=4$ (page 125) both the right and left-hand limits are getting "close" to each other and the limit value, 8, before they start their erratic behavior. Thus, it may be possible to avoid displaying results greatly affected by the round-off error if we cause the program to end when the right and left-hand limits get "close". As before, we will assume that the two values are "close" when their difference is within an arbitrarily chosen tolerance or precision. For our program this can be accomplished by adding the following line, as before the value of 0.0001 is chosen arbitrarily.

IF ABS(D2-D1)<.0001 THEN .....

Rewriting our program to include this test yields the following program.

Revised Program:

```
100 REM ** FIRST DERIVATIVE-2 **
110 REM ** FUNCTION **
120 DEF FNF(X)=.....
130 REM ** INPUT **
140 PRINT "AT WHAT VALUE DO YOU WANT"
150 PRINT "TO TAKE THE DERIVATIVE?"
160 INPUT A
170 PRINT
180 REM ** HEADINGS **
190 PRINT "H";TAB(9);"LEFT-HAND LIM.";
    TAB(25);"RIGHT-HAND LIM."
200 REM ** MAIN PROGRAM **
210 REM ** LOOP FOR COMPUTING LIMIT **
220 FOR I=1 TO 20
230 LET H=(.3)^I
240 LET D1=(FNF(A-H)-FNF(A))/(-H)
250 LET D2=(FNF(A+H)-FNF(A))/H
260 REM ** OUTPUT **
270 PRINT "(.3^)";I;TAB(9);D1;TAB(25);D2
280 IF ABS(D2-D1)<.0001 THEN 300
290 NEXT I
300 END
```

Running DERIVATIVE-2 for our test data $f(x) = x^2 - 9$ at $x=4$:

```
120 DEF FNF(X)=X*X-9
```

```
RUN
AT WHAT VALUE DO YOU WANT
TO TAKE THE DERIVATIVE?
?4
```

H	LEFT-HAND LIM.	RIGHT-HAND LIM.
(.3)^1	7.7	8.30000003
(.3)^2	7.91	8.09000002
(.3)^3	7.97299999	8.02700028
(.3)^4	7.99190023	8.00809926
(.3)^5	7.99757126	8.00242946
(.3)^6	7.99927139	8.00072778
(.3)^7	7.99976708	8.00022699
(.3)^8	7.99992038	8.00009072
(.3)^9	7.99982574	7.99982574

Considering round-off error we would give the value of the limit as 8.

Example 1: Find $f'(5)$ if $f(x) = 1/x$

```
120 DEF FNF(X)=1/X
```

```
RUN
AT WHAT VALUE DO YOU WANT
TO TAKE THE DERIVATIVE?
?5
```

H	LEFT-HAND LIM.	RIGHT-HAND LIM.
(.3)^1	-.0425531916	-.0377358491
(.3)^2	-.0407331975	-.0392927306
(.3)^3	-.0402171745	-.0397851615
(.3)^4	-.0400649043	-.0399353024
(.3)^5	-.0400194556	-.0399805547

Thus, $f'(5) \approx -.04$ We cannot guarantee accuracy after the second decimal place.

Example 2: Find $f'(3)$ if $f(x) = \sqrt{x+1}$.

```
120 DEF FNF(X)=SQR(X+1)

RUN
AT WHAT VALUE DO YOU WANT
TO TAKE THE DERIVATIVE?
?3
```

H	LEFT-HAND LIM.	RIGHT-HAND LIM.
$(.3)^1$	.254871977	.245480454
$(.3)^2$	.24142229	.248609349
$(.3)^3$	.250423292	.249579583
$(.3)^4$	.250126626	.249873732
$(.3)^5$	.250037882	.24996238

Thus, $f'(3) \approx .25$

Example 3: Find $f'(2)$ if

$$f(x) = (1/\sqrt{x})+(1/x^2)+(1/x^4)$$

```
120 DEF FNF(X)=1/SQR(X)+1/(X*X)+1/(X*X*X*X)

RUN
AT WHAT VALUE DO YOU WANT
TO TAKE THE DERIVATIVE?
?2
```

H	LEFT-HAND LIM.	RIGHT-HAND LIM.
$(.3)^1$	-.710364453	-.451519377
$(.3)^2$	-.591359008	-.517219395
$(.3)^3$	-.563077525	-.540926362
$(.3)^4$	-.555118385	-.548475583
$(.3)^5$	-.552774824	-.550782254
$(.3)^6$	-.55207563	-.551477744
$(.3)^7$	-.551865049	-.551688324
$(.3)^8$	-.551795495	-.551745813

Thus, $f'(x) \approx -.55$

Exercise Set 7.1

1. Determine the actual value of the derivative for examples 1 - 3 above. Compare these values with the approximations given by the computer and comment on the accuracy of the computer approximations.

In exercises 2 - 9 use program FIRST DERIVATIVE - 2 to approximate the derivative of the function at the point "a". If possible, determine the actual value of the derivative and compare the value to the computer approximation.

2. $f(x) = 3x^2 - 5x + 2$; $a=2$

3. $f(x) = \sqrt{2x + 3}$; $a=11$

4. $f(x) = \dfrac{1}{x + 1}$; $a=1$

5. $f(x) = x^4 - 1/x^2$; $a=3$

6. $f(x) = \sin x$; $a=0$

7. $f(x) = \sqrt{(x+2)/(x+3)}$; $a=4$

8. $f(x) = \sqrt{3 + \sqrt{x}}$; $a=1$

9. $f(x) = (\cos^3 x)/x^2$; $a=3.14$

*
10. Run program FIRST DERIVATIVE-2 for $f(x) = |x|$ and $a=0$. Interpret and explain the output. Geometrically interpret and explain the output.

**

11.   Left and right-handed derivatives (or one-sided derivatives) are defined as follows.

$$\lim_{h \to 0^+} \frac{f(a+h)-f(a)}{h} \qquad \text{Right-hand derivative}$$

$$\lim_{h \to 0^-} \frac{f(a+h)-f(a)}{h} \qquad \text{Left-hand derivative}$$

a.   Modify program FIRST DERIVATIVE-2 to compute one-sided derivatives. You will need two different programs, one for each sided derivative.

b.   Use one of the programs from part a to compute the right-hand derivative of $f(x) = \sqrt{x}$ at a=0. Does the derivative of $f(x) = \sqrt{x}$ exist at 0? Explain your answer.

c.   Can you determine if the derivative of a function exists if you know the values of both the right and left-hand derivatives? Explain your answer.

***

12.   As we saw in this section, the accuracy to which we could compute the derivative was limited by computer round-off error in calculating f(a+h)-f(a) and f(a-h)-f(a). One method which generally yields more accurate results is based upon the fact that if the derivative exists the right and left-hand derivatives must be equal, and as we pass to the limit the limit value is usually very close to the average of the right and left-hand derivative. Thus,

$$f'(a) = \lim_{h \to 0} \frac{1}{2}\left(\frac{f(a+h)-f(a)}{h} + \frac{f(a-h)-f(a)}{-h}\right)$$

$$= \lim_{h \to 0} \frac{f(a+h)-f(a-h)}{2h}$$

a.   Modify program FIRST DERIVATIVE-2 to approximate the derivative using (f(a+h)-f(a-h))/2h instead of the normal difference quotient.

b.   Run the program from part a for $f(x) = x^2-9$; a=4. Compare the results with those obtained using FIRST DERIVATIVE-2. Which program produces more accurate results?

c. Explain why this method may yield a more accurate result.
<u>Hint</u>: Compare f(a+h)-f(a) and f(a+h)-f(a-h).

7.2 THE SECOND DERIVATIVE

Programming Problem: Approximate the second derivative of a function at a point.

Output: As with our program in section 7.1 the output will consist of a table of values which approximate the right and left-hand limit as h approaches 0. The form of our table will be:

> H   LEFT-HAND LIM.   RIGHT-HAND LIM.

To set the column headings we will need the statement:

PRINT "H";TAB(9);"LEFT-HAND LIM.";TAB(25);"RIGHT-HAND LIM."

Input: A function f(x) and the point at which we will take the derivative.

```
        DEF FNF(X)=.....
        PRINT "AT WHAT POINT DO YOU WANT TO TAKE"
        PRINT "THE SECOND DERIVATIVE?"
        INPUT A
```

Strategy: The procedure in calculus for finding the second derivative of a function is to first symbolically determine the first derivative, that is, find a function which represents the first derivative, and then take the derivative of this function. Finally, we substitute in the value at which we wish to evaluate the second derivative. However, this method will not work on a computer. The computer can approximate the value of the first derivative, but it will not find a function which represents the derivative. What we will do, therefore, is to develop a single expression similar to the difference quotient which becomes the second derivative when we take the limit as h approaches 0, providing, of course, that the limit exists.

Using the definition of the second derivative and then applying the definition of the first derivative to f'(a+h) and f'(a) we obtain the following expression for the second derivative.

$$f''(a) = \lim_{h \to 0} \frac{f'(a+h) - f'(a)}{h}$$

132

$$=\lim_{h\to0}\frac{\lim_{k\to0}\frac{f(a+h+k)-f(a+h)}{k}-f'(a)}{h}$$

$$=\lim_{h\to0}\frac{\lim_{k\to0}\frac{f(a+h+k)-f(a+k)}{k}-\lim_{k\to0}\frac{f(a+k)-f(a)}{k}}{h}$$

Letting h=k results in

$$f''(a)\approx\lim_{h\to0}\frac{f(a+2h)-2f(a+h)+f(a)}{h^2}$$

Incorporating the test used in program FIRST DERIVATIVE-2 to minimize round-off error, we should be able to approximate the second derivative by the following statements.

```
FOR I=1 TO 20
LET H=(.3)^I
LET D1=(FNF(A-2*H)-2*FNF(A-H)+FNF(A))/(-H*-H)
LET D2=(FNF(A+2*H)-2*FNF(A+H)+FNF(A))/(H*H)
PRINT "(.3)^";I;TAB(9);D1;TAB(25);D2
IF ABS(D2-D1)<.0001 THEN .....
NEXT I
```

Program:

```
100 REM ** SECOND DERIVATIVE **
110 REM ** FUNCTION **
120 DEF FNF(X)=.....
130 REM ** INPUT **
140 PRINT "AT WHAT VALUE DO YOU WANT TO TAKE"
150 PRINT "THE SECOND DERIVATIVE?"
160 INPUT A
170 PRINT
180 REM ** HEADINGS **
190 PRINT "H";TAB(9);"LEFT-HAND LIM.";TAB(25);
    "RIGHT-HAND LIM."
200 REM ** MAIN PROGRAM **
210 REM ** LOOP FOR COMPUTING LIMIT **
220 FOR I=1 TO 20
230 LET H=(.3)^I
240 LET D1=(FNF(A-2*H)-2*FNF(A-H)+FNF(A))/(-H*-H)
250 LET D2=(FNF(A+2*H)-2*FNF(A+H)+FNF(A))/(H*H)
260 REM ** OUTPUT **
270 PRINT "(.3^)";I;TAB(9);D1;TAB(25);D2
280 IF ABS(D2-D1)<.0001 THEN 300
290 NEXT I
300 END
```

Test Data:

If $f(x)=x^3-3$, then $f''(4)=24$.

```
120 DEF FNF(X)=X*X*X-3

RUN
AT WHAT VALUE DO YOU WANT TO TAKE
THE SECOND DERIVATIVE?
?4
```

H	LEFT HAND LIM.	RIGHT-HAND LIM.
(.3)^1	22.1999998	25.7999983
(.3)^2	23.4600018	24.5399975
(.3)^3	23.8379537	24.1618343
(.3)^4	23.9515555	24.0498973
(.3)^5	23.9936731	24.0189084
(.3)^6	24.0015241	24.0015241

Therefore, our program seems to be working.

Example 1:

Find $f''(5)$ if $f(x) = \sqrt{x^2 + 1}$

```
120 DEF FNF(X)=SQR(X*X+1)

RUN
```
**AT WHAT VALUE DO YOU WANT TO TAKE
THE SECOND DERIVATIVE?**
?5

H	LEFT HAND LIM.	RIGHT-HAND LIM.
$(.3)^1$	9.04625485E-03	6.39250292E-03
$(.3)^2$	7.95059697E-03	7.16598645E-03
$(.3)^3$	7.66111427E-03	7.42604793E-03
$(.3)^4$	7.5345696E-03	7.5345696E-03

The actual answer is .007542928.

Example 2:

Find $f''(2)$ if $f(x) = x^4 - 4x^2 + 2$.

```
120 DEF FNF(X)=X*X*X*X-4*X*X+2

RUN
```
**AT WHAT VALUE DO YOU WANT TO TAKE
THE SECOND DERIVATIVE?**
?2

H	LEFT HAND LIM.	RIGHT-HAND LIM.
$(.3)^1$	26.86	55.6600005
$(.3)^2$	35.7933983	44.4334021
$(.3)^3$	38.7142158	41.3062347
$(.3)^4$	39.6122322	40.3895976
$(.3)^5$	39.8819426	40.1160783
$(.3)^6$	39.9820949	40.0311634
$(.3)^7$	40.1407888	40.2770903
$(.3)^8$	36.9961187	38.7269312
$(.3)^9$	50.4820333	86.5406284
$(.3)^{10}$	213.680564	400.651058
$(.3)^{11}$	0	0

As illustrated by this output, using

$$\lim_{h \to 0} \frac{f(a+2h) - 2f(a+h) + f(a)}{h^2}$$

135

to find the second derivative can yield results which are affected by round-off error. Even having the test in line 200 has not prevented the round-off error from affecting our results. Based upon past experience, we would probably estimate f''(2) to be 40, which is the actual answer, since the left-hand and right-hand limits appeared to be approaching this value before the erratic behavior began. Again, we must always interpret computer results carefully.

Example 3: The elevation of a balloon above the ground is given by $s(t)=3t^2+(3/t)$ feet. Use programs FIRST DERIVATIVE-2 and SECOND DERIVATIVE to determine the velocity and acceleration of the balloon after 1 second.

Running FIRST DERIVATIVE-2 yields:

```
120 DEF FNF(X)=3*X*X+3/X

RUN
AT WHAT VALUE DO YOU WANT
TO TAKE THE DERIVATIVE?
?1
```

H	LEFT HAND LIM.	RIGHT-HAND LIM.
(.3)^1	.814285713	4.5923077
(.3)^2	2.43329671	3.51770642
(.3)^3	2.8357523	3.15987056
(.3)^4	2.95120165	3.04840482
(.3)^5	2.98540242	3.0145624
(.3)^6	2.99562397	3.00437253
(.3)^7	2.99868835	3.00132007
(.3)^8	2.99962237	3.00038889
(.3)^9	2.99984002	3.00012392
(.3)^10	3.00015546	3.00015546

Running SECOND DERIVATIVE yields:

```
120 DEF FNF(X)=3*X*X+3/X
```

RUN
AT WHAT VALUE DO YOU WANT TO TAKE
THE SECOND DERIVATIVE?
?1

H	LEFT HAND LIM.	RIGHT-HAND LIM.
(.3)^1	27.4285714	8.88461535
(.3)^2	14.0407399	10.6649048
(.3)^3	12.5184931	11.5429372
(.3)^4	12.148614	11.8569103
(.3)^5	12.0438372	11.9567756
(.3)^6	12.0147816	11.9832376
(.3)^7	11.9945311	11.9166445
(.3)^8	12.115688	12.115688

Thus, the velocity of the balloon after 1 second is approximately
3 ft/sec and the acceleration approximately 12 ft/sec^2.

Ch. 7   DERIVATIVES

Exercise Set 7.2

In exercises 1-6 use program SECOND DERIVATIVE to estimate the
second derivative at the indicated point. Check the accuracy of
the results by determining the actual value of the second
derivative.

1. $f(x)=6x^2+2x-1$; $a=2$

2. $f(x)=1/x$; $a=1$

3. $f(x)=x/(1+x)$; $a=2$

4. $f(x)=\sqrt{x} - \sqrt{2x}$; $a=1$

5. $f(x)=\cos 2x$; $a=3.14$

6. $f(x)=\sin^2 x$; $a=0$

In problems 7 and 8 use programs FIRST DERIVATIVE-2 and SECOND
DERIVATIVE to approximate the velocity and acceleration of a
particle at the given instance of time if its position is given by
the function s(t).

7. $s(t)=16t^2+3t+1$; $t=2$

8. $s(t)=4\sin 3t$; $t=1/4$

*
9.   In exercise 12 of Exercise Set 7.1 we used the expression

$$\frac{f(a+h)-f(a-h)}{2h}$$

to calculate the derivative.   Although we will not derive it,   the
expression

$$\frac{f(a+h)-2f(a)+f(a-h)}{h^2}$$

generally   yields   a   more   accurate   estimation   of   the   second
derivative. Modify program SECOND DERIVATIVE to estimate the
second derivative using this expression. Run this program for
x^2-4x+2 with a=2.   Compare your results with example 2 of this
section.   Which program produced the more accurate result?

CHALLENGE ACTIVITY

In section 7.2 we derived a formula for approximating the second derivative of a function. A similar procedure will result in an expression for the third derivative.

$$f'''(a)=\lim_{k\to 0} \frac{f''(a+k)-f''(a)}{k}$$

Replacing the second derivative with the expression found in section 7.2 yields the third derivative approximation to be the limit as k->0 of the following expression.

$$\frac{\lim_{h\to 0}\frac{f(a+k+2h)-2f(a+k+h)+f(a+k)}{h}-\lim_{h\to 0}\frac{f(a+2h)-2f(a+h)+f(a)}{h}}{k}$$

Letting k=h

$$f'''(a) \approx \lim_{h\to 0} \frac{f(a+3h)-2f(a+2h)+f(a+h)-(f(a+2h)-2f(a+h)-f(a))}{h^3}$$

Simplifying,

$$f'''(a) \approx \lim_{h\to 0} \frac{f(a+3h)-3f(a+2h)+3f(a+h)-f(a)}{h^3}$$

a.   Using the expression above, write a computer program to approximate the third derivative of a function. Test your program with $f(x)=4x^4+2x^3-3x^2+2x-1$ at x=2.

b.   Discuss the accuracy of the results of your program.

c.   Determine an expression to compute the 4th derivative of a function similar to the ones we have for the second and third derivatives.

d.   Write a computer program which will approximate the 4th derivative of a function using the expression developed in part c.

CHAPTER 8
Polynomial Functions

In Chapter 7 the value of the derivative was found by using the definition, that is, approximating the limit of the difference quotient. Unfortunately, as illustrated in that chapter, the use of the difference quotient resulted in severe round-off error which we were unable to avoid. However, if we restrict ourselves to polynomial functions, several alternate methods for finding the derivative employing the computer are available.

8.1 Method 1--Direct Evaluation

Programming Problem: Approximate the value of the derivative of a polynomial function at a point a by direct substitution of x=a into the derivative function.

Output: The approximation of the value of the derivative at a point. To clarify the output we will include the following message:

THE VALUE OF THE DERIVATIVE AT _____ IS _____.

Input: A polynomial function and the point at which the derivative is to be evaluated. Unlike earlier programs, the function will not be entered via a DEF statement but by giving its degree and coefficients. This is necessary since we will need the individual coefficients in order to determine the derivative. We will use an array of one dimension to store the coefficients. For a discussion of arrays refer to Appendix D.

Strategy: The evaluation of the derivative by direct substitution requires determining the function which is the derivative and evaluating this function at the given point.   Now, for a polynomial function we know that the derivative always exists and is another polynomial function whose degree is one less than the original polynomial and whose coefficients are related to the original polynomial by the formula

$$\frac{d}{dx}\ A_{n+1}\ x^{n+1}\ =\ A_{n+1}\ (n+1)x^n$$

If we let the array A contain the coefficients of the original polynomial, then we can use the above formula to compute the coefficients of the derivative which we will store in an array B. As for evaluating the derivative we can use the same process we would use without the computer, namely, sum up the values of the individual terms after substituting in the value of x at which the derivative is being evaluated.

Program:

```
100 REM ** POLYNOMIAL DERIVATIVE-1 **
110 DIM A(50),B(50)
120 REM ** ENTERING DEGREE **
130 PRINT "WHAT IS THE DEGREE OF THE"
140 PRINT "POLYNOMIAL? DO NOT EXCEED 50."
150 PRINT
160 INPUT N
170 REM ** TESTING FOR INPUT ERROR **
180 IF N>50 THEN 130
190 PRINT
200 REM ** ENTERING THE COEFFICIENTS **
210 PRINT "ENTER THE COEFFICIENTS ONE AT"
220 PRINT "A TIME.  START WITH THE HIGHEST"
230 PRINT "POWER OF X AND INCLUDE 0"
240 PRINT "FOR MISSING POWERS OF X."
250 PRINT
260 FOR I=N TO 0 STEP -1
270 INPUT A(I)
280 NEXT I
290 PRINT
300 REM ** FINDING COEFFICIENTS OF **
310 REM ** THE DERIVATIVE POLYNOMIAL **
320 LET M=N-1
330 FOR I=M TO 0 STEP -1
340 LET B(I)=A(I+1)*(I+1)
350 NEXT I
360 REM **ENTERING VALUE AT WHICH **
370 REM ** THE DERIVATIVE IS EVALUATED **
380 PRINT "AT WHAT VALUE IS THE DERIVATIVE"
390 PRINT "TO BE EVALUATED?"
400 PRINT
410 INPUT X
420 REM ** EVALUATING THE DERIVATIVE **
430 LET S=0
440 FOR I=0 TO M
450 LET S=S+B(I)*X^I
460 NEXT I
470 PRINT
480 REM ** OUTPUT **
490 PRINT "THE VALUE OF THE DERIVATIVE"
500 PRINT "AT ";X;" IS ";S
510 END
```

143

Test Data: Find f'(2) where

$$f(x)=x^4-2x^3+4x^2-3x+4$$

$$f'(x)=4x^3-6x^2+8x-3$$

$$f'(2)=4(2)^3-6(2)^2+8(2)-3$$
$$=32-24+16-3$$
$$=21$$

```
RUN
WHAT IS THE DEGREE OF THE
POLYNOMIAL? DO NOT EXCEED 50.

?4

ENTER THE COEFFICIENTS ONE AT
A TIME. START WITH THE HIGHEST
POWER OF X AND INCLUDE 0
FOR MISSING POWERS OF X.

?1
?-2
?4
?-3
?4

AT WHAT VALUE IS THE DERIVATIVE
TO BE EVALUATED?

?2

THE VALUE OF THE DERIVATIVE
AT 2 IS 21
```

It appears that our program is working, but it is cumbersome to use. For example, if we wanted to evaluate the derivative at several different points, each time we run the program we would have to reenter the coefficients. To avoid this repeated entry of coefficients several statements could be added to the end of the program which would allow the user to loop back to compute the derivative at another point.

```
510 PRINT
520 PRINT "EVALUATE THE DERIVATIVE AGAIN?(Y/N)"
530 INPUT F$
540 IF F$="Y" THEN 380
550 END
```

Adding these lines results in a much more flexible program.   Note
the use of the character string in line 530.

Revised Program:

```
100 REM ** POLYNOMIAL DERIVATIVE-2 **
110 DIM A(50),B(50)
120 REM ** ENTERING DEGREE **
130 PRINT "WHAT IS THE DEGREE OF THE"
140 PRINT "POLYNOMIAL? DO NOT EXCEED 50."
150 PRINT
160 INPUT N
170 REM ** TESTING FOR INPUT ERROR **
180 IF N>50 THEN 130
190 PRINT
200 REM ** ENTERING THE COEFFICIENTS **
210 PRINT "ENTER THE COEFFICIENTS ONE AT"
220 PRINT "A TIME.  START WITH THE HIGHEST"
230 PRINT "POWER OF X AND INCLUDE 0"
240 PRINT "FOR MISSING POWERS OF X."
250 PRINT
260 FOR I=N TO 0 STEP -1
270 INPUT A(I)
280 NEXT I
290 PRINT
300 REM ** FINDING COEFFICIENTS OF **
310 REM ** THE DERIVATIVE POLYNOMIAL **
320 LET M=N-1
330 FOR I=M TO 0 STEP -1
340 LET B(I)=A(I+1)*(I+1)
350 NEXT I
360 REM **ENTERING VALUE AT WHICH **
370 REM ** THE DERIVATIVE IS EVALUATED **
380 PRINT "AT WHAT VALUE IS THE DERIVATIVE"
390 PRINT "TO BE EVALUATED?"
400 PRINT
410 INPUT X
420 REM ** EVALUATING THE DERIVATIVE **
430 LET S=0
440 FOR I=0 TO M
450 LET S=S+B(I)*X^I
460 NEXT I
```

```
470 PRINT
480 REM ** OUTPUT **
490 PRINT "THE VALUE OF THE DERIVATIVE"
500 PRINT "AT ";X;" IS ";S
510 PRINT
520 PRINT "EVALUATE THE DERIVATIVE AGAIN?(Y/N)"
530 INPUT F$
540 IF F$="Y" THEN 380
550 END
```

Example 1: The position of a particle is given by the equation $s(t)=16t^3-4t^2+2t+8$. Determine the velocity of the particle at .25 sec, .5 sec, 1 sec, and 2 sec.

Since the velocity is the first derivative of the position function, we can use program POLYNOMIAL DERIVATIVE-2 to determine the velocity at the given times.

```
RUN
WHAT IS THE DEGREE OF THE
POLYNOMIAL? DO NOT EXCEED 50.

?3

ENTER THE COEFFICIENTS ONE AT
A TIME. START WITH THE HIGHEST
POWER OF X AND INCLUDE 0
FOR MISSING POWERS OF X.

?16
?-4
?2
?8

AT WHAT VALUE IS THE DERIVATIVE
TO BE EVALUATED?

?.25

THE VALUE OF THE DERIVATIVE
AT .25 IS 3
```

```
EVALUATE THE DERIVATIVE AGAIN?(Y/N)
?Y
AT WHAT VALUE IS THE DERIVATIVE
TO BE EVALUATED?

?.5

THE VALUE OF THE DERIVATIVE
AT .5 IS 10

EVALUATE THE DERIVATIVE AGAIN?(Y/N)
?Y
AT WHAT VALUE IS THE DERIVATIVE
TO BE EVALUATED?

?1

THE VALUE OF THE DERIVATIVE
AT 1 IS 42

EVALUATE THE DERIVATIVE AGAIN?(Y/N)
?Y
AT WHAT VALUE IS THE DERIVATIVE
TO BE EVALUATED?

?2

THE VALUE OF THE DERIVATIVE
AT 2 IS 178

EVALUATE THE DERIVATIVE AGAIN?(Y/N)
?N
```

Summarizing,	t	velocity
	0.25	3
	0.5	10
	1.0	42
	2.0	178

Example 2: Approximate all points on the curve

$$y = .25x^4 - .42x^3 + .02x^2 + .12x - .47$$

where there are horizontal tangent lines.

Since a horizontal line has a slope of 0, we need to approximate the values of x so that the derivative, which is the slope of the tangent line, has value 0. Although our program does not find the zeros of the derivative, we can still use it to help us.

```
RUN
WHAT IS THE DEGREE OF THE
POLYNOMIAL? DO NOT EXCEED 50.

?4

ENTER THE COEFFICIENTS ONE AT
A TIME. START WITH THE HIGHEST
POWER OF X AND INCLUDE 0
FOR MISSING POWERS OF X.

?.25
?-.42
?.02
?.12
?-.47

AT WHAT VALUE IS THE DERIVATIVE
TO BE EVALUATED?

?0

THE VALUE OF THE DERIVATIVE
AT 0 IS .12

EVALUATE THE DERIVATIVE AGAIN?(Y/N)
?Y
AT WHAT VALUE IS THE DERIVATIVE
TO BE EVALUATED?

?1

THE VALUE OF THE DERIVATIVE
AT 1 IS -.0999999996
```

From the Intermediate Value Theorem we know that the derivative must have a root between 0 and 1. By repeated application of this theorem we can arrive at a good approximation of the value of x where the horizontal tangent occurs.

EVALUATE THE DERIVATIVE AGAIN?(Y/N)
?Y
AT WHAT VALUE IS THE DERIVATIVE
TO BE EVALUATED?

?.5

THE VALUE OF THE DERIVATIVE
AT .5 IS -.0499999999

EVALUATE THE DERIVATIVE AGAIN?(Y/N)
?Y
AT WHAT VALUE IS THE DERIVATIVE
TO BE EVALUATED?

?.4

THE VALUE OF THE DERIVATIVE
AT .4 IS -1.60000002e-03

EVALUATE THE DERIVATIVE AGAIN?(Y/N)
?Y
AT WHAT VALUE IS THE DERIVATIVE
TO BE EVALUATED?

?.39675

THE VALUE OF THE DERIVATIVE
AT .39675 IS -1.46680619E-05

EVALUATE THE DERIVATIVE AGAIN?(Y/N)
?Y
AT WHAT VALUE IS THE DERIVATIVE
TO BE EVALUATED?

?.39671

THE VALUE OF THE DERIVATIVE
AT .39671 IS 4.83495353E-06

EVALUATE THE DERIVATIVE AGAIN?(Y/N)
?N

To four place accuracy there is a horizontal tangent line to the curve at x=.3967 We can now determine any other point where there is a horizontal tangent line by using synthetic division to factor and the quadratic formula.

$$f'(x) = x^3 - 1.26x^2 + .06x + .12$$

$$= (x - .3967)(x^2 - .8633x - .28247)$$

Applying the quadratic formula,

$$x = \frac{.8633 \pm \sqrt{(.8633)^2 - 4(-.28247)}}{2} = \frac{.8633 \pm (1.063)}{2}$$

Therefore, if x is in the set {.3967, .9632, -.0999} we will have a horizontal tangent to the curve.

Exercise Set 8.1

In exercises 1-3 use POLYNOMIAL DERIVATIVE-2 to find the value of
the derivative at the indicated point. Check your results by
actually finding the derivative and evaluating at the point.

1. $y=3x^5+2x^4-3x^3-4x^2+x-2$ at $x=1,-1,3$

2. $y=.75x^4-1.26x^2+.47x-.5$ at $x=.15,.25,-.3$

3. $y=1.47x^8-7.21x^4+3.21x^2-5.1$ at $x=0,1,-.37$

4.   The position of a particle at any instance of time is given
by $s(t)=.15t^4-.4t^3+.05t$. Use program POLYNOMIAL DERIVATIVE-2
to find the velocity of the particle for $t=.3,.6,.9,1.2,1.5,2.0$
Also find one time at which the object is at rest.

5.   Use program POLYNOMIAL DERIVATIVE-2 to approximate all points
on the curve $y=.25x^4-.33x^3-.02x^2+.04x+1.4$ where there are
horizontal tangent lines.

*
6. If we enter the degree and coefficients of the first
derivative into program POLYNOMIAL DERIVATIVE-2, what information
would the output contain? Use this idea to approximate the
acceleration of an object whose position is given by
$s(t)=.14t^5-t^4+.2t^3-.6t^2-.1$ for $t=.1,.2,-.1$

**
7. Modify program POLYNOMIAL DERIVATIVE-2 to output not only the
value of the derivative for different values of x but also the
function which represents the derivative. For example, if
$f(x)=3x^3+4x^2+2x-1$ the output should have a line similar to the
one below.

THE DERIVATIVE IS: 9x^2+8x+2

***

8. Modify   program   POLYNOMIAL   DERIVATIVE-2   to   evaluate   the derivative of a function of the form

$$f(x) = a_0 x^{k_0} + a_1 x^{k_1} + \ . \ . \ . \ + a_n x^{k_n}$$

where $k_0, k_1, . \ . \ ., k_n$ are any rational numbers,   at   x=a.   Perform this   modification   and test your program with $f(x) = 3x^{-2} + 4x^{.25}$ at x=2.   <u>Hint</u>: Use an additional array to store the exponents.

8.2 METHOD 2--REPLACING THE DIFFERENCE QUOTIENT

Since the derivative is a limit it is subject to all the results related to limits. Of particular interest is the idea replacing the difference quotient with another function which has the same limit. For polynomials, this idea results in a rather unexpected method for evaluating the derivative. Of equal importance is the fact that this method is subject to minimal round-off error.

By definition the derivative of f(x) at x=a is given by:

$$f'(a) \;=\; \lim_{h \to 0} \frac{f(a+h)-f(a)}{h}$$

However, letting h=x-a results in

$$f'(a) \;=\; \lim_{x \to a} \frac{f(x)-f(a)}{x-a}$$

since a+h=a+x-a=x and h->0 implies x->a. Now to replace the difference quotient with a new function we need to determine an expression equivalent to (f(x)-f(a))/(x-a) everywhere except possibly at a. Toward this end we use the Division Algorithm for Polynomials which states:

> If a polynomial f(x) is divided by a polynomial d(x), called the divisor, then there exists polynomials q(x), called the quotient, and r(x), called the remainder, such that f(x)=d(x)q(x)+r(x) where the degree of r(x) is less than the degree of d(x).

Applying the theorem with d(x)=(x-a) results in the expression

$$f(x) \;=\; (x-a)q(x)+r(x)$$

Also, the Remainder Theorem for Polynomials states:

> If a polynomial f(x) is divided by x-a, then the remainder is f(a).

153

Therefore, the polynomial $f(x)$ can be expressed as

$$f(x) = (x-a)q(x)+f(a)$$

where $q(x)$ is a polynomial. Solving for $q(x)$ yields

$$q(x) = \frac{f(x)-f(a)}{x-a}$$

But the right-hand side is nothing more than the difference quotient which we needed to replace! Returning to the definition of the derivative and replacing $(f(x)-f(a)/(x-a)$ we have

$$f'(a) = \lim_{x->a} \frac{f(x)-f(a)}{x-a} = \lim_{x->a} q(x)$$

However, since $q(x)$ is a polynomial,

$$f'(a) = \lim_{x->a} q(x) = q(a)$$

RESULT: For a polynomial $f(x)$, the value of $f'(a)$ is equal to the value of the quotient polynomial obtained when $f(x)$ is divided by $x-a$, evaluated at a.

Example 1: If

$$f(x)=4x^3 +6x^2 -2x+4$$

find $f'(3)$.

Using synthetic division to divide $f(x)$ by $x-3$,

$$\begin{array}{r|rrrr} 3 & 4 & 6 & -2 & 4 \\ & & 12 & 54 & 156 \\ \hline & 4 & 18 & 52 & \boxed{160} \end{array} = f(3) \text{ by the Remainder Theorem}$$

$$q(x)=4x^2 +18x+52$$

8.2 METHOD 2--REPLACING THE DIFFERENCE QUOTIENT

Thus, $f'(3)=q(3)=4(3)^2+18(3)+52=142$

To check, $f'(x)=12x^2+12x-2$ Note that $f'(x) \neq Q(x)$

Then, $f'(3)=12(3)+12(3)-2=108+36-2=142$

    An efficient method for finding q(a) also follows from the Remainder Theorem for Polynomials--the remainder when q(x) is divided by x-a is q(a). Again we can use synthetic division to determine the remainder.

```
3│    4    18    52
           12    90
      4    30  │142 = q(3)=f'(3)
```

As a practical matter we can combine the two applications of synthetic division.

```
3│    4     6    -2     4
           12    54   146
3│    4    18    52   160  = f(3)
           12    90
      4    30  │142  = q(3) = f'(3)
```

Programming Problem: Determine the value of the derivative of a polynomial function at x=a by the method described above.

Output: The value of the function and its derivative at the point.

Input: Our input will be similar to that in section 5.1 except that it will be more convenient to store the coefficients in reverse order, the coefficient of the highest power of x in A(0), the coefficient of the next highest in A(1) and so on.

Strategy: To write this program we must fully understand how the individual entries in the synthetic division are being determined. Below is a table for the general polynomial $f(x)=a_0 x^n + a_1 x^{n-1} + . . . + a_{n-1} x + a_n$ at the point x=p.

$$
\begin{array}{c|ccccccc}
p & a_0 & a_1 & a_2 & \cdots & a_{n-1} & & a_n \\
 & & pc_0=b_1 & pc_1=b_2 & \cdots\, pc_{n-2}=b_{n-1} & & & pc_{n-1}=b_n \\
\hline
p & a_0=c_0 & a_1+b_1=c_1 & a_2+b_2=c_2 & \cdots\, a_{n-1}+b_{n-1}=c_{n-1} & & & a_n+b_n=c_n \\
 & & pe_0=d_1 & pe_1=d_2 & pe_{n-2}=d_{n-1} & & & \\
\hline
 & a_0=e_0 & c_1+d_1=e_1 & c_2+d_2=e_2 & c_{n-1}+d_{n-1}=e_{n-1}=q(a) & = & f'(a) \\
\end{array}
$$

Since the original coefficients were stored in an array in the input part of the program in descending powers of x. This is the same order as the table above, and, thus, the array corresponds to the first line. If we call the first array A, we can use arrays B, C, D, and E to correspond to the remaining lines of the synthetic division table and complete the values for these arrays by the relations:

$$B(I)=P*C(I-1);\ I=1,2,\ldots,N$$
$$C(I)=A(I)+B(I);\ I=1,2,\ldots,N;C(0)=A(0)$$
$$D(I)=P*E(I-1);\ I=1,2,\ldots,N-1$$
$$E(I)=C(I)+D(I);\ I=1,2,\ldots,N-1,E(0)=A(0)$$

As in program POLYNOMIAL DERIVATIVE-2 we will include a problem to evaluate the derivative at several values of x.

8.2 METHOD 2--REPLACING THE DIFFERENCE QUOTIENT

Program:

```
100 REM ** POLYNOMIAL DERIVATIVE-3 **
110 DIM A(50),B(50),C(50),D(50),E(50)
120 REM ** ENTERING DEGREE **
130 PRINT "WHAT IS THE DEGREE OF THE"
140 PRINT "POLYNOMIAL? DO NOT EXCEED 50."
150 PRINT
160 INPUT N
170 REM ** TESTING FOR INPUT ERROR **
180 IF N>50 THEN 130
190 PRINT
200 REM ** ENTERING THE COEFFICIENTS **
210 PRINT "ENTER THE COEFFICIENTS ONE AT"
220 PRINT "A TIME.  START WITH THE HIGHEST"
230 PRINT "POWER OF X AND INCLUDE 0"
240 PRINT "FOR MISSING POWERS OF X."
250 PRINT
260 FOR I=0 TO N
270 INPUT A(I)
280 NEXT I
290 PRINT
300 REM **ENTERING VALUE AT WHICH **
310 REM ** THE DERIVATIVE IS EVALUATED **
320 PRINT "AT WHAT VALUE IS THE DERIVATIVE"
330 PRINT "TO BE EVALUATED?"
340 PRINT
350 INPUT X
360 REM ** COMPUTING THE ARRAYS **
370 LET C(0)=A(0)
380 LET E(0)=A(0)
390 FOR I=1 TO N
400 LET B(I)=P*C(I-1)
410 LET C(I)=B(I)+A(I)
420 NEXT I
430 FOR I=1 TO (N-1)
440 LET D(I)=P*E(I-1)
450 LET E(I)=C(I)+D(I)
460 NEXT I
470 REM ** PRINTING VALUES **
480 PRINT
490 PRINT "F(";P;")=";C(N)
500 PRINT "F'(";P;")=";E(N-1)
510 PRINT
520 PRINT "EVALUATE THE DERIVATIVE AGAIN?(Y/N)"
530 INPUT F$
540 IF F$="Y" THEN 290
550 END
```

Test Data: Find f(a) and f'(a) for

$$f(x)=5x^5+6x^4-4x^3-2x^2+2x-4, \ a=-1 \ \text{and} \ a=0$$

Now, $f'(x)=25x^4+24x^3-12x^2-4x+2$

Then, $f(-1)=-3$, $f'(-1)=-5$, $f(0)=-4$, $f'(0)=2$

```
RUN
WHAT IS THE DEGREE OF THE
POLYNOMIAL? DO NOT EXCEED 50.

?5

ENTER THE COEFFICIENTS ONE AT
A TIME. START WITH THE HIGHEST
POWER OF X AND INCLUDE 0
FOR MISSING POWERS OF X.

?5
?6
?-4
?-2
?2
?-4

AT WHAT VALUE IS THE DERIVATIVE
TO BE EVALUATED?

?-1

F(-1)=-3
F'(-1)=-5

EVALUATE THE DERIVATIVE AGAIN?(Y/N)
?Y

AT WHAT VALUE IS THE DERIVATIVE
TO BE EVALUATED?

?0

F(0)=-4
F'(0)=2

EVALUATE THE DERIVATIVE AGAIN?(Y/N)
?N
```

Example 2: Find the equation of the tangent and normal lines to

$$y = x^6 - .27x^5 + .21x^4 - 1.26x^3 - .67x^2 + .12x - .37 \text{ at } x = .26$$

We can use the program POLYNOMIAL DERIVATIVE-3 to find the slope of the tangent and the value of the function at $x=.26$.

```
RUN
WHAT IS THE DEGREE OF THE
POLYNOMIAL? DO NOT EXCEED 50.

?6

ENTER THE COEFFICIENTS ONE AT
A TIME. START WITH THE HIGHEST
POWER OF X AND INCLUDE 0
FOR MISSING POWERS OF X.

?1
?-.27
?.21
?-1.26
?-.67
?.12
?-.37

AT WHAT VALUE IS THE DERIVATIVE
TO BE EVALUATED?

?.26

F(.26)=-.405289992
F'(.26)=-.468204511

EVALUATE THE DERIVATIVE AGAIN?(Y/N)
?N
```

$$\begin{aligned}
\text{Tangent line:}\quad & y-y_1 = m(x-x_1) \\
& y-(-.4053) = (-.4682)(x-.26) \\
& y+.4053 = -.4682x+.1217 \\
& .4682x+y = -.2836
\end{aligned}$$

$$\begin{aligned}
\text{Normal line:}\quad & m = -1/(-.4682) = 2.1358 \\
\\
& y-(-.4053) = 2.1358(x-.26) \\
& y+.4053 = 2.1358x-.5553 \\
& 2.1358x-y = .9606
\end{aligned}$$

Ch. 8 POLYNOMIAL FUNCTIONS

Exercise Set 8.2

In exercises 1-3 use program POLYNOMIAL DERIVATIVE-3 to determine the value of the function and the first derivative at the indicated points.

1. $f(x)=.57x^3-2x^2+.47x-.53$; $x=-.12,.1,3$

2. $f(x)=x^4-.33x^3+1.67x^2-1.3x-1$; $x=-1.1,.17,2$

3. $f(x)=x^5+.25x^4+2.2x^3-4.2x^2+2.1x+4$; $x=-.3,.1,.6$

In exercises 4-6 find the equations of the tangent and normal lines at the given point using program POLYNOMIAL DERIVATIVE-3.

4. $y=x^3-.2x^2-1.3x-5.1$; $x=.15$

5. $y=x^4-.31x^3+2.1x^2+2.6x+1.5$; $x=2.6$

6. $y=x^5-.68x^4-.32x^3+.17x^2-.2.9x-4.3$; $x=-3.4$

*
7. The population of a town in January,1985, was 25,648. The predicted population t years later is given by

$$p(t)=25648(-.006t^4+.21t^3+.16t^2+.32t+1).$$

a) Find the initial growth rate.

b) Find the predicted growth rate and population in 1993.

c) Will the town ever achieve zero population growth? If so, when?

8.2 METHOD 2--REPLACING THE DIFFERENCE QUOTIENT

**

8. Consider the following.

$$f(x)=(x-a)q(x)+f(a)$$
$$f'(x)=q(x)+(x-a)q'(x)$$
$$f''(x)=q'(x)+q'(x)+(x-a)q''(x) \quad \text{Using Product Rule}$$
Thus, $f''(a)=2q'(a)$ since $a-a=0$

Thus, the second derivative of f at "a" can be obtained by evaluating the derivative of the quotient polynomial. Extend program POLYNOMIAL DERIVATIVE-3 to compute f"(a) by evaluating q'(a) using synthetic division. Test your program with $f(x)=3x^3+4x^2-5$ at $x=2$.

CHALLENGE ACTIVITY

A rational function is a function of the form $f(x)=p(x)/q(x)$ where $p(x)$ and $q(x)$ are polynomials. Write a computer program which will evaluate the derivative of a rational function by combining the theorem for the derivative of a quotient with one of the methods for finding the derivative of a polynomial function developed in this chapter. Then run your program for the following rational functions at the indicated value of x.

a. $f(x)=1/x$; $x=3$

b. $f(x)=(x-1)/(x+1)$; $x=2$

c. $f(x)=(x^2+5x-6)/(x^4-5x^2+6)$; $x=1$

Expand the program from above so that the derivative function of a rational function is determined. To simplify the output list the numerator and denominator separately. For example, if $f(x)=(x^2+2)/(x+1)$ your output should be:

NUMERATOR: X^2+2X-2
DENOMINATOR: (X+1)^2

CHAPTER 9
Applications Of The Derivative

The derivative is a useful tool for solving problems. Specifically, in this chapter we will investigate the application of the derivative to locating roots of functions and determining maximum and minimum values of a function on a closed interval.

9.1 ROOTS OF FUNCTIONS

A frequently occurring problem in mathematics is to determine a solution to equations of the form $f(x)=0$. However, only in a relatively few cases, such as easily factorable polynomials or quadratic equations, can these solutions be exactly determined. Usually the solutions or roots must be approximated to within a given tolerance or precision. One effective method for doing this was developed by Sir Issac Newton in the seventeenth century. Known as Newton's Method, successive approximations of the roots are obtained using the iterative formula

$$x_{n+1} = x - \frac{f(x_n)}{f'(x_n)} \qquad n=1,2,3,\ldots$$

given an initial guess for the root of the differentiable function $f(x)$.

The iteration formula is obtained from the idea that at a point $(x_0, f(x_0))$ on a curve the tangent line to the curve at that point is a good approximation to the curve in the vicinity of the point. Thus, we can approximate a root of the function $f(x)$ by the zero of the tangent line. Now at the point $(x_0, f(x_0))$ the slope of the tangent line is given by $f'(x_0)$ and the equation of the tangent line by

$$y - f(x_0) = f'(x_0)(x - x_0)$$

163

or

$$y=f'(x_0)(x-x_0)+f(x_0)$$

Assuming $f'(x_0) \neq 0$, we can easily solve for the zero of the tangent line, which we call x_1.

$$f'(x_0)(x_1-x_0)+f(x_0)=0$$

$$f'(x_0)(x_1-x_0)=-f(x_0)$$

$$x_1-x_0=-f(x_0)/f'(x_0)$$

$$x_1=x_0 - f(x_0)/f'(x_0)$$

Repeating or iterating the process results in Newton's Method.

Geometrically, Newton's Method is illustrated in the figure below.

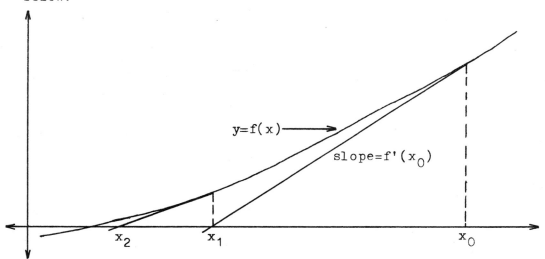

Programming Problem: Approximate a root of a function by Newton's Method.

Output: A table which lists the iteration number and successive approximations of the root followed by a message summarizing the results of the iterative process.

Input: The function f(x), its derivative, the desired precision to which we wish to calculate the root and the initial guess for the root.

Strategy: As the framework for our program we can use the following procedure.

1. Use the formula

$$x_{n+1} = x_n - \frac{f(x_n)}{f'(x_n)}$$

to determine the next estimate of the root if $f'(x_n) = 0$.

2. Check to see if successive approximations are within the desired precision. If so, output the results. Otherwise, go back to step 1.

3. Finally, since we want the process to be performed several times, we should enclose the statements in a FOR-NEXT loop. We will perform a maximum of 20 iterations. Often Newton's Method will converge, if it is going to converge, in fewer than 20 iterations.

Program:

```
100 REM ** NEWTON'S METHOD **
110 REM ** THE FUNCTION **
120 DEF FNF(X)=.....
130 REM ** THE DERIVATIVE **
140 DEF FNG(X)=.....
150 REM ** INPUT **
160 PRINT "ENTER THE INITIAL GUESS FOR THE "
170 PRINT "ROOT OF THE FUNCTION."
180 INPUT X0
190 PRINT
200 PRINT "ENTER THE DESIRED PRECISION."
210 INPUT P
220 PRINT
230 REM ** HEADINGS **
240 PRINT "ITERATION","APPROXIMATION"
250 REM ** ITERATION **
260 LET X1=X0
270 FOR I=1 TO 20
280 IF FNG(X1)=0 THEN 390
290 LET X2=X1-(FNF(X1)/FNG(X1))
300 PRINT " ";I,X2
310 REM ** TESTING ROOTS FOR PRECISION **
320 IF ABS(X2-X1)<P THEN 420
330 LET X1=X2
340 NEXT I
350 PRINT
360 PRINT "AFTER 20 ITERATIONS NO ROOT WAS"
370 PRINT "FOUND TO WITHIN A PRECISION OF";P;"!!!"
380 GOTO 470
390 PRINT "THE DERIVATIVE AT X=";X1;" EQUALS"
400 PRINT "ZERO.  THE PROCESS HAS BEEN TERMINATED!!!"
410 GOTO 470          can change to  go back to beginning  to reenter
420 PRINT
430 PRINT "AT X= ";X2;" A ROOT OF THE"
440 PRINT "FUNCTION TO WITHIN A PRECISION OF ";P
450 PRINT "HAS BEEN FOUND.  THE FUNCTIONAL VALUE"
460 PRINT "IS ";FNF(X2)
470 END
```

Test Data:

$f(x)=x^2+2x-3=(x-1)(x+3)$ has roots 1 and -3.

```
120 DEF FNF(X)=X*X+2*X-3

140 DEF FNG(X)=2*X+2
```

```
RUN
ENTER THE INITIAL GUESS FOR THE
ROOT OF THE FUNCTION.
?0

ENTER THE DESIRED PRECISION.
?.00001

ITERATION          APPROXIMATION
    1              1.5
    2              1.05
    3              1.00060976
    4              1.00000009
    5              1.00000004

AT X= 1.00000004 A ROOT OF THE
FUNCTION TO WITHIN A PRECISION OF 1E-05
HAS BEEN FOUND. THE FUNCTIONAL VALUE
IS 8.7544322E-08
```

Choosing a different initial guess leads to the second root.

```
RUN
ENTER THE INITIAL GUESS FOR THE
ROOT OF THE FUNCTION.
?-10

ENTER THE DESIRED PRECISION.
?.00001

ITERATION          APPROXIMATION
    1              -5.72222223
    2              -3.78464052
    3              -3.11054582
    4              -3.00289508
    5              -3.00000209
    6              -3

AT X= -3 A ROOT OF THE
FUNCTION TO WITHIN A PRECISION OF 1E-05
HAS BEEN FOUND. THE FUNCTIONAL VALUE
IS 0
```

.714286
-3.00001

Ch. 9   APPLICATIONS OF THE DERIVATIVE

Example 1: Find a root of

$$f(x)=x^5+x^4-\sqrt{3x+1}\ -1$$

using the program NEWTON'S METHOD.

```
120 DEF FNF(X)=X*X*X*X*X+X*X*X*X-SQR(3*X+1)-1

140 DEF FNG(X)=5*X*X*X*X+4*X*X*X-3/(2*SQR(3*X+1))

RUN
ENTER THE INITIAL GUESS FOR THE
ROOT OF THE FUNCTION.
?1.5

ENTER THE DESIRED PRECISION.
?.00001

ITERATION          APPROXIMATION
    1              1.25608239
    2              1.13252723
    3              1.10165801
    4              1.09992429
    5              1.09991897

AT X= 1.09991897 A ROOT OF THE
FUNCTION TO WITHIN A PRECISION OF 1E-05
HAS BEEN FOUND. THE FUNCTIONAL VALUE
IS 1.39698386E-09
```

The root is approximately 1.09991897

Example 2: Approximate the point of intersection of $y=\sin x+\cos x$ and $y=\tan x$ with $0 \le x \le \pi/2$.

If the two functions have a point of intersection, then $\sin x+\cos x=\tan x$, or equivalently, $\sin x+\cos x-\tan x=0$. Thus, the x-coordinate will be a root of $f(x)=\sin x+\cos x-\tan x$.

```
120 DEF FNF(X)=SIN(X)+COS(X)-TAN(X)

140 DEF FNG(X)=COS(X)-SIN(X)-1/(COS(X)*COS(X))
```

```
RUN
ENTER THE INITIAL GUESS FOR THE
ROOT OF THE FUNCTION.
?1

ENTER THE DESIRED PRECISION.
?.00001

ITERATION          APPROXIMATION
    1                 .952871167
    2                 .948991063
    3                 .948968021
    4                 .948968029

AT X= .94896802 A ROOT OF THE
FUNCTION TO WITHIN A PRECISION OF 1E-05
HAS BEEN FOUND. THE FUNCTIONAL VALUE
IS 0
```

To find the ordinate value of the point of intersection we substitute for x in each equation.

y=sin(.94896802)+cos(.94896802) y=tan(.94896802)
  =.81281477+.58252222   =1.3953369
  =1.3953369

Therefore, the point of intersection is approximately (.94896802,1.3953369).

Example 3: Run the program NEWTON'S METHOD to find roots of $f(x)$=sin x using 1.1655611, 1.1655614, and 1.1655617 as initial guesses of the root. Note: The output for this example may vary depending upon the computer being used.

```
120 DEF FNF(X)=SIN(X)

140 DEF FNG(X)=COS(X)

RUN
ENTER THE INITIAL GUESS FOR THE
ROOT OF THE FUNCTION.
?1.1655611

ENTER THE DESIRED PRECISION.
?.00001
```

ITERATION	APPROXIMATION
1	-1.16556073
2	1.16555869
3	-1.16554762
4	1.16548746
5	-1.16516061
6	1.1633868
7	-1.15381583
8	1.10374606
9	-.879355586
10	.328722395
11	-.0123755985
12	6.31827788E-07
13	1.47004631E-12

AT X= 1.47004631E-12 A ROOT OF THE
FUNCTION TO WITHIN A PRECISION OF 1E-05
HAS BEEN FOUND. THE FUNCTIONAL VALUE
IS 0

Running the program again for the second guess of the root:

```
RUN
ENTER THE INITIAL GUESS FOR THE
ROOT OF THE FUNCTION.
?1.1655614

ENTER THE DESIRED PRECISION.
?.00001
```

ITERATION	APPROXIMATION
1	-1.16556236
2	1.16556755
3	-1.16559576
4	1.16574907
5	-1.16658271
6	1.17112796
7	-1.19628307
8	1.34782849
9	-3.06255312
10	-3.14175766
11	-3.141592665
12	-3.14159265

AT X= -3.14159265 A ROOT OF THE
FUNCTION TO WITHIN A PRECISION OF 1E-05
HAS BEEN FOUND. THE FUNCTIONAL VALUE
IS -7.3145904E-10

Finally, running the program for the third guess for the root:

```
RUN
ENTER THE INITIAL GUESS FOR THE
ROOT OF THE FUNCTION.
?1.1655617

ENTER THE DESIRED PRECISION.
?.00001

ITERATION           APPROXIMATION
    1                -1.16556399
    2                 1.1655764
    3                -1.16564388
    4                 1.16601063
    5                -1.16800658
    6                 1.17894002
    7                -1.24104009
    8                 1.68077905
    9                10.736426
   10                 6.96440357
   11                 6.15372529
   12                 6.28391344
   13                 6.28318531
   14                 6.28318531

AT X= 6.28318531 A ROOT OF THE
FUNCTION TO WITHIN A PRECISION OF 1E-05
HAS BEEN FOUND. THE FUNCTIONAL VALUE
IS 0
```

The values of the initial estimation of the root differ by .0000006, yet three different roots were obtained: $0, -\pi$, and 2π!! For more on this example see exercise 14 in exercise set 9.1

Example 4: Using an initial guess of 2.1 run the program NEWTON'S METHOD for the function $f(x)=(x-1)/x^2$.

```
120 DEF FNF(X)=(X-1)/(X*X)

140 DEF FNG(X)=(-2*X*X+2*X)/(X*X*X*X)

RUN
```
**ENTER THE INITIAL GUESS FOR THE
ROOT OF THE FUNCTION.**
?2.1

ENTER THE DESIRED PRECISION.
?.00001

ITERATION	APPROXIMATION
1	3.15
2	4.725
3	7.0875
4	10.63125
5	15.946875
6	23.9203125
7	35.8804687
8	53.8207031
9	80.7310547
10	121.096582
11	181.644873
12	272.467309
13	408.700964
14	613.051446
15	919.577169
16	1379.36575
17	2069.04863
18	3103.57295
19	4655.35942
20	6983.03913

**AFTER 20 ITERATIONS NO ROOT WAS
FOUND TO WITHIN A PRECISION OF 1E-05!!!**

But $f(x)=(x-1)/x^2$ has a root at 1!

As illustrated by the test data and examples 1 and 2, Newton's Method appears to produce a sequence of values which converge rapidly to the root. However, as shown in the last two examples, Newton's Method can lead to unexpected results. Such factors as the initial guess for the root, concavity and points of inflection for the curve can lead to sequences of values which either do not converge to a root or produce unusual results as in example 3. Several of these situations are illustrated graphically below.

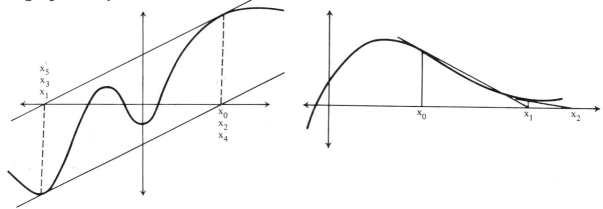

Oscillation due to a point of inflection at x=0 and the choice of the initial guess x_0.

Values heading away from the root due to the choice of the initial guess x_0 to the wrong side of the point $(x,f(x))$ where a local maximum occurs.

An additional difficulty with Newton's Method arises when the function has no root. The method will produce a sequence of values which does not converge, but we cannot detect whether the lack of convergence is due to the initial guess. the shape of the curve, or the fact that there is no root. Generating sequences of values which do not converge for several different initial guesses may lead us to believe that there is no root, but Newton's Method provides no way to be sure. A graphic illustration of Newton's Method in the case of no root is shown below.

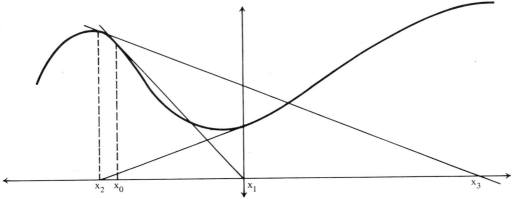

Newton's Method and a function with no root.

Because of difficulties associated with Newton's Method for approximating roots other methods for finding roots have been developed. One of simplest is the Bisection Method. The basic idea of this method is to determine an interval whose length is less than the desired precision for the root and contains the root. Then the midpoint of the interval is used as an approximation of the root. The desired interval is found by halving previous intervals, always choosing the half which contains the root. To decide which half of the interval to choose we apply the Intermediate Value Theorem. A root will lie in the half of the interval which contains the endpoint at which the functional value has a different sign from the sign of the functional value at the midpoint of the interval.

Example 5: The function

$$f(x)=x^3-2x^2+1$$

has a root in the interval [-1,0]. Determine in which half of the interval the root lies.

$$f(-1)=(-1)^3-2(-1)^2+1=-2 \qquad \text{and} \qquad f(0)=0^3-2(0)^2+1=1$$

Now, the midpoint of [-1,0] is -.5 and

$$\begin{aligned} f(-.5)&=(-.5)^3-2(-.5)^2+1 \\ &=-.125-2(.25)+1 \\ &=.375 \end{aligned}$$

Thus, the root lies in the half interval [-1,-.5] since the sign of f(-1) is different from f(-.5)

Programming Problem: Approximate a root of a function using the Bisection Method.

Output: A message giving the approximation of the root, its precision, and the functional value at the value of the approximation.

Input: The function f(x), the precision to which the root is to be found, and an initial interval containing the root.

Strategy: The Bisection Method can be subdivided into the following tasks.

1. Check to see if a root occurs at one of the endpoints.

2. For the current interval [L,R] compute the midpoint M=(L+R)/2.

3. Compare the sign of f(M) with f(L) and f(R). If f(M) and f(L) have different signs choose the new interval [L,M]. Otherwise, let the new interval be [M,R].

4. Compare the length of the new interval with the precision. If the length is less than the value of the desired precision, approximate the root by the midpoint. Otherwise, go back to the first task.

Program:

```
100 REM ** BISECTION METHOD **
110 REM ** FUNCTION **
120 DEF FNF(X)=.....
130 REM ** INPUT **
140 PRINT "ENTER THE LEFT-ENDPOINT OF THE"
150 PRINT "INTERVAL CONTAINING THE ROOT'
160 INPUT L
170 PRINT
180 PRINT "ENTER THE RIGHT-ENDPOINT OF THE"
190 PRINT "INTERVAL CONTAINING THE ROOT"
200 INPUT R
210 PRINT
220 PRINT "ENTER THE PRECISION FOR THE ROOT"
230 INPUT P"240 PRINT
250 REM ** TESTING THE ENDPOINTS **
260 IF FNF(L)<>0 THEN 290
270 LET M=L
280 GOTO 460
290 IF FNF(R)<>0 THEN 330
300 LET M=R
310 GOTO 460
320 REM ** FINDING MIDPOINT **
330 LET M=(L+R)/2
340 REM ** COMPARING SIGN OF F(M) AND F(L) **
350 IF SGN(FNF(M))=SGN(FNF(L)) THEN 390
360 REM ** NEW INTERVAL [L,M] **
370 LET R=M
380 GOTO 420
390 REM ** NEW INTERVAL [M,R] **
400 LET L=M
410 REM ** TESTING LENGTH AGAINST PRECISION **
420 IF ABS(R-L)>=P THEN 260
430 REM ** OUTPUT **
440 LET M=(L+R)/2
450 PRINT
460 PRINT "A ROOT OF F(X) OCCURS AT"
470 PRINT M;" TO WITHIN A PRECISION OF"
480 PRINT P;".  THE VALUE OF THE FUNCTION AT"
490 PRINT "THIS VALUE OF X IS ";FNF(M)
500 END
```

Test Data:

$f(x)=x^2+2x-3$ has a root at 1.

```
120 DEF FNF(X)=X*X*+2*X-3
```

```
RUN
```
ENTER THE LEFT-ENDPOINT OF THE
INTERVAL CONTAINING THE ROOT.
```
?-1
```

ENTER THE RIGHT-ENDPOINT OF THE
INTERVAL CONTAINING THE ROOT.
```
?3
```

ENTER THE PRECISION FOR THE ROOT.
```
?.00001
```

A ROOT OF F(X) OCCURS AT
1 TO WITHIN A PRECISION OF
1E-05. THE VALUE OF THE FUNCTION AT
THIS VALUE OF X IS 0

Example 6: Use the program BISECTION METHOD to approximate a root of the function in example 4.

```
120 DEF FNF(X)=(X-1)/(X*X)
RUN
```
ENTER THE LEFT-ENDPOINT OF THE
INTERVAL CONTAINING THE ROOT.
```
?-1
```

ENTER THE RIGHT-ENDPOINT OF THE
INTERVAL CONTAINING THE ROOT.
```
?5
```

ENTER THE PRECISION FOR THE ROOT.
```
?.00001
```

A ROOT OF F(X) OCCURS AT
1.00000095 TO WITHIN A PRECISION OF
1E-05. THE VALUE OF THE FUNCTION AT
THIS VALUE OF X IS 9.53672498E-07

Notice the improvement when using the Bisection Method in place of Newton's Method for this particular example.

Ch. 9   APPLICATIONS OF THE DERIVATIVE

Exercise Set 9.1

In exercises 1-3 use the program NEWTON'S METHOD to find a root of the following functions using the initial guess.

1. $f(x)=x^3-2x^2+5x-6$; initial guess=1

2. $f(x)=x^2+\sin x$; initial guess=-1

3. $f(x)=x^3+\sqrt{x^2+1}-\cos x$; initial guess=-3

4.   Find all the roots of $f(x)=2x^3-5x^2+4x-3$

In exercises 5 and 6 find all points of intersection of the two curves.

5. $y=x^3+7x^2-4x+5$ and $y=x$

6. $y=(x-2)^{1/2}+x$ and $y=(x-3)/(x^2+4)$

In exercises 7-10 use program BISECTION METHOD to approximate a root of the given function on the indicated interval.

7. $f(x)=x^3+2x^2-x+4$; $[-4,1]$

8. $f(x)=x^3-\sin x - 1$; $[0,3]$

9. $f(x)=\ln(x+1)-x+1$; $[0,3]$

10. $f(x)=2-xe^x$; $[0,3]$

*
11.   What is a disadvantage of the Bisection Method?

12.   Graph the function used in example 4 and graphically explain why using Newton's Method failed to locate the root at 1.
<u>Hint</u>: What happens to the graph when x=2?

13.   Try to find a root of $f(x)=2x^2+8x+8.1$ using the program NEWTON'S METHOD.   If you cannot find a root, explain what is wrong.

**
14.   In example 3 we found roots of $f(x)=\sin x$ using as initial estimates of the root 1.1655611, 1.1655614, and 1.1655617.

        a.   Show that these values are approximations to a root of $f(x)=2x-\tan x$.

        b.   Let x_0 be a root of $g(x)=2x-\tan x$. Using this value, x_0, as the initial guess for the root, write out the first six values of the sequence of approximations for a root of $f(x)=\sin x$ in terms of x_0.
        <u>Hint</u>: $2x_0=\tan x_0$ and $x_1=x_0-((\sin x_0)/(\cos(x_0)))$.

15.   In section 8.2 f'(a) and f(a) were found for a polynomial $f(x)$ by using synthetic division.   Using this method to find f'(a) and f(a), write a program to approximate the root of a polynomial by Newton's Method.

16.   In Newton's Method the successive approximations, x_n, are located by using the tangent line to the curve at $(x_n,f(x_n))$. If, instead, a secant line through two points, $(x_n,f(x_n))$ and $(x_{n-1},f(x_{n-1}))$, is used to locate the next approximation, the iterative formula

$$x_{n+1}=x_n-f(x_n)\left(\frac{x_n-x_{n-1}}{f(x_n)-f(x_{n-1})}\right)$$

is obtained. Write a program to approximate the root of a function using this formula.   Use as test data the function $f(x)=x^2-3x-4$ which has roots -1 and 4.   This method for approximating roots is called the Secant Method.   Is there ay advantage to the Secant Method when compared to Newton's Method and the Bisection Method?

Ch. 9   APPLICATIONS OF THE DERIVATIVE

17.   An equation in two variables $f(x,y)=0$ defines y implicitly as a function of x.   Write a program which will determine the value of y given a value of x for an implicit function $f(x,y)=0$.   Test your program with $f(x,y)=x^2 y+3xy^3-1=0$ for $x=2$.
<u>Hint</u>: DEF FNF(Y)=X*X*Y+3*X*Y*Y*Y-1 and use Newton's Method to compute a root of FNF(Y).

9.2 MAXIMUM AND MINIMUM VALUES

Besides finding roots, the derivative can also be useful in determining the location of maximum and minimum values of a function. We know that on a closed interval a continuous function must attain both a maximum and minimum value. In addition, we also know that these extreme values occur at the endpoints of the interval or when the derivative is undefined or equal to zero. We will use all of these facts in our program.

Programming Problem: Determine the extrema of a function on a closed interval.

Output: A message giving the information:

    AT X=____ THE MAXIMUM IS ____

    AT X=____ THE MINIMUM IS ____

Input: The function, its derivative, the x value of the endpoints of the closed interval, and values of x at which the derivative is undefined. We must input the values of x at which the derivative is undefined since the computer will terminate execution of the program if the evaluation of a function is undefined.

Strategy: The extrema of the function must occur at the endpoints of the interval or when the derivative is undefined or equal to zero. Thus, our program must accomplish the following tasks.

1. Input and store the x-values of the endpoints of the interval, and points at which the derivative is undefined in an array, C.

2. Search the interval for values of x at which the derivative is equal to zero and store in C.

3. Evaluate the function for values of x at the endpoints, where the derivative is undefined, and where the derivative is equal to zero and store the function values in an array V.

4. Select the maximum and minimum function values.

Probably the most difficult part of our program is task 2, searching the interval for values of x at which the derivative is equal to zero. Unfortunately, finding the exact value of the roots of the derivative is often impossible, and the best that can be achieved is an approximation. One method for approximating the roots is the Bisection Method which we introduced in the previous section. The premise of the Bisection Method is that if f(x) is continuous on some interval [L,R] with f(L) and f(R) having different signs, then a root of f(x) is in the interval [L,R]. By comparing the sign of f((L+R)/2) with that of f(L) and f(R) a smaller interval containing the root can be determined.   By continuing the process the length of the interval can be made as small as we like, and, hence, the location of the root can be determined to the desired precision. Since the actual number of values where the derivative is zero on the given interval is unknown, an accurate search for roots will be necessary.

We will begin our search by dividing the original interval into 100 parts. Although this number of divisions is arbitrarily chosen, we can be reasonably confident that all roots will be located. It is possible that we may miss roots; however, accuracy must be weighed against the amount of time spent on the search. In the case of a large interval we may want to use a number greater than 100.   In general, if we store the endpoints of the interval in C(1) and C(2), then the length of each division is D=(C(2)−C(1))/100 and the interval is subdivided as shown below.

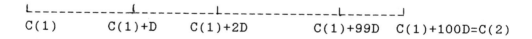

C(1)   C(1)+D   C(1)+2D   C(1)+99D   C(1)+100D=C(2)

Having subdivided our interval, we will check to see if there is a root in each subinterval. If so, we will use the Bisection Method to approximate the root. A flowchart for this portion, finding values at which the derivative is zero, of the program is given below.

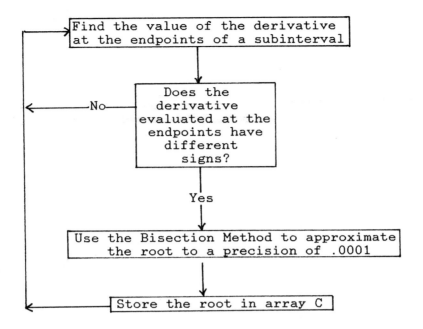

Program:

```
100 REM ** MAX-MIN **
110 REM ** THE FUNCTION **
120 DEF FNF(X)=.....
130 REM ** THE DERIVATIVE **
140 DEF FNG(X)=.....
150 DIM C(101),V(101)
160 REM ** INPUT **
170 PRINT "ENTER THE LEFT-HAND ENDPOINT"
180 PRINT "OF THE INTERVAL."
190 INPUT C(1)
200 PRINT
210 PRINT "ENTER THE RIGHT-HAND ENDPOINT"
220 "OF THE INTERVAL."
230 INPUT C(2)
240 PRINT
250 PRINT "ENTER HOW MANY TIMES THE DERIVATIVE"
260 PRINT "IS UNDEFINED ON THE INTERVAL."
270 INPUT N
```

```
280 IF N=0 THEN 350
290 LET T=N+2
300 PRINT "ENTER THE VALUE(S) AT WHICH THE DERIVATIVE"
310 PRINT "IS UNDEFINED ONE AT A TIME."
320 FOR I=3 TO T
330 INPUT C(I)
340 NEXT I
350 REM ** FINDING LENGTH OF A **
360 REM ** SUBDIVISION **
370 LET D=(C(2)-C(1))/100
380 REM ** SEARCHING THE INTERVAL **
390 REM ** FOR VALUES AT WHICH THE **
400 REM ** DERIVATIVE IS ZERO **
410 LET T=N+2
420 FOR I=1 TO 100
430 REM ** LEFT-HAND ENDPOINT **
440 LET L=C(1)+D*(I-1)
450 REM ** VALUE OF DERIVATIVE **
460 LET Y1=FNG(L)
470 REM ** RIGHT-HAND ENDPOINT **
480 LET R=C(1)+I*D
490 REM ** VALUE OF DERIVATIVE **
500 LET Y2=FNG(R)
510 REM ** TESTING FOR DIFFERENT SIGNS **
520 IF SGN(Y1)=SGN(Y2) THEN 690
530 REM ** BISECTION TO FIND VALUE **
540 IF FNG(L)<>0 THEN 570
550 LET T=T+1
560 LET C(T)=L
570 IF FNG(R)<>0 THEN 610
580 LET T=T+1
590 LET C(T)=R
600 GOTO 690
610 LET M=(L+R)/2
620 IF SGN(FNG(M))=SGN(FNG(L)) THEN 650
630 LET R=M
640 GOTO 660
650 LET L=M
660 IF ABS(R-L)>=.00001 THEN 540
670 LET T=T+1
680 LET C(T)=(L+R)/2
690 NEXT I
700 REM ** FINDING THE FUNCTIONAL VALUES **
710 FOR I=1 TO T
720 LET V(I)=FNF(C(I))
730 NEXT I
740 REM ** FINDING MAX AND MIN **
750 LET A=C(1)
760 LET B=C(1)
770 LET MAX=V(1)
```

```
780 LET MIN=V(1)
790 FOR I=2 TO T
800 IF V(I)<=MAX THEN 840
810 LET MAX=V(I)
820 LET A=C(I)
830 GOTO 870
840 IF V(I)>(MIN) THEN 870
850 LET B=C(I)
860 LET MIN=V(I)
870 NEXT I
880 REM ** OUTPUT **
890 PRINT
900 PRINT "AT X=";A;" THE MAXIMUM IS ";MAX
910 PRINT "AT X=";B;" THE MINIMUM IS ";MIN
920 END
```

Test Data: For the function

$$f(x)=x^3-3x^2-9x \text{ on } [-2,6]$$

Minimum: $f(3)=-27$ Maximum: $f(6)=54$

```
120 DEF FNF(X)=X*X*X-3*X*X-9*X

140 DEF FNG(X)=3*X*X-6*X-9

RUN
ENTER THE LEFT-HAND ENDPOINT
OF THE INTERVAL.
?-2

ENTER THE RIGHT-HAND ENDPOINT
OF THE INTERVAL.
?6

ENTER HOW MANY TIMES THE DERIVATIVE
IS UNDEFINED ON THE INTERVAL.
?0

AT X=6 THE MAXIMUM IS 54
AT X=3.00000488 THE MINIMUM IS -27
```

Ch. 9   APPLICATIONS OF THE DERIVATIVE

Example: At any time t the height of an object, relative to its rest position, attached to a spring is given by

$$h(t)=.47\sin(1.5t)+.29\cos(.8t)$$

What is the maximum and minimum height of the object?

$$h'(t)=.705\cos(1.5t)-.232\sin(.8t)$$

The period of the function is $2\pi/.8=2.5\pi \approx 7.86$

Thus, the maximum or minimum height will be obtained on the interval [0,7.86].

```
120 DEF FNF(X)=.47*SIN(1.5*X)+.29*COS(.8*X)

140 DEF FNG(X)=.705*COS(1.5*X)-.232*SIN(.8*X)

RUN
ENTER THE LEFT-HAND ENDPOINT
OF THE INTERVAL.
?0

ENTER THE RIGHT-HAND ENDPOINT
OF THE INTERVAL.
?7.86

ENTER HOW MANY TIMES THE DERIVATIVE
IS UNDEFINED ON THE INTERVAL.
?0

AT X=.901218275 THE MAXIMUM IS .676614503
AT X=3.25454565 THE MINIMUM IS -.712309866
```

Exercise Set 9.2

In exercises 1-4 determine the maximum and minimum value of the function $f(x)$ on the given interval by using program MAX-MIN.

1. $f(x)=3x^4-2.5x^3+1.7x^2+.9x-2$ on $[0,4]$

2. $f(x)=3.5\sin x +6.2\cos(1.4x)$ on $[0,6.3]$

3. $f(x)=\dfrac{x^3-2x+1}{x^4-3x^2+15}$ on $[-2,2]$

4. $f(x)=\dfrac{\cos x}{\sin x}$ on $[-\pi/2,\pi/2]$

5. The height of an object, relative to its rest position, attached to a spring is given at any time t by $h(t)=.53\sin(.25t)-1.2\cos(.6t)$. Determine the maximum and minimum height reached by the object.

**

6. An advertising flyer is to have a border of 1.25 inches at the top and bottom and 1 inch on each side. If 84 square inches is needed for the printed material, what are the dimensions of the page having the smallest area? Use program MAX-MIN.

7. The program MAX-MIN can be used to aid in the graphing of a function by dividing the domain into a series of closed intervals. Use this method to aid in the graphing of

$$y=\frac{3x^3-2x^2}{x+2}$$

CHALLENGE ACTIVITY

A rational function is a function of the form $f(x)=p(x)/q(x)$, where $p(x)$ and $q(x)$ are polynomials. Applying quotient and product rules for derivatives, we can obtain expressions for $f'(x)$ and $f''(x)$.

$$f'(x) = \frac{p'(x)q(x)-p(x)q'(x)}{[q\,(x)]^2}$$

$$f''(x)=\frac{p''(x)q^3(x)-p(x)q^2(x)q''(x)-2q'(x)p'(x)q^2(x)+2p(x)(q'(x))^2q(x)}{[q(x)]^4}$$

Write a program which, for a given value of x, namely "a", will output the following information for a rational function.

1.   The point $(a,f(a))$.

2.   A message "INCREASING","DECREASING", or "CRITICAL POINT", depending on whether $f'(a)$ is positive, negative, or zero.

3.   If "a" is a critical point, a message "MAX","MIN", or "NO MAX OR MIN", depending on the type extrema of the point.

4.   A message "CONCAVE UP","CONCAVE DOWN", or "POSSIBLE INFLECTION POINT", depending on whether $f''(a)$ is positive, negative, or zero.

Your program should define the rational function by entering its numerator and denominator separately. The values of $f'(a)$ and $f''(a)$ should be calculated using the formula above after calculating $p(a)$, $p'(a)$, $p''(a)$, $q(a)$, $q'(a)$, and $q''(a)$ by either method presented in Chapter 8.

CHAPTER 10
The Integral

Besides the derivative, the calculus has one other major concept--the integral. The definite integral, as we know, is defined in terms of a summation. Fortunately, a computer is perfectly suited to handle such sums and can be a useful tool in the investigation of the definite integral and its applications.

10.1 APPROXIMATING THE DEFINITE INTEGRAL

Programming Problem: Approximate the value of the definite integral

$$\int_a^b f(x) \, dx$$

Output: A table listing the approximation and the number of subintervals used to obtain the approximation followed by a message summarizing the results.

Input: The function which is being integrated and the limits of integration.

Strategy: If the definite integral of a function $f(x)$ exists, then it can be approximated by evaluating

$$\sum_{i=1}^n f(x_i) \triangle x$$

for specified values of x_i and N. This sum is easily calculated in BASIC by using a FOR-NEXT loop. The only decisions that need to be made concern the selection of the x_i's and N.

189

Since there is no way to pre-determine a value of N to guarantee the approximation to a given precision on a computer, finding an approximation for a single value of N would be of little value. We would never know how close the approximation may be to the actual value. Therefore, instead of selecting a single value for N, we will start with N=10 and perform the following steps:

1. Approximate the definite integral for the current value of N.

2. Compare the last two approximations and stop if they are within .0001 of each other. Note that this precision is arbitrarily chosen and should be changed if greater precision is desired.

3. If the last two approximations are not within .0001 of each other, double the value of N and start over.

As for the selection of the xi's, we will choose the midpoint of each subinterval. The midpoint can be found by taking the average of the left and right-hand endpoints of the subinterval.

If the length of each subinterval is DX, then

LE (left-hand endpoint) = A+(I-1)*DX

RE (right-hand endpoint)= A+I*DX

M (midpoint)= (LE+RE)/2

```
L----------1-----------1--------     -1----------1-------J
A          A+DX        A+2DX         A+(I-1)DX A+I*DX    A+N*DX=B
```

Program:

```
100 REM ** INTEGRAL **
110 REM ** FUNCTION **
120 DEF FNF(X)=.....
130 REM ** INPUT **
140 PRINT "ENTER THE LOWER LIMIT OF INTEGRATION."
150 INPUT A
160 PRINT "ENTER THE UPPER LIMIT OF INTEGRATION."
170 INPUT B
180 REM ** HEADINGS **
190 PRINT
200 PRINT " N";TAB(10);"APPROXIMATION"
210 REM ** INITIALIZATION **
220 LET N=10
230 LET S2=0
240 LET S1=0
250 REM ** FINDING DELTA X **
260 LET DX=(B-A)/N                    finding delta x
270 REM ** LOOP TO APP. INTEGRAL **
280 FOR I=1 TO N
290 LET LE=A+(I-1)*DX
300 LET RE=A+I*DX
310 LET M=(LE+RE)/2
320 LET S1=S1+FNF(M)*DX
330 NEXT I
340 REM ** OUTPUT TO TABLE **
350 PRINT N;TAB(10);S1
360 REM ** TESTING TO END PROGRAM **
370 IF ABS(S1-S2) < .0001 THEN 410
380 LET S2=S1
390 LET N=2*N
400 GO TO 240
410 PRINT
420 PRINT "THE APPROXIMATION OF THE INTEGRAL"
430 PRINT "FROM ";A;" TO ";B;" IS ";S1
440 END
```

Test Data:

$$\int_{0}^{3} x^2 \ dx = x^3/3 \ \Big|_{0}^{3} = \ 9$$

```
120 DEF FNF(X)=X*X
```

```
RUN
ENTER THE LOWER LIMIT OF INTEGRATION.
?0
ENTER THE UPPER LIMIT OF INTEGRATION.
?3

N        APPROXIMATION
10       8.97750001
20       8.994375
40       8.99859375
80       8.99964845
160      8.99991212
320      8.99997804

THE APPROXIMATION OF THE INTEGRAL
FROM 0 TO 3 IS 8.99997804
```

Example 1: Approximate

$$\int_0^2 \sqrt{2x^2+3} \ dx \text{ using program INTEGRAL.}$$

```
120 DEF FNF(X)=SQR(2*X*X+3)

RUN
ENTER THE LOWER LIMIT OF INTEGRATION.
?0
ENTER THE UPPER LIMIT OF INTEGRATION.
?2

N        APPROXIMATION
10       4.65777188
20       4.65927976
40       4.65965668
80       4.65975095

THE APPROXIMATION OF THE INTEGRAL
FROM 0 TO 2 IS 4.6597509
```

Using the Fundamental Theorem of Calculus to find the exact value
of the integral results in the value 4.6597823 to seven decimal
places.   Thus, our program is accurate.

Example 2: Approximate the area in the first quadrant bounded by
the curves $y=\sqrt{1+x^2}$, $x=0$ and $y=x^2$.

The point of intersection can be determined by substitution.

$$\sqrt{1+x^2} = x^2$$
$$1+x^2 = x^4$$
$$x^4 - x^2 - 1 = 0$$

Squaring both sides

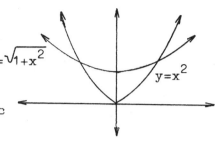

$y = \sqrt{1+x^2}$

$y = x^2$

Let $u = x^2$

$$u^2 - u - 1 = 0$$

Substituting

$$u \in \frac{1 \pm \sqrt{1+4}}{2} = \frac{1 \pm \sqrt{5}}{2}$$

Using Quadratic Formula

But $u = x^2$. Thus, $x^2 = \dfrac{1 \pm \sqrt{5}}{2}$

However, $1 - \sqrt{5}$ is negative, so $x^2 = \dfrac{1 + \sqrt{5}}{2}$

Therefore, $x = \pm \dfrac{\sqrt{1 + \sqrt{5}}}{\sqrt{2}} \approx \pm 1.27201965$

Note that the point of intersection could also be found by using one of the methods for root approximation discussed in section 9.1 Thus, the area is given by the integral

$$\text{Area} = \int_{0}^{1.27201965} (\sqrt{1+x^2} - x^2)\, dx$$

and can be approximated by the program INTEGRAL.

```
120 DEF FNF(X)=SQR(1+X*X)-X*X
```

```
RUN
```
ENTER THE LOWER LIMIT OF INTEGRATION.
```
?0
```
ENTER THE UPPER LIMIT OF INTEGRATION.
```
?1.27201965
```

N	APPROXIMATION
10	.874851061
20	.873962312
40	.873740106
80	.873684554

THE APPROXIMATION OF THE INTEGRAL FROM 0 TO 1.27201965 IS .873684554

Therefore, area ≈ .87368 square units.

Example 3: Use program INTEGRAL to approximate the distance in meters travelled by a particle during the first 1 1/2 seconds if its velocity at any time t is given by the function $v(t) = .87\sin^2(.4t) + .29\cos^2(.2t)$

The distance travelled is given by the integral

$$\int_0^{1.5} (.87\sin^2(.4t) + .29\cos^2(.2t))\,dt$$

```
120 DEF FNF(X)=.87*SIN(.4*X)*SIN(.4*X)
            +.29*COS(.2*X)*COS(.2*X)
```

RUN
ENTER THE LOWER LIMIT OF INTEGRATION.
?0
ENTER THE UPPER LIMIT OF INTEGRATION.
?1.5

N	APPROXIMATION
10	.567613144
20	.567818292
40	.567869557

THE APPROXIMATION OF THE INTEGRAL FROM 0 TO 1.5 IS .567869557

Therefore, distance ≈ .56787 m/sec.

Example 4: Of particular interest in both theoretical and applied statistics is the normal distribution since it occurs frequently in applications. The probability density function associated with this distribution is

$$f(x) = \frac{1}{s\sqrt{2\pi}} \exp(.5((x-m)/s)^2)$$

where m is the mean and s is the standard deviation of the distribution.

10.1 APPROXIMATING THE DEFINITE INTEGRAL

If a certain characteristic of a population has a normal distribution, then the probability that a member of that population, chosen at random, will have a value for that characteristic between two values a and b is given by the following formula:

$$\text{Probability } (a<x<b) = \int_a^b f(x)\ dx$$

Using this property of the normal distribution solve the following problem.

A radar unit is used to measure automobile speeds during rush hour. The speeds are normally distributed with a mean of 57 mph and a standard deviation of 4.67 Determine the probability that a car is travelling at a speed between 52 and 60 mph.

Since m=57 and s=4.67, the probability distribution function is

$$f(x) = \frac{1}{4.67\ \sqrt{2\pi}}\ \exp(.5((x-57)/4.67)^2)$$

and

$$\text{Probability } (52<x<60) = \int_{52}^{60} f(x)\ dx$$

The best approximation to π on a computer is found by using 4 tan^{-1}(1). Then approximating the integral using program INTEGRAL yields the following output.

```
120 DEF FNF(X)=EXP(-.5*((X-57)/4.67)^2)/(4.67*SQR(8*ATN(1)))

RUN
ENTER THE LOWER LIMIT OF INTEGRATION.
?52
ENTER THE UPPER LIMIT OF INTEGRATION.
?60
```

N	APPROXIMATION
10	.598082786
20	.59766979
40	.597566724
80	.597540968

**THE APPROXIMATION OF THE INTEGRAL
FROM 52 TO 60 IS .597540968**

The probability that a car is travelling at a speed between 52 mph
and 60 mph is approximately .5975

Example 5: Run program INTEGRAL several times to illustrate the
following theorem.   Use $f(x)=x^2 \sqrt{1+x}$, a=0, b=4 and c=3.

> **Theorem:** If f is continuous on any interval
> containing a, b, and c, then
>
> $$\int_a^b f(x)\ dx = \int_a^c f(x)\ dx + \int_c^b f(x)\ dx$$
>
> regardless of the order of a, b and c.

Running program INTEGRAL for limits of integration 0 to
4:

```
120 DEF FNF(X)=X*X*SQR(1+X)

RUN
ENTER THE LOWER LIMIT OF INTEGRATION.
?0
ENTER THE UPPER LIMIT OF INTEGRATION.
?4
```

```
N       APPROXIMATION
10      42.2962177
20      42.403609
40      42.4304456
80      42.4371545
160     42.4388311
320     42.4392503
640     42.4393551
1280    42.4393814
```

**THE APPROXIMATION OF THE INTEGRAL
FROM 0 TO 4 IS 42.4393814**

Running program INTEGRAL for limits of integration 0 to
3:

```
RUN
```
ENTER THE LOWER LIMIT OF INTEGRATION.
```
?0
```
ENTER THE UPPER LIMIT OF INTEGRATION.
```
?3

N       APPROXIMATION
10      16.0989244
20      16.1390204
40      16.1490411
80      16.15146
160     16.1521722
320     16.1523288
640     16.1523679
```

**THE APPROXIMATION OF THE INTEGRAL
FROM 0 TO 3 IS 16.1523679**

Running program INTEGRAL for limits of integration 3 to
4:

```
RUN
```
ENTER THE LOWER LIMIT OF INTEGRATION.
```
?3
```
ENTER THE UPPER LIMIT OF INTEGRATION.
```
?4
```

N	APPROXIMATION
10	26.2840023
20	26.2862574
40	26.2868212
80	26.2869621
160	26.2869974

THE APPROXIMATION OF THE INTEGRAL FROM 3 TO 4 IS 26.2869974

Now, from the above,

$$\int_a^b f(x)\ dx = \int_0^4 f(x)\ dx = 42.4393814$$

$$\int_a^c f(x)\ dx = \int_0^3 f(x)\ dx = 16.1523679$$

$$\int_c^b f(x)\ dx = \int_3^4 f(x)\ dx = 26.2869974$$

$$\int_a^c f(x)\ dx + \int_c^b f(x)\ dx = 16.1523679 + 26.2869974$$

$$= 42.4393653$$

Thus,

$$\int_a^b f(x)\ dx = \int_a^c f(x)\ dx + \int_c^b f(x)\ dx$$

taking into account the precision of the approximations.

Exercise Set 10.1

In exercises 1-4 approximate the definite integral using the program INTEGRAL.

1. $\int_{0}^{2} \sqrt{4-x^2}\ dx$

2. $\int_{1}^{3} dx/x$

3. $\int_{0}^{1} dx/\sqrt{1+x^2}$

4. $\int_{0}^{\pi/2} \ln(\tan x)\ dx$

<u>Hint</u>: $F(X)=LOG(TAN(X))$ and $\pi/2=1.570796327$.

In exercises 5-8 approximate the definite integral using the program INTEGRAL. Also, find the exact value of the integral and compare the results to the approximation.

5. $\int_{2}^{5} x^3-2x^2+1\ dx$

6. $\int_{0}^{2} (x-1)\sqrt{x^2-2x}\ dx$

7. $\int_{0}^{3} \cos 4x\ dx$

8. $\displaystyle\int_1^2 x \sin x \, dx$

Hint: $d(\sin x - x \cos x)/dx = x \sin x$.

*
In exercises 9-10 approximate the area bounded by the indicated curves.

9. $y = \sqrt{1+x}, \quad y = (x+1)^2$

10.   Above the line $x+y=4$ and inside the circle $x^2+y^2=16$.

In exercises 11-12 determine the distance travelled by an object for the stated time interval if its velocity is given at any instance of time by the function $v(t)$.

11. $v(t) = 3(t+2)^{3/2} + t; \quad 1 \le t \le 4$

12. $v(t) = .2\sin(.4t) + .3\cos(.3t); \quad 0 \le t \le 1$

13.   If grades in a calculus course are normally distributed with a mean of 81.6 and a standard deviation of 8.6, what is the probability that a student chosen at random from the class will have a grade between 80 and 90?

14.   To correct the orbit of a certain satellite its thrusters need to be fired for $2.43 \pm .07$ seconds. From past experience it is known that if a signal is sent to the satellite instructing it to fire its thrusters for x seconds, then the actual firing time will be normally distributed with mean x and standard deviation .04 What is the probability that the orbit will be corrected with the first signal transmitted? Hint: The probability distribution function has mean 2.43 and standard deviation .04 .

15.   Redo example 5 with first c=-1 and then c=7 to show that c does not need to be between a and b in the statement of the theorem.

**

16. Write a program to approximate the average value of f(x) on [a,b]. By definition

$$f_{avg} = 1/(b-a) \int_a^b f(x) \, dx$$

Test your program with f(x)=x³ on [0,5].

17. Find the total distance travelled by an object in its first 4 seconds if its velocity is given by v(t)=2 cos πt. Be careful!

***

18. In program INTEGRAL the value of x_i was chosen as the midpoint of the subinterval. However, the definition of the definite integral places no restrictions on x_i except that it must be some value in the ith subinterval.

   a. Modify program INTEGRAL so that the approximation will use the value of x_i which is
      (1) left-hand endpoint of the subinterval.
      (2) right-hand endpoint of the subinterval.
      (3) a random value on the subinterval. <u>Hint</u>: Consider the expression X=A+(I-1)*DX+RND*DX where RND is the random number generator for your computer.

      Note that you will have three different programs when finished.

   b. Test the three programs of part a by approximating

$$\int_0^2 \sqrt{2x^2 + 3} \, dx$$

   c. Compare the approximations of part b with the approximation obtained in example 1 and state a conclusion concerning the efficiency (number of subdivisions needed) of using the various methods for choosing x_i.

   d. Geometrically try to explain your conclusion formed in part c.

19.   The graph of $x^2/a^2 + y^2/b^2 = 1$ is an ellipse as shown in the figure below. Since the graph is symmetric with respect to both the x and y axes, the area enclosed by the curve is four times the area enclosed in the first quadrant (shaded region). Run program INTEGRAL several times for different values of a and b and make a conjecture about a formula for the area of an ellipse not involving an integral.

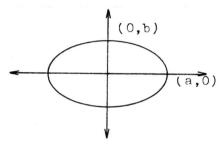

20.   Find an expression for

$$\int_0^1 (\ln 1/x)^n \, dx$$

which does not involve an integral. <u>Hint</u>: See exercise 19.

10.2 FURTHER APPLICATIONS OF THE DEFINITE INTEGRAL

In this section we will be concerned with modifying program INTEGRAL in order to approximate volumes, arc lengths, and surface areas.

Programming Problem: Approximate the volume of a figure of revolution using the disk (washer) method if the figure is revolved about the x-axis.

Output: The output will be similar to that of program INTEGRAL except we will use headings N and VOLUME.

Input: Two functions which give the outer and inner radius and the limits of integration.

Strategy: Since the figure is being revolved about the x-axis, the thickness of a typical disk (washer) is given by $\triangle x$. Thus, the volume can be approximated by the sum

$$\sum_{i=1}^{n} \pi((\text{outer radius})^2 - (\text{inner radius})^2)\triangle x_i$$

f(x) - outer radius

g(x) - inner radius

The program INTEGRAL can be easily modified to approximate this sum in place of

$$\sum_{i=1}^{n} f(x_i) \triangle x_i$$

Program:

```
100 REM ** VOLUME-DISK **
110 REM ** OUTER RADIUS **
120 DEF FNF(X)=.....
130 REM ** INNER RADIUS **
140 DEF FNG(X)=.....
150 REM ** INPUT **
160 PRINT "ENTER THE LOWER LIMIT OF INTEGRATION."
170 INPUT A
180 PRINT "ENTER THE UPPER LIMIT OF INTEGRATION."
190 INPUT B
200 REM ** HEADINGS **
210 PRINT
220 PRINT " N";TAB(10);"VOLUME"
230 REM ** INITIALIZATION **
240 LET P=4*ATN(1)                         approximating π
250 LET N=10
260 LET S2=0
270 LET S1=0
280 REM ** FINDING DELTA X **
290 LET DX=(B-A)/N
300 REM ** LOOP TO APP.   INTEGRAL **
310 FOR I=1 TO N
320 LET LE=A+(I-1)*DX
330 LET RE=A+I*DX
340 LET M=(LE+RE)/2
350 LET R1=FNF(M)*FNF(M)                   squaring outer radius
360 LET R2=FNG(M)*FNG(M)                   squaring inner radius
370 LET S1=S1+P*(R1-R2)*DX
380 NEXT I
390 REM ** OUTPUT TO TABLE **
400 PRINT N;TAB(10);S1
410 REM ** TESTING TO END PROGRAM **
420 IF ABS(S1-S2) < .0001 THEN 410
430 LET S2=S1
440 LET N=2*N
450 GOTO 240
460 PRINT
470 PRINT "THE APPROXIMATION OF THE VOLUME"
480 PRINT "FROM ";A;" TO ";B;" IS ";S1
490 END
```

Test Data: The plane region bounded by $y=x$, $y=0$, and $x=1$ revolved about the x-axis is a cone of radius one and height one. Using the formula for the volume of a cone, $1/3\pi r^2 h$, the volume for this figure of revolution will be $1/3 \ \pi \approx 1.047197551$

```
    outer radius:   y=x
    inner radius:   y=0
```

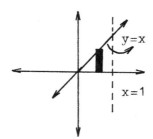

```
120 DEF FNF(X)=X

140 DEF FNG(X)=0

RUN
ENTER THE LOWER LIMIT OF INTEGRATION.
?0
ENTER THE UPPER LIMIT OF INTEGRATION.
?1
```

N	VOLUME
10	1.04457956
20	1.04654305
40	1.04703393
80	1.04715665
160	1.04718733

THE APPROXIMATION OF THE VOLUME
FROM 0 TO 1 IS 1.04718733

Thus, volume $\approx$ 1.0471 cubic units.

Example 1: Approximate the volume of the figure obtained by revolving the plane region bounded by $y=\cos x$, $y=\tan x$, and $x=0$ about the x-axis.

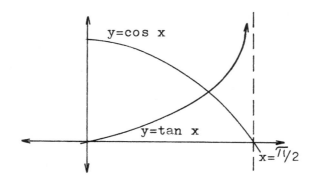

The point of intersection can be determined by substitution or by one of the methods introduced in section 9.1. Here we will use substitution.

$$\cos x = \tan x$$
$$\cos x = (\sin x)/(\cos x)$$
$$\cos^2 x = \sin x$$
$$1 - \sin^2 x = \sin x$$
$$\sin^2 x + \sin x - 1 = 0$$

Using the Quadratic Formula,

$$\sin x = (-1 \pm \sqrt{5})/2$$

Thus, $\sin x = (-1 + \sqrt{5})/2 \approx .618033988$

Then, $x \approx \sin^{-1}(.618033988)$

$$x \approx .666239432$$

Outer radius: y=cos x
Inner radius: y=tan x

Running program INTEGRAL to approximate the volume:

    120 DEF FNF(X)=COS(X)

    140 DEF FNG(X)=TAN(X)

    RUN
    ENTER THE LOWER LIMIT OF INTEGRATION.
    ?0
    ENTER THE UPPER LIMIT OF INTEGRATION.
    ?.666239432

N	VOLUME
10	1.43505219
20	1.43352259
40	1.43313973
80	1.43304398

THE APPROXIMATION OF THE VOLUME
FROM 0 TO .666239432 IS 1.43304398

The volume is approximately 1.4330 cubic units.

Programming Problem: Approximate the volume of a figure of revolution using the shell method if the figure is revolved about the y-axis.

Output: A table similar to that in program VOLUME-DISK and a message summarizing the results.

10.2 FURTHER APPLICATIONS OF THE DEFINITE INTEGRAL

Input: Two functions which will be used to determine the inner and outer length and the limits of integration.

Strategy: Since the figure is being revolved about the y-axis, the average radius is x_i, and the width is $\triangle x_i$. Thus, the volume can be approximated by the sum

$$\sum_{i=1}^{n} 2\pi \ (x_i \ )(\text{outer length} - \text{inner length}) \ \triangle x_i$$

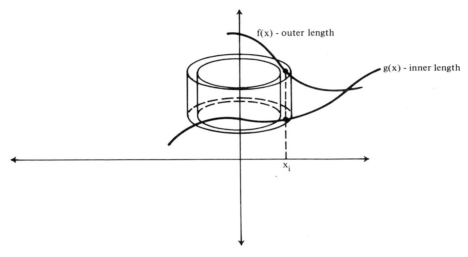

f(x) - outer length

g(x) - inner length

x_i

The program VOLUME-DISK can be modified to approximate the sum.

Program:

```
100 REM ** VOLUME-SHELL **
110 REM ** OUTER RADIUS **
120 DEF FNF(X)=.....
130 REM ** INNER RADIUS **
140 DEF FNG(X)=.....
150 REM ** INPUT **
160 PRINT "ENTER THE LOWER LIMIT OF INTEGRATION."
170 INPUT A
180 PRINT "ENTER THE UPPER LIMIT OF INTEGRATION."
190 INPUT B
200 REM ** HEADINGS **
210 PRINT
220 PRINT " N";TAB(10);"VOLUME"
230 REM ** INITIALIZATION **
240 LET P=4*ATN(1)
250 LET N=10
260 LET S2=0
270 LET S1=0
280 REM ** FINDING DELTA X **
290 LET DX=(B-A)/N
300 REM ** LOOP TO APP.  INTEGRAL **
310 FOR I=1 TO N
320 LET LE=A+(I-1)*DX
330 LET RE=A+I*DX
340 LET M=(LE+RE)/2
350 LET L=FNF(M)-FNG(M)
360 LET S1=S1+2*P*M*L*DX
370 NEXT I
380 REM ** OUTPUT TO TABLE **
390 PRINT N;TAB(10);S1
400 REM ** TESTING TO END PROGRAM **
410 IF ABS(S1-S2) < .0001 THEN 450
420 LET S2=S1
430 LET N=2*N
440 GOTO 270
450 PRINT
460 PRINT "THE APPROXIMATION OF THE VOLUME"
470 PRINT "FROM ";A;" TO ";B;" IS ";S1
480 END
```

Test Data: If the plane region bounded by $y=x$, $y=1$, and $x=0$ is revolved about the y-axis, the resulting figure will be a cone of radius one and height one. The volume of this cone is approximately 1.047197551

```
        outer length:  y=1
        inner length:  y=x
```

```
120 DEF FNF(X)=1

140 DEF FNG(X)=X

RUN
```
ENTER THE LOWER LIMIT OF INTEGRATION.
```
?0
```
ENTER THE UPPER LIMIT OF INTEGRATION.
```
?1
```

N	VOLUME
10	1.05243354
20	1.04850655
40	1.0475248
80	1.04727936
160	1.04721801

THE APPROXIMATION OF THE VOLUME
FROM 0 TO 1 IS 1.04721801

Thus, volume$\approx$1.0472 cubic units.

Example 2: Approximate the volume of the figure obtained by revolving the plane region bounded by $y=2^x$, $y=2x^2$ and $x=0$ about the y-axis.

```
        outer length:  y=2ˣ
        inner length:  y=2x²
```

By observation the point of intersection of the two curves is (1,2).

```
120 DEF FNF(X)=2^X

140 DEF FNG(X)=2*X*X
```

```
RUN
```
ENTER THE LOWER LIMIT OF INTEGRATION.
```
?0
```
ENTER THE UPPER LIMIT OF INTEGRATION.
```
?1
```

N	VOLUME
10	1.91968318
20	1.91258618
40	1.91081221
80	1.91036874
160	1.91025787
320	1.91023015

**THE APPROXIMATION OF THE VOLUME
FROM 0 TO 1 IS 1.91023015**

Thus, volume $\approx$ 1.9102 cubic units.

Another application of the integral is finding the length of a curve or arc length. The program INTEGRAL can easily be modified to find arc length.

Programming Problem: Approximate the length of a curve.

Output: A table similar to those used earlier in this section, except the headings will be "N" and "ARC LENGTH" and a message summarizing the results.

Input: A function which describes the curve and the limits over which the arc length is to be approximated.

Strategy: The arc length of a curve is given by the integral

$$\int_a^b \sqrt{1+(dy/dx)^2}\; dx$$

which gives the Riemann sum

$$\sum_{i=1}^{n} \sqrt{1 + (\triangle y/\triangle x)^2}\; \triangle x$$

10.2 FURTHER APPLICATIONS OF THE DEFINITE INTEGRAL

Bringing the $\triangle$x inside the square root yields

$$\sum_{i=1}^{n} \sqrt{(\triangle x)^2 + (\triangle y)^2}$$

We will modify program INTEGRAL to approximate this sum.

Program:

```
100 REM ** ARC LENGTH **
110 REM ** FUNCTION **
120 DEF FNF(X)=.....
130 REM ** INPUT **
140 PRINT "ENTER THE LOWER LIMIT OF INTEGRATION."
150 INPUT A
160 PRINT "ENTER THE UPPER LIMIT OF INTEGRATION."
170 INPUT B
180 REM ** HEADINGS **
190 PRINT
200 PRINT " N";TAB(10);"ARC LENGTH"
210 REM ** INITIALIZATION **
220 LET N=10
230 LET S2=0
240 LET S1=0
250 REM ** FINDING DELTA X **
260 LET DX=(B-A)/N
270 REM ** LOOP TO APP.  INTEGRAL **
280 FOR I=1 TO N
290 LET LE=A+(I-1)*DX
300 LET RE=A+I*DX
310 LET DY=FNF(RE)-FNF(LE)
320 LET S1=S1+SQR(DX*DX+DY*DY)
330 NEXT I
340 REM ** OUTPUT TO TABLE **
350 PRINT N;TAB(10);S1
360 REM ** TESTING TO END PROGRAM **
370 IF ABS(S1-S2) < .0001 THEN 410
380 LET S2=S1
390 LET N=2*N
400 GOTO 240
410 PRINT
420 PRINT "THE APPROXIMATION OF THE ARC LENGTH"
430 PRINT "FROM ";A;" TO ";B;" IS ";S1
440 END
```

Test Data: If $y=-x+1$, then the length of the curve from x=0 to x=1 using the distance formula is $\sqrt{2} \approx 1.414213562$.

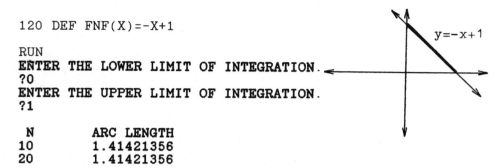

```
120 DEF FNF(X)=-X+1

RUN
ENTER THE LOWER LIMIT OF INTEGRATION.
?0
ENTER THE UPPER LIMIT OF INTEGRATION.
?1

   N        ARC LENGTH
  10        1.41421356
  20        1.41421356

THE APPROXIMATION OF THE ARC LENGTH
FROM 0 TO 1 IS 1.41421356
```

Example 3: Approximate the length of the hypocycloid

$$y^{2/3} + x^{2/3} = 1$$

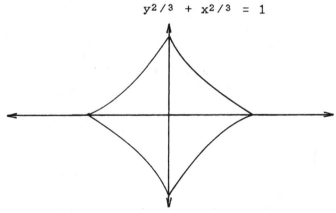

As we can see from the graph, the length of the curve will be four times the length of the curve in Quadrant I. Using the program ARC LENGTH with $y=(1-x^{2/3})^{3/2}$, we can approximate the length of the curve in Quadrant I.

```
120 DEF FNF(X)=(1-(X*X)^(2/3))^1.5

RUN
ENTER THE LOWER LIMIT OF INTEGRATION.
?0
ENTER THE UPPER LIMIT OF INTEGRATION.
?1

    N       ARC LENGTH
    10      1.49717413
    20      1.49894348
    40      1.49959976
    80      1.49984664
    160     1.49994069

THE APPROXIMATION OF THE VOLUME
FROM 0 TO 1 IS 1.49994069
```

Therefore, the length of the hypocycloid is approximately 4(1.49994069)=5.99976276

Finally, the program INTEGRAL can be modified to find surface area of a surface of revolution.

Programming Problem: Approximate the area of the surface obtained by revolving a plane curve about the x-axis.

Output: A table giving the number of subdivisions and the approximation of the surface area and a message summarizing the results.

Input: A function which describes the curve and the limits of integration.

Strategy: Since we are approximating the surface area for a curve revolved about the x-axis, the average radius will be $f(x_i)$ and the surface area will be approximated by the Riemann Sum

$$\sum_{i=1}^{n} 2\pi f(x_i) \sqrt{(\triangle x_i)^2+(\triangle y_i)^2}$$

$f(x_i)$

213

Program:

```
100 REM ** SURFACE AREA **
110 REM ** FUNCTION **
120 DEF FNF(X)=.....
130 REM ** INPUT **
140 PRINT "ENTER THE LOWER LIMIT OF INTEGRATION."
150 INPUT A
160 PRINT "ENTER THE UPPER LIMIT OF INTEGRATION."
170 INPUT B
180 REM ** HEADINGS **
190 PRINT
200 PRINT " N";TAB(10);"SURFACE AREA"
210 REM ** INITIALIZATION **
220 LET N=10
230 LET P=4*ATN(1)
240 LET S2=0
250 LET S1=0
260 REM ** FINDING DELTA X **
270 LET DX=(B-A)/N
280 REM ** LOOP TO APP.  INTEGRAL **
290 FOR I=1 TO N
300 LET LE=A+(I-1)*DX
310 LET RE=A+I*DX
320 LET M=(LE+RE)/2
330 LET R=FNF(M)
340 LET DY=FNF(RE)-FNF(LE)
350 LET L=SQR(DX*DX+DY*DY)
360 LET S1=S1+2*P*R*L
370 NEXT I
380 REM ** OUTPUT TO TABLE **
390 PRINT N;TAB(10);S1
400 REM ** TESTING TO END PROGRAM **
410 IF ABS(S1-S2) < .0001 THEN 410
420 LET S2=S1
430 LET N=2*N
440 GOTO 250
450 PRINT
460 PRINT "THE APPROXIMATION OF THE SURFACE AREA"
470 PRINT "FROM ";A;" TO ";B;" IS ";S1
480 END
```

Test Data: The surface area of a cone is given by $\pi r l$ where l is the slant height. If the curve $y=x$ for $0<x<1$, see figure below, is revolved about the x-axis, then a cone of radius 1 and slant height $\sqrt{2}$ will be generated. The surface area will then be $\pi(1)(2) \approx 4.442882938$

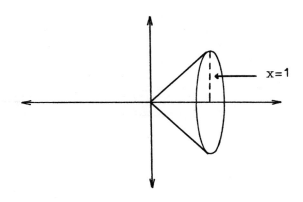

```
120 DEF FNF(X)=X

RUN
ENTER THE LOWER LIMIT OF INTEGRATION.
?0
ENTER THE UPPER LIMIT OF INTEGRATION.
?1

    N       SURFACE AREA
    10      4.44288294
    20      4.44288294

THE APPROXIMATION OF THE SURFACE AREA
FROM 0 TO 1 IS 4.44288294
```

Example 4: Approximate the surface area of the figure formed by revolving the hypocycloid in Example 3 about the x-axis.

The surface area will be twice the surface area of rotating that portion of the curve in Quadrant I about the x-axis. Also, since our function is complicated, the precision used in line 410 of the program will be changed to 0.0002 to reduce the time needed to run the program. <u>NOTE: Although computer time will vary by machine, this program may take over an hour to run.</u>

```
120 DEF FNF(X)=((1-(X*X)^(2/3)))^1.5
```

RUN
ENTER THE LOWER LIMIT OF INTEGRATION.
?0
ENTER THE UPPER LIMIT OF INTEGRATION.
?1

N	SURFACE AREA
10	3.66130108
20	3.7256247
40	3.75202847
80	3.76273442
160	3.76704227
320	3.76876724
640	3.76945579
1280	3.7697301
2560	3.76983925
5120	3.76988263

THE APPROXIMATION OF THE SURFACE AREA
FROM 0 TO 1 IS 3.76988263

Thus, the surface area of the hypocycloid is approximately
2(3.76988263)=7.53976526

10.2 FURTHER APPLICATIONS OF THE DEFINITE INTEGRAL

Exercise Set 10.2

In exercises 1-6 approximate the volume of the three-dimensional figure obtained by revolving the given plane region about the indicated axis by using program VOLUME-DISK or VOLUME-SHELL.

1. The region bounded by $y=x+\sqrt{x}$, $y=0$, and $x=3$ about the x-axis.

2. The region bounded by $y=(x-1)/(x+1)$, $x=0$, and $y=0$ about the x-axis.

3. The region bounded by $y=\tan x$, $x=0$, and $y=5$ about the y-axis.

4. The region bounded by $x=y^2-1$, $x=5$, and $y=0$ about the x-axis.

5. The region bounded by $y=\cos x$, $y=\sec x$, and $0<x<\pi/4$ about the y-axis.

6. The region bounded by $y=4\cos^2 x$, $y=\sin x$, and $y=0$ about the x-axis.

In exercises 7-8 approximate the length of the curve on the specified interval using program ARC LENGTH.

7. $y=\tan x$, $0 \leq x \leq \pi/3$

8. $y=x^{1/3}/(1-x^2)$, $-1/2 \leq x \leq 1/2$

In exercises 9-10 approximate the area of the surface obtained by revolving the given curve on the specified interval about the x-axis by using program SURFACE AREA.

9. $y=\tan x$, $0 \leq x \leq \pi/3$

10. $y=x^{1/3}+x^{1/2}$, $0 \leq x \leq 1$

*
11.   Approximate the volume of the torus obtained by revolving the circle $x^2+(y-2)^2=1$ about the x-axis by using program VOLUME-DISK.

12.   Approximate the surface area of the torus in exercise 11 using program SURFACE AREA.

13.   Use program ARC LENGTH to approximate the length of the ellipse $x^2/81 + y^2/144 = 1$.

14.   Approximate the volume of the figure obtained by revolving the ellipse of exercise 13 about the y-axis by using program VOLUME-SHELL.

15.   Approximate the surface area of the figure in exercise 14 using program SURFACE AREA.

**
16.   Write a program which will approximate the volume of the figure obtained when a plane region is revolved about the line x=a. As test data revolve the region bounded by $y=\sqrt{x}$, the x-axis, and x=1 about the line x=3.   Volume ≈ 10.05309649

17.   In program ARC LENGTH the approximation was based upon using the Riemann Sum

$$\sum_{i=1}^{n} \sqrt{(\triangle x)^2+(\triangle y)^2}$$

Modify this program to approximate the length of the curve using the Riemann Sum

$$\sum_{i=1}^{n} \sqrt{1+(\triangle y/\triangle x)^2}\ \triangle x$$

Compute $\triangle y/\triangle x$ instead of bringing $\triangle x$ inside the radical. Run your program for the curves which were used as test data for the program ARC LENGTH and example 3. Compare the results with those obtained using program ARC LENGTH. Decide which program (if either) is more accurate. The exact answer to example 3 is 6

10.2 FURTHER APPLICATIONS OF THE DEFINITE INTEGRAL

CHALLENGE ACTIVITY

Write a computer program which will approximate the moment of area (density equals 1) about the x and y-axis and the centroid of a plane region. Your input will include the curves which bound the region above and below and the limits of integration. Your output should be similar to that below.

```
MOMENT(X=0);_____
MOMENT(Y=0);_____
AREA;_____
X-BAR:_____  Y-BAR:_____
```

Test your program using the region bounded by $y=x^2$, the x-axis, and $x=2$.

Answers: M(x=0):4, M(y=0):16/5, Area:8/3, x:3/2, y:6/5.

Then use your program to approximate the moment of area about the x and y axes and the centroid of the region bounded by $y=x^{1/3}+x^{1/2}$ and $y=2x^2$.

CHAPTER 11
Numerical Integration

The method that we have been using in our programs to approximate definite integrals is known as the midpoint rule. However, there are several alternative methods for approximating definite integrals. In this chapter we will investigate several of these methods as well as the error incurred by such approximations.

11.1 TRAPEZOIDAL AND SIMPSON'S METHODS

In this section we will discuss two of the more commonly used methods to approximate definite integrals--the Trapezoidal Rule and Simpson's Rule.

Trapezoidal Rule

The definite integral

$$\int_a^b f(x) \, dx$$

can be approximated by the Reimann Sum

$$\sum_{i=1}^n f(x_i) \, \triangle x_i$$

Geometrically this sum can be interpreted as estimating the area under the curve $f(x)$ by the sum of the areas of rectangles.

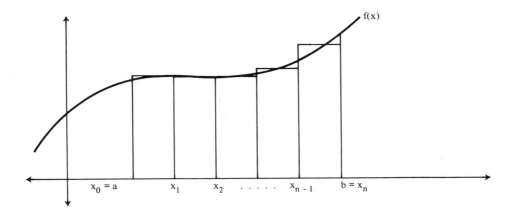

However, figures other than rectangles can also be used to estimate the area under the curve. As its name implies, the Trapezoidal Rule uses the sum of the areas of trapezoids to approximate the area.

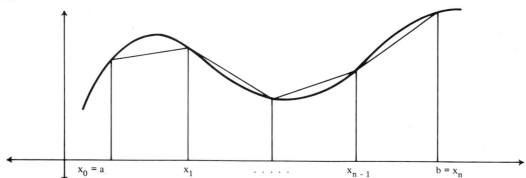

To develop an expression for approximating the definite integral by a sum of the area of trapezoids we will assume that $f(x)$ is continuous and positive on $[a,b]$ and that

$$\int_a^b f(x) \, dx$$

represents the area bounded by the curves $y=f(x)$, $y=0$, $x=a$ and $x=b$.

We begin by subdividing the interval [a,b] into n subdivisions of equal length $\triangle x = (b-a)/n$ with consecutive endpoints

$$a = x_0, x_1, x_2, \ldots, x_{n-1}, x_n = b$$

Now the area of a trapezoid with bases b_1 and b_2 and height h is given by

$$A = (1/2)h(b_1 + b_2)$$

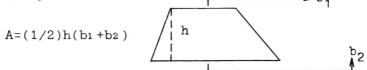

Applying this formula to a typical trapezoid in an interval $[x_{i-1}, x_i]$ we have

Area of ith trapezoid $= (1/2)\triangle x(f(x_{i-1}) + f(x_i))$

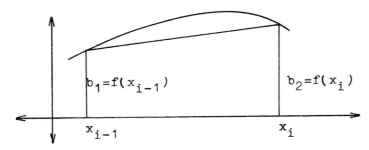

Thus, repeating the process for all the trapezoids we obtain

$$\int_a^b f(x)\,dx \approx (\triangle x/2)[f(x_0) + f(x_1)] + (\triangle x/2)[f(x_1 + f(x_2)]$$

$$+ (\triangle x/2)[(f(x_2) + f(x_3)] + \ldots$$

$$+ (\triangle x/2)[f(x_{n-2}) + f(x_{n-1})] + (\triangle x/2)[f(x_{n-1}) + f(x_n)]$$

Simplifying, we obtain an approximation of the integral.

$$\int_a^b f(x)\,dx \approx (\triangle x/2)[f(x_0) + 2f(x_1) + 2f(x_2 + \ldots + 2f(x_{n-1}) + f(x_n)]$$

Example 1: Approximate

$$\int_0^3 \sqrt{9-x^2}\ dx$$

using the Trapezoidal Rule with n=6.

With n=6, $\triangle x=(3-0)/6 = .5$, and the interval is subdivided as follows.

$$\begin{array}{ccccccc} \vdash & \text{-----} & \vdash & \text{-----} & \vdash & \text{-----} & \dashv \\ 0 & .5 & 1 & 1.5 & 2.0 & 2.5 & 3 \end{array}$$

Thus, using the formula developed above for the Trapezoidal Rule

$$\int_0^3 \sqrt{9-x^2}\ dx \approx (.5/2)[\sqrt{9-0^2} + 2\sqrt{9-(.5)^2} + 2\sqrt{9-1^2}$$

$$+ 2\sqrt{9-(1.5)^2} + 2\sqrt{9-2^2} + 2\sqrt{9-(2.5)^2}$$

$$+ \sqrt{9-3^2}\ ]$$

$$\approx .25[\sqrt{9} + 2\sqrt{8.75} + 2\sqrt{8} + 2\sqrt{6.75}$$

$$+ 2\sqrt{5} + 2\sqrt{2.75} + \sqrt{0}\ ]$$

$$\approx .25(3+5.9160798+5.6568542+5.1961524$$

$$+4.472136+3.3166248+0)$$

$$\approx .25(27.557847)$$

$$\approx 6.8894618$$

Programming Problem: Approximate the value of a definite integral using the Trapezoidal Rule.

Output: The output for this program will be a table giving the number of subdivisions and the corresponding approximation followed by a summary of the results.

Ch. 11   NUMERICAL INTEGRATION

Input: The function being integrated and the limits of integration.

Strategy: As shown above, the Trapezoidal Rule for approximating the definite integral is

$$\int_a^b f(x)\ dx \approx [f(x_0)+2f(x_1)+2f(x_2)+\ldots+2f(x_{n-1})+f(x_n)](\triangle x/2)$$

where $\triangle x=(b-a)/n$ and $a=x_0,x_1,\ldots,x_n=b$ are endpoints of subintervals. However, for the purposes of writing a computer program it is more useful to rewrite the formula in terms of a summation, obtaining the following equivalent expression.

$$\int_a^b f(x)\ dx \approx [f(x_0)+f(x_n)+\sum_{i=1}^{n-1} 2f(x_i)](\triangle x/2)$$

Now, the summation

$$\sum_{i=1}^{n-1} 2f(x_i)$$

can be computed using a FOR-NEXT loop. Adding $f(x_0)$ and $f(x_n)$ can be accomplished by initializing the variable used to accumulate the sum to $f(a)+f(b)$ instead of zero. As for the x_i's, since they are the endpoints of the subintervals, they can be generated using the formula

$$A+I*DX$$

where DX is the width of each subinterval.

```
L_____l_____l_/      /_____l_____J
A=X0          X1=A+DX        X2=A+2DX        XN-1=A+(N-1)DX   XN=B
```

Finally, as discussed in section 10.1, we will not compute a single approximation but a sequence of approximations starting with N=10 and doubling N until two successive approximations are within .0001 of each other. Again, the value .0001 is arbitrarily chosen and can be changed if a different precision is required.

Program

```
100 REM ** TRAPEZOIDAL RULE **
110 REM ** FUNCTION **
120 DEF FNF(X)=.....
130 REM ** INPUT **
140 PRINT "ENTER THE LOWER LIMIT OF INTEGRATION."
150 INPUT A
160 PRINT "ENTER THE UPPER LIMIT OF INTEGRATION."
170 INPUT B
180 REM ** HEADINGS **
190 PRINT
200 PRINT " N";TAB(10);"APP.  TRAP.  RULE"
210 REM ** INITIALIZATION **
220 LET N=10
230 LET S2=0
240 LET S1=FNF(A)+FNF(B)
250 REM ** FINDING DELTA X **
260 LET DX=(B-A)/N
270 REM ** LOOP TO APP.  INTEGRAL **
280 FOR I=1 TO (N-1)
290 REM ** FINDING NEXT X-I **
300 LET X=A+I*DX
310 LET S1=S1+2*FNF(X)
320 NEXT I
330 REM ** MULTIPLYING BY DELTA-X/2 **
340 LET S1=S1*DX/2
350 REM ** OUTPUT TO TABLE **
360 PRINT N;TAB(10);S1
370 REM ** TESTING TO END PROGRAM **
380 IF ABS(S1-S2) < .0001 THEN 420
390 LET S2=S1
400 LET N=2*N
410 GOTO 240
420 PRINT
430 PRINT "THE APPROXIMATION OF THE INTEGRAL"
440 PRINT "FROM ";A;" TO ";B;" IS ";S1
450 END
```

Test Data: The integral

$$\int_0^3 \sqrt{9-x^2}\ dx$$

represents the area of that portion of the circle of radius 3 in the first quadrant.

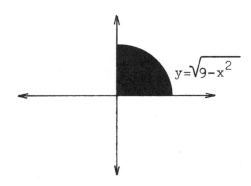

Thus,

$$\int_0^3 \sqrt{9-x^2}\ dx\ =\ 1/4\ \pi(3)^2\ \approx\ 7.068583471$$

```
120 DEF FNF(X)=SQR(9-X*X)

RUN
ENTER THE LOWER LIMIT OF INTEGRATION.
?0
ENTER THE UPPER LIMIT OF INTEGRATION.
?3

N          APP. TRAP. RULE
10         6.98516624
20         7.03904598
40         7.05813241
80         7.06488705
160        7.06727634
320        7.06812129
640        7.06842005
1280       7.06852571
2560       7.06856311

THE APPROXIMATION OF THE INTEGRAL
FROM 0 TO 3 IS 7.06856311
```

Example 2: An elliptic integral of the first kind has the form

$$F(k,a)\ =\ \int_0^a\ dx/\ \sqrt{1-k^2\ \sin^2 x}$$

where $k^2 < 1$ and a is an angle.

226

Use program TRAPEZOIDAL RULE to approximate $F(1/2, \pi/6)$ for $\pi/6 \approx .523598775$.

```
120 DEF FNF(X)=1/SQR(1-.25*SIN(X)*SIN(X))
```

```
RUN
ENTER THE LOWER LIMIT OF INTEGRATION.
?0
ENTER THE UPPER LIMIT OF INTEGRATION.
?.523598775
```

```
N       APP. TRAP. RULE
10      .529455875
20      .529435438
```

THE APPROXIMATION OF THE INTEGRAL
FROM 0 TO .523598775 IS .529435438

Thus, $F(1/2, \pi/6) \approx .5294$.

Simpson's Rule

In approximating the definite integral by the midpoint rule we approximated $f(x)$ on each subinterval by a horizontal line or a polynomial of degree 0. Using the Trapezoidal Rule we approximated $f(x)$ on each subdivision by a straight line or polynomial of degree 1. Simpson's Rule takes this process a step further and approximates $f(x)$ by polynomials of degree 2 as illustrated in the figure below.

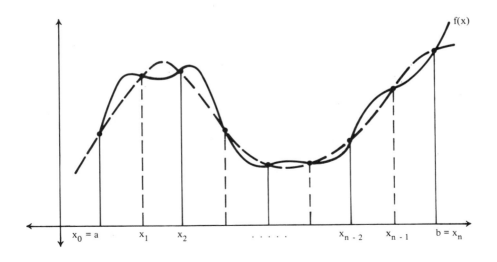

Simpson's Rule: Let the interval [a,b] be divided into n subdivisions of uniform length $\triangle x=(b-a)/n$ with endpoints

$$a=x_0, x_1, \ldots, x_{n-1}, x_n=b.$$

Then if n is even and f(x) is continuous on [a,b]

$$\int_a^b f(x) \ dx \approx (\triangle x/3)[f(x_0)+4f(x_1)$$
$$+2f(x_2)+4f(x_3) +\ldots$$
$$+2f(x_{n-2})+4f(x_{n-1})+f(x_n)].$$

Example 3: Approximate

$$\int_0^3 \sqrt{9-x^2} \ dx$$

using Simpson's Rule with n=6.

With n=6, $\triangle x=(3-0)/6=.5$ and the endpoints of the subdivision are

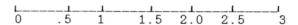

Thus,

$$\int_0^3 \sqrt{9-x^2} \ dx \approx (.5/3)[\sqrt{9-0^2} + 4\sqrt{9-(.5)^2} + 2\sqrt{9-1^2}$$
$$+4\sqrt{9-(1.5)^2} + 2\sqrt{9-2^2} + 4\sqrt{9-(2.5)^2}$$
$$+\sqrt{9-(3)^2}]$$

$$\approx (.5/3)[\sqrt{9} + 4\sqrt{8.75} + 2\sqrt{8} + 4\sqrt{6.75} + 2\sqrt{5}$$

$$+4\sqrt{2.75} + \sqrt{0}]$$

$$\approx (.5/3)[3+11.83216+5.6568542+10.392305$$

$$+4.472136+6.6332496+0]$$

$$\approx (.5/3)[41.986704]$$

$$\approx 6.997784$$

Notice that the approximation using Simpson's Rule is better for n=6 than the Trapezoidal Rule. For a comparison see example 1 and the test data for the program TRAPEZOIDAL RULE.

Programming Problem: Approximate the value of a definite integral using Simpson's Rule.

Output: The output will again be a table giving the number of subdivisions and the corresponding approximation followed by a summary of the results.

Input: The function being integrated and the limits of integration.

Strategy: From above the approximation of a definite integral using Simpson's Rule is given by

$$\int_{a}^{b} f(x)\ dx \approx (\triangle x/3)[f(x_0)+4f(x_1)+2f(x_2)+4f(x_3)+\ldots+4f(x_{n-1})+f(x_n)]$$

where n is even and $a=x_0, x_1, \ldots, x_{n-1}, x_n=b$ are the endpoints of the subdivisions. Noting that except for $f(x_0)$ and $f(x_n)$, $f(x_i)$ is multiplied by 4 if "i" is odd and 2 if "i" is even, we can rewrite this formula as we did for the Trapezoidal Rule to obtain an equivalent expression involving summations.

$$\int_a^b f(x)\ dx \approx (\triangle x/3)[f(x_0)+f(x_n)+ \sum_{\substack{i=1 \\ \text{odd}}}^{n-1} 4f(x_i)+ \sum_{\substack{i=2 \\ \text{even}}}^{n-2} 2f(x_i)]$$

The odd and even under the summation means that only odd or even values are used for i.

Except for the fact that we have two summations, this expression is similar to that of the Trapezoidal Rule. Thus, by expanding the program TRAPEZOIDAL RULE to include both sums we will be able to compute the Simpson's Rule approximation. Finally, notice that by starting with N=10 and doubling it each time, the condition that N is even will be satisfied. We will again use the technique of computing approximations until the difference between two successive approximations is less than 0.0001

Program:

```
100 REM ** SIMPSON'S RULE **
110 REM ** FUNCTION **
120 DEF FNF(X)=.....
130 REM ** INPUT **
140 PRINT "ENTER THE LOWER LIMIT OF INTEGRATION."
150 INPUT A
160 PRINT "ENTER THE UPPER LIMIT OF INTEGRATION."
170 INPUT B
180 REM ** HEADINGS **
190 PRINT
200 PRINT " N";TAB(10);"APP.  SIMP.  RULE"
210 REM ** INITIALIZATION **
220 LET N=10
230 LET S2=0
240 LET S1=FNF(A)+FNF(B)
250 REM ** FINDING DELTA X **
260 LET DX=(B-A)/N
270 REM ** LOOP TO APP.  INTEGRAL **
280 REM ** TERMS MULTIPLIED BY 4 **
290 FOR I=1 TO (N-1) STEP 2
300 REM ** FINDING NEXT X-I **
310 LET X=A+I*DX
320 LET S1=S1+4*FNF(X)
330 NEXT I
340 REM ** TERMS MULTIPLIED BY 2 **
350 FOR I=2 TO (N-2) STEP 2
360 REM ** FINDING NEXT X **
370 LET X=A+I*DX
380 LET S1=S1+2*FNF(X)
```

```
390 NEXT I
400 REM ** MULTIPLYING BY DELTA-X/3 **
410 LET S1=S1*DX/3
420 REM ** OUTPUT TO TABLE **
430 PRINT N;TAB(10);S1
440 REM ** TESTING TO END PROGRAM **
450 IF ABS(S1-S2) < .0001 THEN 490
460 LET S2=S1
470 LET N=2*N
480 GOTO 240
490 PRINT
500 PRINT "THE APPROXIMATION OF THE INTEGRAL"
510 PRINT "FROM ";A;" TO ";B;" IS ";S1
520 END
```

Test Data: From the test data for the Trapezoidal Rule

$$\int_0^3 \sqrt{9-x^2}\ dx \approx 7.068583471$$

```
120 DEF FNF(X)=SQR(9-X*X)
```

RUN
ENTER THE LOWER LIMIT OF INTEGRATION.
?0
ENTER THE UPPER LIMIT OF INTEGRATION.
?3

N	APP. SIMP. RULE
10	7.03576836
20	7.05700589
40	7.06449455
80	7.06713859
160	7.06807278
320	7.06840294
640	7.06851966
1280	7.06856093

THE APPROXIMATION OF THE INTEGRAL
FROM 0 TO 3 IS 7.06856093

Comparing the approximations for

$$\int_0^3 \sqrt{9-x^2}\ dx$$

using the Trapezoidal Rule and Simpson's Rule shows that Simpson's Rule results in more accurate approximations for every value of N.

Example 4: Illustrate that the statement "Simpson's Rule gives the exact value of

$$\int_a^b f(x)\ dx$$

if f(x) is a polynomial of degree 3 is true.

We will use the program SIMPSON'S RULE to approximate

$$\int_0^2 2x^3+3x^2-5x+1\ dx$$

$$\int_0^2 2x^3+3x^2-5x+1\ dx = x^4/2+x^3-(5/2)x^2+x \ \Big|_0^2 = 8$$

```
120 DEF FNF(X)=2*X*X*X+3*X*X-5*X+1

RUN
ENTER THE LOWER LIMIT OF INTEGRATION.
?0
ENTER THE UPPER LIMIT OF INTEGRATION.
?2

   N        APP. SIMP. RULE
   10          8
   20          8

THE APPROXIMATION OF THE INTEGRAL
FROM 0 TO 2 IS 8
```

Notice that for N=10 and N=20 the approximation gives the exact value.

Example 5: Illustrate that Simpson's Rule need not give an exact value for a polynomial of degree 4.

We will use the program SIMPSON'S RULE to approximate

$$\int_0^2 5x^4 - 8x^3 + 6x^2 + 3x + 9 \ dx$$

$$\int_0^2 5x^4 - 8x^3 + 6x^2 + 3x + 9 \ dx \ = \ x^5 - 2x^4 + 2x^3 + (3/2)x^2 + 9x \ \Big|_0^2$$

$$= \ 40$$

```
120 DEF FNF(X)=5*X*X*X*X-8*X*X*X+6*X*X+3*X+9
```

RUN
ENTER THE LOWER LIMIT OF INTEGRATION.
?0
ENTER THE UPPER LIMIT OF INTEGRATION.
?2

N	APP. SIMP. RULE
10	40.0021333
20	40.0001333
40	40.0000083
80	40.0000005

THE APPROXIMATION OF THE INTEGRAL
FROM 0 TO 2 IS 40.0000005

Unlike example 4 we do not obtain the exact value for the integral.

Ch. 11   NUMERICAL INTEGRATION

Exercise Set 11.1

In exercises 1-4 use programs TRAPEZOIDAL RULE and SIMPSON'S RULE to approximate each definite integral. Also compute the exact value of the integral. Compare the exact value with both approximations and make a conjecture concerning the efficiency and accuracy of both methods.

1. $\int_{0}^{\pi/4} \tan x \, dx$

2. $\int_{0}^{\pi} \sin^2 x \, dx$

3. $\int_{3}^{5} xe^x \, dx$

4. $\int_{0}^{1} dx/(x^2-4)$

*
5.   The gamma and exponential probability distributions are quite important in probability theory since they describe such random phenomena as lengths of waiting time, life of electron tubes, and time intervals between successive breakdowns of electronic systems and accidents.

Gamma: $f(t) = \begin{cases} \dfrac{m}{(r-1)!} (mt)^{r-1} e^{-mt} & ; t \geq 0 \\ 0 & ; t \leq 0 \end{cases}$

Exponential: $f(t) = \begin{cases} me^{-mt} & ; t \geq 0 \\ 0 & ; t \leq 0 \end{cases}$

a.   Show that the exponential probability distribution is a special case of the gamma probability distribution.

b.   Use program SIMPSON'S RULE to approximate the following definite integral which gives the probability [$1 \leq t \leq 5$] for a gamma probability distribution with r=3 and m=2.5.

$$\int_{1}^{5} 7.8125t \ e^{-2.5t} dt$$

c.   Use program TRAPEZOIDAL RULE to approximate the following definite integral which gives the probability [$0 \leq t \leq 1.5$] for an exponential probability distribution with m=1.75.

$$\int_{0}^{1.5} 1.75 \ e^{-1.75t} \ dt$$

6.   Another important function in probability and statistics is the error function erf(x).

$$erf(x) = 2/\sqrt{\pi} \int_{0}^{x} e^{-t*t} \ dt$$

Using program SIMPSON'S RULE approximate erf(1.5).

7.   Use program TRAPEZOIDAL RULE to approximate the value of $J_0(2.1)$ where $J_0(x)$ is the Bessel function.

$$J_0(x) = 1/\pi \int_{0}^{\pi} \cos(x \sin y) \ dy$$

8. An elliptic integral of the third kind is of the form

$$P(k,n,a) = \int_0^a \frac{dx}{(1+n\,\sin^2 x)\,\sqrt{1-k^2\sin^2 x}}$$

for $k^2 < 1$ and n an integer.

Use program SIMPSON'S RULE to approximate $P(1/2,\ 5,\ \pi/3)$.

**

9. Write a program to approximate the value of a definite integral using the Corrected Trapezoidal Rule given below. Include in your input the values of f'(a) and f'(b). Test your program using

$$\int_0^3 \sqrt{16-x^2}\ dx$$

and compare your results to the approximations obtained using the Trapezoidal Rule and Simpson's Rule. Which method appears to be "better"? Explain.

Corrected Trapezoidal Rule: Let f(x) be differentiable on [a,b]. If [a,b] is subdivided into n subdivisions of the same length $\triangle x = (b-a)/n$, then

$$\int_a^b f(x)\ dx \approx (\triangle x/2)[f(x_0)+2f(x_1)+\ldots+2f(x_{n-1})+f(x_n)]$$

$$+[f'(a)-f'(b)]\ (\triangle x)^2/12$$

***
10.   Same as exercise 9 except use Weddle's Rule.

Weddle's Rule: Let f be a continuous function on the interval [a,b]. If [a,b] is divided into n subdivisions, where n is a multiple of 6, with uniform length $\triangle x=(b-a)/n$, and the points x_i are the endpoints of the subdivisions, then

$$\int_a^b f(x)dx \approx (3\triangle x/10)[f(x_0)+5f(x_1)+f(x_2)+6f(x_3)+f(x_4)$$
$$+5f(x_5)+\overset{2}{f}(x_6)+5f(x_7)+f(x_8)+6f(x_9)$$
$$+f(x_{10})+5f(x_{11})+\overset{2}{f}(x_{12})+ \ldots$$
$$+\overset{2}{f}(x_{n-6})+5f(x_{n-5})+f(x_{n-4})+6f(x_{n-3})$$
$$+f(x_{n-2})+5f(x_{n-1})+f(x_n)]$$

$2f(x_6)$

$+2f(x_{12})$

$2f(x_n-6)$

→ <u>where</u> n must be a multiple of 6.
<u>Note</u>: for problem # 4

11.2 Discretization Error

As pointed out many times in this text calculations involving the computer are subject to round-off error which is unavoidable. However, numerical estimation of definite integrals are subject to an additional type of error which is due entirely to the type of method used. This error results from the fact that we approximate the curve between the points $(x_i, f(x_i))$ and $(x_{i+1}, f(x_{i+1}))$ by a polynomial. Since this results in using only a finite or discrete number of functional values of the function being integrated, the error which is generated is called discretization error.

Unlike round-off error, discretization error is not due to the device (computer) being used to perform the calculation but to the method employed and can thus be analyzed. For the Trapezoidal Rule with N subdivisions a bound on the discretization error in

approximating $\int f(x)\ dx$ is:

$$E_N \leq \frac{K(b-a)^3}{12N^2}$$

where $K = \max |f''(x)|$ on $[a,b]$. (For a derivation see a text on numerical analysis.) Notice that as N increases the discretization error decreases. However, an increasing number of subdivisions does increase the number of computations which can cause an increase in the total round-off error. Generally there is a balance between the two types of error, and there is an optimal number of subdivisions to choose in order to minimize the total error. Finally, it should be pointed out that the bound given above for the discretization error resulting from using the Trapezoidal Rule can be improved.

For example, suppose that we are estimating

$$\int_a^b f(x)\ dx$$

by the trapezoidal rule with N subdivisions whose endpoints are $x_0, x_1, \ldots, x_n$. There are two ways to view the trapezoidal rule--estimating an integral by using N trapezoids or estimating N integrals each by one trapezoid. The latter approach is basically the same as rewriting the integral as:

$$\int_a^b f(x) \, dx = \int_{x_0}^{x_1} f(x) \, dx + \int_{x_1}^{x_2} f(x) \, dx + \ldots + \int_{x_{n-1}}^{x_n} f(x) \, dx$$

Since each integral on the right-hand side of the equation is being approximated by a single trapezoid, we can apply the error formula, $K(b-a)^3/12N^2$, for the trapezoidal rule with $N=1$. Thus,

$$E_n \le \frac{k_1(x_1-x_0)^3}{12(1)^2} + \ldots + \frac{k_i(x_i-x_{i-1})^3}{12(1)^2} + \ldots + \frac{k_n(x_n-x_{n-1})^3}{12(1)^2}$$

where $k_i = \max|f''(x)|$, $x_{i-1} \le x \le x_i$. However, since for the trapezoidal rule the endpoints are evenly spaced, $x_i - x_{i-1} = (b-a)/N = \triangle x$ for each subdivision, and

$$E_n = \frac{k_1(\triangle x)^3}{12} + \ldots + \frac{k_i(\triangle x)^3}{12} + \ldots + \frac{k_n(\triangle x)^3}{12}$$

$$= \frac{(\triangle x)^3}{12} \sum_{i=1}^{n} k_i$$

Example 1: Estimate the discretization error which results when

$$\int_0^1 7-6x^5 \, dx$$

is approximated by the trapezoidal rule with $N=10$ by the method above and by the formula $E_n \approx k(b-a)^3/12N^2$. Compare the results.

Since $f(x)=7-6x^5$, $f''(x)=-120x^3$ and $|f''(x)|=120x^3$ which is an increasing function. Thus, the $\max|f''(x)|$ will occur at the right-hand endpoint of whatever interval x is restricted.

Estimating E_{10} using $((\triangle x)^3/12) \sum_{i=1}^{10} k_i$:

The interval [0,1] is being divided into 10 subintervals as shown below and $\triangle x = .1$

$$\llcorner ----\llcorner ---\llcorner ---\llcorner ---\llcorner ---\llcorner ---\llcorner ---\llcorner ---\llcorner ---\llcorner ---\lrcorner$$
$$0 \quad .1 \quad .2 \quad .3 \quad .4 \quad .5 \quad .6 \quad .7 \quad .8 \quad .9 \quad 1$$

Thus,

$$E_n \leq \frac{(.1)^3}{12} [120(.1)^3 + 120(.2)^3 + \ldots + 120(.9)^3 + 120(1)^3$$

$$\leq \frac{(.1)^3}{12} (363)$$

$$\leq .03025$$

Estimating E_{10} using $\frac{k(b-a)^3}{12N^2}$:

$$E_{10} \leq \frac{120(1-0)^3}{12(10)^2}$$

$$\leq .1$$

Note that a much smaller estimation of the error is obtained by the new method.

Programming Problem: Estimate the discretization error for the trapezoidal rule by the method described above.

Output: A message giving the limits of integration, the estimated error and the number of subdivisions used in the calculations.

Input: The second and third derivative of the function being integrated, the limits of integration, and the number of subdivisions.

Strategy: To estimate the discretization error by the formula

$$E_n = ((\triangle x)^3/12) \sum_{i=1}^{n} k_i$$

we must be able to approximate max $|f''(x)|$ for each subdivision $[x_{i-1}, x_i]$. Since we know that the maximum or minimum of a function on a closed interval will occur at either the endpoints of the interval or a critical point and the max $|f''(x)|$ will occur at either a maximum or minimum of $f''(x)$, (see figure below) we can approximate the max $|f''(x)|$ on each subinterval by the following steps:

1. Determine the endpoints of the subinterval.

2. Find the critical points of $f''(x)$ on the subinterval.

3. Evaluate $|f''(x)|$ at each critical point and the endpoints of the interval.

4. Choose the maximum.

The problem of finding critical points of a function on a closed interval by a computer was discussed and programmed in section 9.2 of this text. Simplifying that program by assuming that $f''(x)$ is defined at all points on the interval [a,b], we can use that portion of the program MAX-MIN involved with determining critical points.

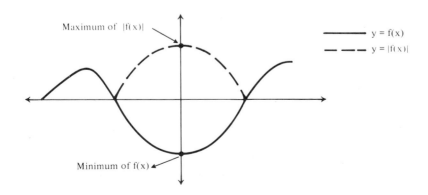

Maximum of $|f(x)|$

$y = f(x)$
$y = |f(x)|$

Minimum of $f(x)$

Program:

```
100 REM ** ERROR ESTIMATION **
110 REM ** TRAPEZOIDAL RULE **
120 REM ** SECOND DERIVATIVE **
130 DEF FNF(X)=.....
140 REM ** THIRD DERIVATIVE **
150 DEF FNG(X)=.....
160 DIM C(101),V(101)
170 REM ** INPUT **
180 PRINT "ENTER THE LOWER LIMIT OF INTEGRATION."
190 INPUT A
200 PRINT
210 PRINT "ENTER THE UPPER LIMIT OF INTEGRATION."
220 INPUT B
230 PRINT
240 PRINT "ENTER THE NUMBER OF SUBDIVISIONS"
250 PRINT "TO BE USED IN ESTIMATING THE ERROR."
260 INPUT N
270 REM ** MAIN PROGRAM **
280 LET DX=(B-A)/N
290 REM ** E USED TO KEEP SUM OF MAX **
300 REM ** VALUES OF ABS(F''(X)) **
310 LET E=0
320 REM ** LOOP FOR FINDING SUM **
330 REM ** OF THE MAXIMUM OF ABS(F''(X)) **
340 FOR J=1 TO N
350 LET C(1)=A+(J-1)*DX
360 LET C(2)=A+J*DX
370 REM ** FINDING THE LENGTH OF A **
380 REM ** SUBDIVISION **
390 LET LET D=(C(2)-C(1))/100
400 REM ** SEARCHING THE INTERVAL **
410 REM ** FOR VALUES AT WHICH THE **
420 REM ** DERIVATIVE IS ZERO **
430 LET T=2
440 FOR I=1 TO 100
450 REM ** LEFT-HAND ENDPOINT **
460 LET L=C(1)+D*(I-1)
470 REM ** VALUE OF THE DERIVATIVE **
480 LET Y1=FNG(L)
490 REM ** RIGHT-HAND ENDPOINT **
500 LET R=C(1)+I*D
510 REM ** VALUE OF DERIVATIVE **
520 LET Y2=FNG(R)
530 REM ** TESTING FOR DIFFERENT SIGNS **
```

```
540 IF SGN(Y1)=SGN(Y2) THEN 710
550 REM ** BISECTION TO FIND VALUE **
560 IF FNG(L)<>0 THEN 590
570 LET T=T+1
580 LET C(T)=L
590 IF FNG(R)<>0 THEN 630
600 LET T=T+1
610 LET C(T)=R
620 GOTO 710
630 LET M=(L+R)/2
640 IF SGN(FNG(M))=SGN(FNG(L)) THEN 670
650 LET R=M
660 GOTO 680
670 LET L=M
680 IF ABS (R-L)>=.00001 THEN 560
690 LET T=T+1
700 LET C(T)=(L+R)/2
710 NEXT I
720 REM ** FINDING THE FUNCTIONAL VALUES **
730 REM ** OF ABS(F''(X)) **
740 FOR I=1 TO T
750 LET V(I)=ABS(FNF(C(I)))
760 NEXT I
770 REM ** FINDING MAX AND MIN **
780 LET MAX=V(1)
790 FOR I=2 TO T
800 IF V(I)<=MAX THEN 820
810 LET MAX=V(I)
820 NEXT I
830 LET E=E+MAX
840 NEXT J
850 REM ** APPROXIMATING THE ERROR **
860 LET E=E*DX*DX*DX/12
870 REM ** OUTPUT **
880 PRINT "THE DISCRETIZATION ERROR FOR"
890 PRINT "THE TRAPEZOIDAL RULE ON THE"
900 PRINT "INTERVAL [";A;",";B;"] FOR"
910 PRINT N;" SUBDIVISIONS IS ";E
920 END
```

In searching for the critical points each subinterval has been divided into 100 parts. On many computers this will result in a program which takes an extensive amount of time to run. To speed up the program change 100 to a smaller number in lines 390 and 440.

Test Data: Estimate the discretization error for

$$\int_0^1 7-6x^5 \ dx$$

with N=10.

$$f(x) = 7-6x^5$$
$$f'(x) = -30x^4$$
$$f''(x) = -120x^3$$
$$f'''(x) = -360x^2$$

```
130 DEF FNF(X)=-120*X*X*X
150 DEF FNG(X)=-360*X*X

RUN
ENTER THE LOWER LIMIT OF INTEGRATION.
?0

ENTER THE UPPER LIMIT OF INTEGRATION.
?1

ENTER THE NUMBER OF SUBDIVISIONS
TO BE USED IN ESTIMATING THE ERROR.
?10
THE DISCRETIZATION ERROR FOR
THE TRAPEZOIDAL RULE ON THE
INTERVAL [0,1] FOR
10 SUBDIVISIONS IS .03025
```

The result agrees with that computed in example 1.

Example 2: Determine a value of N so that the discretization error in approximating

$$\int_{-1}^1 x^4 \cos x \ dx$$

by the trapezoidal rule is less than .001

We will run the program ERROR ESTIMATION for different values of N until the estimated error is less than .001

$$f(x) \quad = x^4 \cos x$$
$$f''(x) \quad = (12x^2 - x^4) \cos x - 8x^3 \sin x$$
$$f'''(x) = (24x - 12x^3) \cos x - (36x^2 - x^4) \sin x$$

For N=10:

```
110 DEF FNF(X)=(12*X*X-X*X*X*X)*COS(X)-8*X*X*X*SIN(X)
150 DEF FNG(X)=(24*X-12*X*X*X)*COS(X)-(36*X*X-X*X*X*X)*SIN(X)

RUN
ENTER THE LOWER LIMIT OF INTEGRATION.
?-1

ENTER THE UPPER LIMIT OF INTEGRATION.
?1

ENTER THE NUMBER OF SUBDIVISIONS
TO BE USED IN ESTIMATING THE ERROR.
?10
THE DISCRETIZATION ERROR FOR
THE TRAPEZOIDAL RULE ON THE
INTERVAL [-1,1] FOR
10 SUBDIVISIONS IS .0122424692
```

For N=30:

```
RUN
ENTER THE LOWER LIMIT OF INTEGRATION.
?-1

ENTER THE UPPER LIMIT OF INTEGRATION.
?1

ENTER THE NUMBER OF SUBDIVISIONS
TO BE USED IN ESTIMATING THE ERROR.
?30
THE DISCRETIZATION ERROR FOR
THE TRAPEZOIDAL RULE ON THE
INTERVAL [-1,1] FOR
30 SUBDIVISIONS IS 1.12981365E-03
```

For N=40:

```
RUN
ENTER THE LOWER LIMIT OF INTEGRATION.
?-1

ENTER THE UPPER LIMIT OF INTEGRATION.
?1

ENTER THE NUMBER OF SUBDIVISIONS
TO BE USED IN ESTIMATING THE ERROR.
?40
THE DISCRETIZATION ERROR FOR
THE TRAPEZOIDAL RULE ON THE
INTERVAL [-1,1] FOR
40 SUBDIVISIONS IS 6.20948346E-04
```

For N=40 the estimated discretization error is less than .001 However, it may be possible to choose a smaller value for N between 30 and 40.

Example 3: The Maxwell probability distribution with parameter $\alpha > 0$ is given by

$$f(x) = \begin{cases} \dfrac{4}{\sqrt{\pi}} \ \dfrac{1}{\alpha^3} \ x^2 \exp(-x^2/\alpha^2) & , x > 0 \\ \\ 0 & , x \leq 0 \end{cases}$$

Using the trapezoidal rule approximate the probability that $.5 \leq x \leq 1$ if x has a Maxwell probability distribution with $\alpha = .2$ Also, estimate the discretization error.

By definition the probability that $.4 \leq x \leq .7$ is given by the definite integral

$$\int_{.4}^{.7} \frac{4}{\sqrt{\pi}} \ \frac{1}{.008} \ x^2 \ \exp(-x^2/.04) \ dx$$

We will use the program TRAPEZOIDAL RULE from section 8.1 to approximate the integral. We will use $(4/\sqrt{\pi})(1/.008) \approx 282.0947918$

```
130 DEF FNF(X)=282.0947918*EXP(-X*X/.04)
             *(2-5*X*X/.02+X*X*X*X/.0004)

150 DEF FNG(X)=282.0947918*EXP(-X*X/.04)*(-6*X/.01
             +9*X*X*X/.0004-X*X*X*X*X/.000008)

RUN
```
ENTER THE LOWER LIMIT OF INTEGRATION.
```
?.4
```
ENTER THE UPPER LIMIT OF INTEGRATION.
```
?.7
```

N	APP. TRAP. RULE
10	.0469200277
20	.0462241416
40	.0460500891
80	.0460065711

THE APPROXIMATION OF THE INTEGRAL
FROM .4 TO .7 IS .0460065711

Thus, $Prob(.4 \leq x \leq .7) = .0460065711$

Now to estimate the discretization error we will use program ERROR ESTIMATION.

$$f(x) = \frac{4}{\sqrt{\pi}} \frac{1}{.008} x^2 \exp(-x^2/.04)$$

$$f''(x) = \frac{4}{\sqrt{\pi}} \frac{1}{.008} \exp(-x^2/.04)(2 - \frac{5x^2}{.02} + \frac{x^4}{.0004})$$

$$f'''(x) = \frac{4}{\sqrt{\pi}} \frac{1}{.008} \exp(-x^2/.04)(\frac{-6x}{.01} + \frac{9x^3}{.0004} - \frac{x^5}{.000008})$$

```
130 DEF FNF(X)=282.0947918*EXP(-X*X/.04)
            *(2-5*X*X/.02+X*X*X*X/.0004)

150 DEF FNG(X)=282.0947918*EXP(-X*X/.04)
            *(-6*X/.01+9*X*X*X/.0004 -X*X*X*X*X/.000008)

RUN
```
ENTER THE LOWER LIMIT OF INTEGRATION.
?.4

ENTER THE UPPER LIMIT OF INTEGRATION.
?.7

ENTER THE NUMBER OF SUBDIVISIONS
TO BE USED IN ESTIMATING THE ERROR.
?30
THE DISCRETIZATION ERROR FOR
THE TRAPEZOIDAL RULE ON THE
INTERVAL [.4,.7] FOR
30 SUBDIVISIONS IS 1.08755562E-04

Therefore, the actual probability should be in the interval

.046±.000109 or [.045891,.046109]

neglecting round-off error.

Exercise Set 11.2

In exercises 1-3 use program ERROR ESTIMATION to determine N so that the estimated discretization error will be less than .001 if the integral is being approximated by the trapezoidal rule.

1. $\displaystyle\int_0^1 \cos x \, dx$

2. $\displaystyle\int_{-4}^3 x^2 \sin x^2 \, dx$

3. $\displaystyle\int_0^1 e^x \cos x \, dx$

In exercises 4-7 approximate the definite integral using program TRAPEZOIDAL RULE. Also estimate the discretization error using program ERROR ESTIMATION and give an interval in which the answer should be found. Finally, determine the exact value of the integral and decide if the exact answer is in the interval determined by using TRAPEZOIDAL RULE and ERROR ESTIMATION. Comparing the exact answer to the approximations should give you an idea about the round-off error being generated.

4. $\displaystyle\int_0^2 4x^3 - 3x^2 + 2x - 8 \, dx$

5. $\displaystyle\int_{-1}^2 \cos^2 x \sin x \, dx$

6. $\displaystyle\int_1^3 x \, e^x \, dx$

7. $\displaystyle\int_1^2 x^2 \ln 3x \, dx$

Ch. 11   NUMERICAL INTEGRATION

*
8.   The Cauchy probability distribution function with parameters
$-\infty < A < \infty$ and $B > 0$ is given by

$$f(x) = \frac{1}{\pi B(1 + ((x-A)/B)^2)}$$

Determine an interval for the probability $1 \leq x \leq 2$ if x has a Cauchy
probability distribution with A=.5 and B=2.

***
For exercises 9-11:

a.   Develop an   improved   formula   for   estimation   of   the
discretization error for the indicated rule in a manner similar to
what was done in this section.

b.   Use your formula to compute the estimated error for

$$\int_0^1 7-6x^5 \; dx$$

with N=10.

c.   Write   a   program   to estimate the discretization error using
your formula.

d.   Test your program with

$$\int_0^1 7-6x^5 \; dx \text{ with N=10.}$$

9.   Midpoint Rule: An error estimate for the Midpoint Rule for
the integral of f(x) from a to b is

$$E_N \leq K(b-a)^3 / 24N^2$$

where $K = \max \left| f''(x) \right|$ on [a,b].

10. Simpson's Rule: An error estimate for Simpson's Rule for the integral of f(x) from a to b is

$$E_N \leq K(b-a)^5/180N^4$$

where $K = \max \left| f^{(4)}(x) \right|$ on [a,b].

11. Weddle's Rule: An error estimate for Weddle's Rule for the integral of f(x) from a to b is

$$E_N \leq K(b-a)^7/140N^6$$

where $k = \max \left| f^{(6)}(x) \right|$ on [a,b].

CHALLENGE ACTIVITY

To this point all the methods presented have approximated the value of the integral by means of a summation involving values of the function being integrated and the width of the subdivisions. The only time two different approximations were used was in deciding if the program should terminate. Yet there are several methods for approximating the value of a definite integral which make use of successive approximations. One such method is Romberg integration. Although space does not permit a detailed description of this process (for a complete description see a text on numerical analysis), it basically involves combining successive trapezoidal approximations, doubling the number of trapezoids each time, to obtain a series of approximations which theoretically become more accurate as we proceed. This series of approximations can be conveniently expressed in a table like the one below.

Romberg Integration Table

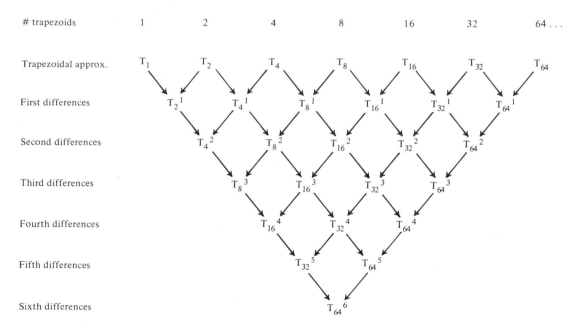

# trapezoids	1	2	4	8	16	32	64 ...
Trapezoidal approx.	T_1	T_2	T_4	T_8	T_{16}	T_{32}	T_{64}
First differences		T_2^1	T_4^1	T_8^1	T_{16}^1	T_{32}^1	T_{64}^1
Second differences			T_4^2	T_8^2	T_{16}^2	T_{32}^2	T_{64}^2
Third differences				T_8^3	T_{16}^3	T_{32}^3	T_{64}^3
Fourth differences					T_{16}^4	T_{32}^4	T_{64}^4
Fifth differences						T_{32}^5	T_{64}^5
Sixth differences							T_{64}^6

Each term in the table, except for the trapezoidal approximations is obtained using the formula

$$T_{2m}{}^i = \frac{4^i \, T_{2m}{}^{i-1} - T_m{}^{i-1}}{4^i - 1}$$

For example,

$$T_{32}{}^2 = \frac{4^2 \, T_{32}{}^1 - T_{16}{}^1}{4^2 - 1}$$

The arrows in the table indicate which two entries are combined to obtain the indicated entry.

Write a program which will output the Romberg integration table going up to 128 trapezoids. Test your program by approximating

$$\int_0^3 \sqrt{9-x^2} \; dx$$

Compare your results with those in section 11.1 for this integral and comment on the efficiency and accuracy of this method.

CHAPTER 12
Antiderivatives And Approximations

Any definite integral may be thought of as an area or an algebraic sum of signed areas. Hence, every continuous function has an antiderivative. The Fundamental Theorem of Calculus gives us a method of evaluating a definite integral when the indefinite integral of the integrand is known. However, there exist many functions, such as 1/x, whose antiderivatives are not expressible as a combination of elementary functions. In this case we express the antiderivative as

$$\int_a^x f(t) \, dt$$

Although the antiderivative is not a combination of simple functions that we know, it is sometimes helpful to have an elementary function which will approximate the antiderivative. In this chapter we will investigate a method for approximating the antiderivative of a function by a polynomial. To derive the polynomial we will need several functional values for various values of x.

12.1 FUNCTIONAL VALUES OF THE ANTIDERIVATIVE

Programming Problem: Approximate the value of a function defined by an integral, that is,

$$F(x) = \int_a^x f(t)\ dt$$

for a given value of x.

Output: An appropriate message giving the value of x and the functional value. Also, we will construct the program so that the function can be evaluated at more than one value of x.

Input: The function which is the integrand, the value of a (the lower limit of integration), and the value at which the function is to be evaluated.

Strategy: Our purpose is almost identical to that of program INTEGRAL of section 10.1 with the difference that we may wish to evaluate the integral more than one time. Also, there is no need to print the table of approximations as in program INTEGRAL but only the final approximation. Thus, by including a test to decide if the program is to be repeated, we can modify the program INTEGRAL to meet our needs.

Program:

```
100 REM ** ANTIDERIVATIVE **
110 REM ** FUNCTION **
120 DEF FNF(T)=.....
130 REM ** INPUT **
140 PRINT "ENTER THE VALUE OF A, THE LOWER"
150 PRINT "LIMIT OF INTEGRATION."
160 INPUT A
170 PRINT
180 PRINT "ENTER THE VALUE AT WHICH THE"
190 PRINT "ANTIDERIVATIVE IS TO BE"
200 PRINT "APPROXIMATED."
210 INPUT X
220 PRINT
230 REM ** INITIALIZATION **
240 LET N=10
250 LET S2=0
260 LET S1=0
270 REM ** FINDING DELTA X **
280 LET DX=(B-A)/N
290 REM ** LOOP TO APP.  INTEGRAL **
300 FOR I=1 TO N
310 LET LE=A+(I-1)*DX
320 LET RE=A+I*DX
330 LET M=(LE+RE)/2
340 LET S1=S1+FNF(M)*DX
350 NEXT I
360 REM ** TESTING TO END PROGRAM **
370 IF ABS(S1-S2) < .0001 THEN 410
380 LET S2=S1
390 LET N=2*N
400 GO TO 260
410 PRINT
420 PRINT "THE APPROXIMATION OF THE ANTIDERIVATIVE"
430 PRINT "AT ";X;" IS ";S1
440 PRINT
450 PRINT "APPROXIMATE THE ANTIDERIVATIVE AT"
460 PRINT "ANOTHER VALUE OF X?(Y/N)"
470 INPUT F$
480 IF F$="Y" THEN 170
490 END
```

Test Data: If

$$F(x) = \int_0^x t^2 + 1 \; dt$$

then $F(3) = 3^3/3 + 3 = 12$

and $F(-2) = -2^3/3 + (-2) = -8/3 - 2 = -14/3 \approx -4.6666667$

```
120 DEF FNF(T)=T*T+1

RUN
ENTER THE VALUE OF A, THE LOWER
LIMIT OF INTEGRATION.
?0

ENTER THE VALUE AT WHICH THE
ANTIDERIVATIVE IS TO BE
APPROXIMATED.
?3

THE APPROXIMATION OF THE ANTIDERIVATIVE
AT 3 IS 11.999978

APPROXIMATE THE ANTIDERIVATIVE AT
ANOTHER VALUE OF X?(Y/N)
?Y

ENTER THE VALUE AT WHICH THE
ANTIDERIVATIVE IS TO BE
APPROXIMATED.
?-2

THE APPROXIMATION OF THE ANTIDERIVATIVE
AT -2 IS -4.66664063

APPROXIMATE THE ANTIDERIVATIVE AT
ANOTHER VALUE OF X?(Y/N)
?N
```

Ch. 12   ANTIDERIVATIVES AND APPROXIMATIONS

Example 1: If the distance travelled, in feet, by a particle for any time x is given by

$$F(x) = \int_{1}^{x} (t+1)/(t^2 + 1) \ dt$$

approximate the average speed of the particle between 4 and 10 seconds.

$$\text{Average Speed} = \frac{\text{total distance travelled}}{\text{total elapsed time}} = \frac{F(10)-F(4)}{10-4}$$

```
120 DEF FNF(T)=(T+1)/(T*T+1)

RUN
ENTER THE VALUE OF A, THE LOWER
LIMIT OF INTEGRATION.
?1

ENTER THE VALUE AT WHICH THE
ANTIDERIVATIVE IS TO BE
APPROXIMATED.
?4

THE APPROXIMATION OF THE ANTIDERIVATIVE
AT 4 IS 1.61042795

APPROXIMATE THE ANTIDERIVATIVE AT
ANOTHER VALUE OF X?(Y/N)
?Y

ENTER THE VALUE AT WHICH THE
ANTIDERIVATIVE IS TO BE
APPROXIMATED.
?10

THE APPROXIMATION OF THE ANTIDERIVATIVE
AT 10 IS 2.64670008

APPROXIMATE THE ANTIDERIVATIVE AT
ANOTHER VALUE OF X?(Y/N)
?N
```

$$\text{Average Speed} = \frac{2.64670008 - 1.61042795}{10 - 4}$$

$$= \frac{1.03627213}{6}$$

$$= .17271202 \text{ ft/sec}$$

Example 2: Use the program ANTIDERIVATIVE to obtain points to plot the function

$$F(x) = \int_0^x dt/\sqrt{1-t^2}$$

Notice that for values of t such that t $\geq$ 1 the function is undefined. Thus, the domain of F(x) is -1<x<1. Actually the domain includes both -1 and 1, but this leads to an improper integral which we are unable to handle at the present time.

Running the program ANTIDERIVATIVE with

    120 DEF FNF(T)=1/SQR(1-T*T)

for different values of x results in the following table. The actual printout is omitted due to its length.

X	F(X)	X	F(X)
-.9	-1.11975519	.1	.100167315
-.7	-0.775372982	.3	.304689414
-.5	-0.523578736	.5	.523578736
-.3	-0.304689414	.7	.775372982
-.1	-0.100167315	.9	1.11975519
0	0		

Plotting these points and sketching a smooth curve should yield a good approximation of the graph. This function is also known as the inverse sine function.

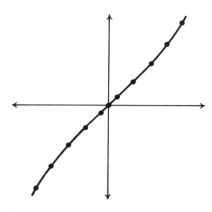

Example 3: Approximate H(-1) if

$$H(x) = \int_x^4 \sin t^2 \, dt$$

Since

$$\int_x^4 \sin t^2 \, dt = -\int_4^x \sin t^2 \, dt$$

we can approximate H(-1) using the program ANTIDERIVATIVE.

```
120 DEF FNF(T)=SIN(T*T)

RUN
ENTER THE VALUE OF A, THE LOWER
LIMIT OF INTEGRATION.
?4

ENTER THE VALUE AT WHICH THE
ANTIDERIVATIVE IS TO BE
APPROXIMATED.
?-1

THE APPROXIMATION OF THE ANTIDERIVATIVE
AT -1 IS -1.05741888

APPROXIMATE THE ANTIDERIVATIVE AT
ANOTHER VALUE OF X?(Y/N)
?N
```

Therefore, H(-1)≈-(-1.057418884)=1.057418884

Example 4: One of the theorems of the calculus states:

If f is a continuous function of the interval
[a,b] and F is the function defined by

$$F(x) = \int_{a}^{x} f(t)\ dt$$

for all x in [a,b], then F'(x)=f(x) on [a,b].

Letting f(t)=t²+2t+1 and a=1, approximate F'(2) by determining
the value of the difference quotient for values of x close to 2.
Compare your answer to f(2), which equals 9, to illustrate the
validity of this theorem.

Using program ANTIDERIVATIVE with

    120 DEF FNF(T)=T*T+2*T+1

we will approximate F for values close to and at 2. Again, due to
the length of the output most of the results will be recorded in a
table.

```
    120 DEF FNF(T)=T*T+2*T+1

    RUN
    ENTER THE VALUE OF A, THE LOWER
    LIMIT OF INTEGRATION.
    ?1

    ENTER THE VALUE AT WHICH THE
    ANTIDERIVATIVE IS TO BE
    APPROXIMATED.
    ?2

    THE APPROXIMATION OF THE ANTIDERIVATIVE
    AT 2 IS 6.33332032

    APPROXIMATE THE ANTIDERIVATIVE AT
    ANOTHER VALUE OF X?(Y/N)
    ?N
```

X	F(X)	F(X)-F(2)	△X	Difference Quotient
1.99	6.24362037	-.08969995	-.01	8.969995
1.999	6.32432335	-.00899697	-.001	8.99697
1.9999	6.33242035	-.00089997	-.0001	8.9997
2.0001	6.33422034	-.00090002	.0001	9.0002
2.001	6.34232328	.00900296	.001	9.00296
2.01	6.42362025	.09029993	.01	9.029993

Although this does not constitute a proof, it does seem reasonable to conclude that F'(2)=9.

Exercise Set 12.1

In exercises 1-3 use program ANTIDERIVATIVE to approximate the value of the function at the given value of x.

1. $F(-1.2)$ if $F(x) = \int_{-5}^{x} (t^2 + 2t - 7)/(t^2 + t + 7)\ dt$

2. $F(\pi/2)$ if $F(x) = \int_{0}^{x} \sin t\ dt$

3. $F(1.576)$ if $F(x) = \int_{1}^{x} \sqrt{t^4 - t + 1}\ dt$

In exercises 4-6 use program ANTIDERIVATIVE to aid in graphing each of the following functions.

4. $F(x) = \int_{0}^{x} dt/(1 + t^2)$ (Inverse tangent function)

5. $F(x) = \int_{2}^{x} dt/t\sqrt{t^2 - 1}$; $x > 2$ (Inverse secant function)

6. $F(x) = \int_{0}^{x} dt/\sqrt{1 + t^2}$ (Inverse hyperbolic sine)

In exercises 7-8 use program ANTIDERIVATIVE to approximate the value of the function at the given value of x.

7. $F(-6)$ if $F(x) = \int_{x}^{0} \sin^3 t\ dt$

8.   F(5) if $F(x) = \int_{x}^{10} dt/(t^2 + \sin t + 2)$

*
9.   Let the function F(x) be defined by the integral

$$\int_{1}^{x} dt/t$$

Use program ANTIDERIVATIVE to:

   a.   Approximate F(3), F(2), and F(6). Make a conjecture about the relation between F(ab), F(a), and F(b).

   b.   Approximate F(5), F(2), and F(2.5). Make a conjecture about the relation between F(a/b), F(a), and F(b).

   c.   Approximate F(3), F(9), F(27), and F(81). Make a conjecture about the relation between $F(a^n)$ and F(a).

   d. Approximate sufficient points to graph the function for x>0. What name is given to this function?

***
   Not all functions which are defined by integrals have the variable as one of the limits of integration. In exercises 10-13 use program INTEGRAL (Section 10.1), ANTIDERIVATIVE or any other program to make a conjecture about the value of the functions.

   10. $F(m,n) = \int_{-\pi}^{\pi} \sin mx \sin nx \, dx$

   11. $F(m,n) = \int_{-\pi}^{\pi} \cos mx \cos nx \, dx$

   12. $F(m,n) = \int_{-\pi}^{\pi} \sin mx \cos nx \, dx$

Be sure to consider the case when m=n.

13. a. Approximate $\lim_{r \to \infty} \int_{0}^{1} \sin rx \, dx$

b. Approximate $\lim_{r \to \infty} \int_{0}^{1} x^2 \sin rx \, dx$

c. Make a conjecture about the value of

$$\lim_{r \to \infty} \int_{a}^{b} f(x) \sin rx \, dx$$

if $f(x)$ is any integrable function of $[a,b]$. This result is known as the Riemann-Lebesgue Lemma and along with the results of exercises 10-12 plays an important part in the theory of Fourier series.

Ch. 12   ANTIDERIVATIVES AND APPROXIMATIONS

12.2 APPROXIMATING A FUNCTION

As mentioned in the introduction of this chapter the antiderivative of a function need not be expressible in terms of elementary functions but only in the more complicated form

$$\int_a^x f(t)\ dt$$

However, at times it may be helpful to have an elementary function which will approximate the antiderivative. In this section we will develop a method which will approximate

$$\int_a^x f(t)\ dt$$

by a polynomial. In fact, the method developed will approximate any function for which several functional values are known by a polynomial. This method is actually polynomial interpolation, and the particular interpolating polynomial which we generate is called the Lagrange interpolating polynomial.

Lagrange Interpolating Polynomial

Suppose that for n+1 different values of x, not necessarily evenly spaced, we know the corresponding values of f(x). Letting the n+1 different values of x be x_0, x_1, ..., x_n, then the nth degree Lagrange interpolating polynomial is given by

$$p_n(x) = f(x_0)L_0(x) + f(x_1)L_1(x) + \ldots + f(x_n)L_n(x)$$

where each $L_i(x)$ is the nth degree polynomial

$$L_i(x) = \frac{(x-x_0)(x-x_1)\ldots(x-x_{i-1})(x-x_{i+1})\ldots(x-x_n)}{(x_i-x_0)(x_i-x_1)\ldots(x_i-x_{i-1})(x_i-x_{i+1})\ldots(x_i-x_n)}$$

Notice that for any $x_j \neq x_i$ one of the factors in the numerator will be zero, and, thus, $L_i(x_j)=0$. For $x=x_i$ the numerator and denominator are equal, and $L_i(x_i)=1$. Thus, the interpolating polynomial has the property that $p_n(x_i)=f(x_i)$. In fact, this property is required by any interpolation scheme. Also note that is n+1 points on the graph of the function are known, then the Lagrange interpolating polynomial will be a polynomial of degree n formed by adding together n+1 polynomials of degree n, $f(x_i)L_i(x)$.

Example 1: Determine the Lagrange interpolating polynomial for a function which passes through the points (0,-5), (1,0) and (2,7).

Since we are given three points, the interpolating polynomial will be of degree 2 and have the form

$$p_2(x)=f(x_0)L_0(x)+f(x_1)L_1(x)+f(x_2)L_2(x)$$

where $x_0=0$, $x_1=1$, and $x_2=2$.

Step 1. Finding $L_0(x)$

$$L_0(x)=\frac{(x-x_1)(x-x_2)}{(x_0-x_1)(x_0-x_2)}=\frac{(x-1)(x-2)}{(0-1)(0-2)}=\frac{x^2-3x+2}{2}$$

Step 2. Finding $L_1(x)$

$$L_1(x)=\frac{(x-x_0)(x-x_2)}{(x_1-x_0)(x_1-x_2)}=\frac{(x-0)(x-2)}{(1-0)(1-2)}=\frac{x^2-2x}{-1}$$

Step 3. Finding $L_2(x)$

$$L_2(x)=\frac{(x-x_0)(x-x_1)}{(x_2-x_0)(x_2-x_1)}=\frac{(x-0)(x-1)}{(2-0)(2-1)}=\frac{x^2-x}{2}$$

Step 4. Finding $p_2(x)$

$$p_2(x) = f(x_0)L_0(x) + f(x_1)L_1(x) + f(x_2)L_2(x)$$

$$= -5\frac{(x^2-3x+2)}{2} + 0\frac{(x^2-2x)}{-1} + 7\frac{(x^2-x)}{2}$$

$$= -2.5x^2 + 7.5x - 5 + 3.5x^2 - 3.5x$$

$$= x^2 + 4x - 5$$

Check: $p_2(0) = -5$, $p_2(1) = 1+4-5 = 0$, $p_2(2) = 4+8-5 = 7$.

Although an interpolating polynomial does pass through the given points on the graph of the function, it may not pass through any other points of the graph--it is only an approximation. Also, the approximation is only valid on the smallest interval containing x_0 through x_n and may be extremely inaccurate outside of this interval. See the graph below.

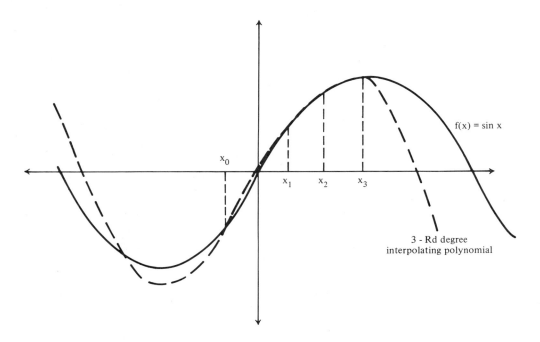

f(x) = sin x

3 - Rd degree
interpolating polynomial

Programming Problem: Approximate the Lagrange interpolating polynomial.

Output: The output will list the power of x and the coefficient for each term of the interpolating polynomial.

Input: The number of points and the actual points which will be used in forming the interpolating polynomial.

Strategy: The most difficult part of the program will be in determining the numerator of the polynomial $L_i(x)$ which is the product of n linear factors of the form $x-x_j$ with $j=0,\ldots,n$ and $j \neq i$. Although there are several different approaches for determining the product of linear factors, in this program we will achieve the product by multiplying each linear factor with the polynomial obtained by multiplying the preceeding linear factors. An example of this method is presented below.

$$(x-1)(x-2)(x-3)(x-4) = (x^2-3x+2)(x-3)(x-4)$$

$$= (x^3-6x^2+11x-6)(x-4)$$

$$= x^4-10x^3+35x^2-50x+24$$

Formalizing the process, we show how the coefficients of a specific power of x is obtained when the polynomial is multiplied by a linear factor. The coefficient of x^k, $0 \leq k \leq j+1$ in the product

$$(A_j x_j + \ldots + A_1 x + A_0)(x-a)$$

is given by $A_{k-1} - aA_k$ if $k>0$ and by $-aA_0$ if $k=0$.

Two final observations are in order before we write the program. In calculating the coefficients we must start with the highest power of x and work down since the new coefficient of the x^k power requires the old coefficient of the x^{k-1} power. Also, before we start multiplying the linear factors, we must decide whether $(x-x_0)$ or $(x-x_1)$ will be the first factor.

Program:

```
100 REM ** INTERPOLATING POLYNOMIAL **
110 DIM F(50),X(50),A(50),C(50)
120 REM ** INPUT **
130 PRINT "ENTER THE NUMBER OF POINTS TO"
140 PRINT "BE USED IN THE INTERPOLATION."
150 PRINT "MAXIMUM NUMBER OF POINTS IS 50."
160 INPUT M
170 IF M>50 THEN 130
180 LET N=M-1                    determining degree of
190 PRINT                            polynomial
200 FOR I=0 TO N
210 LET C(I)=0                    initializing array C
220 PRINT "ENTER POINT ";I+1;":"
230 INPUT X(I),F(I)
240 NEXT I
250 REM ** MAIN PROGRAM **
260 FOR I=0 TO N
270 FOR J=0 TO N                  initilizing the array A for
280 LET A(J)=0                        calculating the next L
290 NEXT J
300 REM ** CALCULATING L FOR A GIVEN I **
310 REM ** CALCULATING DENOMINATOR **
320 LET D=1
330 FOR J=0 TO N
340 IF I=J THEN 360               skipping mult.  by X-X(I)
350 LET D=D*(X(I)-X(J))
360 NEXT J
370 REM ** CALCULATING NUMERATOR **
380 LET A(0)=-X(0)
390 LET A(1)=1
400 LET S=1
410 IF I<>0 THEN 450              deciding which is the first
420 LET A(0)=-X(1)                        linear term
430 LET S=2
440 REM ** MULTIPLYING BY X-X(J) **
450 FOR J=S TO N
460 IF I=J THEN 510               skipping mult. by X(I)-X(I)
470 FOR K=N TO 1 STEP -1
480 LET A(K)=A(K-1)-X(J)*A(K)     finding coeff. of X^K in num.
490 NEXT K
500 LET A(0)=-X(J)*A(0)           finding the constant
510 NEXT J
520 REM ** ADDING NEW POLYNOMIAL **
530 REM ** TO PREVIOUS ONES **
540 FOR K=0 TO N
550 LET T=(F(I)*A(K))/D           finding coeff. of X^K in L
560 LET C(K)=C(K)+T
570 NEXT K
```

```
580 NEXT I
590 REM ** OUTPUT **
600 PRINT
610 PRINT "INTERPOLATING POLYNOMIAL"
620 PRINT
630 FOR I=N TO 0 STEP -1
640 PRINT "COEFF.  OF ";I;"POWER IS ";C(I)
650 NEXT I
660 END
```

Test Data: From example 1 we know that the Lagrange interpolating polynomial for the points $(0,-5)$, $(1,0)$ and $(2,7)$ is x^2+4x-5.

```
RUN
ENTER THE NUMBER OF POINTS TO
BE USED IN THE INTERPOLATION.
MAXIMUM NUMBER OF POINTS IS 50.
?3

ENTER POINT 1:
?0,-5
ENTER POINT 2:
?1,0
ENTER POINT 3:
?2,7

INTERPOLATING POLYNOMIAL

COEFF. OF 2 POWER IS 1
COEFF. OF 1 POWER IS 4
COEFF. OF 0 POWER IS -5
```

Thus, $p_2(x) = x^2+4x-5$

Example 2: Approximate the third degree Lagrange interpolating polynomial for sin x and investigate its accuracy.

Using a calculator to obtain the value of sin x for x=1, 2, 2.5, 3.1415 and program INTERPOLATING POLYNOMIAL we can obtain a third degree interpolating polynomial.

```
RUN
ENTER THE NUMBER OF POINTS TO
BE USED IN THE INTERPOLATION.
MAXIMUM NUMBER OF POINTS IS 50.
?4

ENTER POINT 1:
?1,.841470984
ENTER POINT 2:
?2,.909297427
ENTER POINT 3:
?2.5,.598472143
ENTER POINT 4:
?3.1415,.000092654

INTERPOLATING POLYNOMIAL

COEFF. OF 3 POWER IS .0873630121
COEFF. OF 2 POWER IS -.940147907
COEFF. OF 1 POWER IS 2.27672909
COEFF. OF 0 POWER IS -.582473201
```

Thus, using only four decimal places

$$p_3(x) = .0874x^3 - .9401x^2 + 2.2767x - .5825$$

The table below gives $p_3(x)$ for various values of x.

x	p (x)	sin x (calculator)	Error
0	-.5825	0	.5825
0.5	.33175	0.479425538	.1477
1.5	1.0123	0.997494986	.0148
3.0	.1465	0.141120008	.0054
4.0	-.9237	-.756802495	.1669
10.0	15.5745	-.54402111	16.1185

From our earlier discussion the interpolating polynomial obtained
above is guaranteed to be an approximation only on the interval
[1,3.1415] which is illustrated by the above table.

Example 3: Approximate the antiderivative of f(x)+1/x by a fifth degree polynomial on the interval [.25,4].

Since f(x)=1/x is continuous on [.25,4], we know that its antiderivative can be given by

$$\int_{1}^{x} dt/t$$

Using program ANTIDERIVATIVE to approximate six points on the graph, output not shown, with x values in the interval [.25,4], including the endpoints, and then using program INTERPOLATING POLYNOMIAL a fifth degree polynomial approximating the antiderivative can be obtained.

```
RUN
ENTER THE NUMBER OF POINTS TO
BE USED IN THE INTERPOLATION.
MAXIMUM NUMBER OF POINTS IS 50.
?6

ENTER POINT 1:
?.25,-1.386280629
ENTER POINT 2:
?.5,-.693127652
ENTER POINT 3:
?1.5,.4054506428
ENTER POINT 4:
?2,.693127652
ENTER POINT 5:
?3,1.098589143
ENTER POINT 6:
?4,1.386280629

INTERPOLATING POLYNOMIAL

COEFF. OF 5 POWER IS .0402216314
COEFF. OF 4 POWER IS -.463143983
COEFF. OF 3 POWER IS 2.03908206
COEFF. OF 2 POWER IS -4.40086949
COEFF. OF 1 POWER IS 5.28484441
COEFF. OF 0 POWER IS -2.46252817
```

The antiderivative is approximated on [.25,4] by
$p_5(x)=.0402x^5-.4631x^4+2.0391x^3-4.4009x^2+5.2848x-2.4625$

273

Exercise Set 12.2

In exercises 1-3 use program INTERPOLATING POLYNOMIAL to approximate the Lagrange interpolating polynomial for the function whose graph contains the given point. Also, complete a table similar to that of example 2 to investigate the accuracy of the interpolating polynomial. $F(x)$ is the function used to generate the points.

1. | x | $F(x_i)$ |
    |----|----|
    | -1 | -3 |
    | 0 | 1 |
    | 1 | 3 |

    $F(x) = x^3 - x^2 + 2x + 1$

2. | x | $F(x_i)$ |
    |----|----|
    | -1 | .5 |
    | 0 | 1 |
    | 1 | 2 |
    | 2 | 4 |

    $F(x) = 2^x$

3. | x | $F(x_i)$ |
    |----|----|
    | -1 | .3679 |
    | 0 | 1 |
    | 2 | 7.3891 |
    | 3 | 20.0855 |

    $F(x) = e^x$

4.   Approximate the antiderivative of $f(x) = 1/(1-x^2)$ by a third degree polynomial on the interval $[.1, .5]$.

*

5. a. If $f(x)$ is approximated by the nth degree interpolating polynomial $p_n(x)$ on $[x_0, x_n]$, then for any $\bar{x} \in [x_0, x_n]$ the error in the estimation is given by

$$Error(\bar{x}) = f(\bar{x}) - p_n(\bar{x}) \leq |f^{(n+1)}(a)| \frac{|(\bar{x}-x_0)\ldots(\bar{x}-x_n)|}{(n+1)!}$$

for some $a \in [x_0, x_n]$. Thus, if $f^{(n+1)}(x)$ can be bounded on $[x_0, x_n]$, $|f^{(n+1)}(x)| < M$, then the error could be bounded by

$$\text{Error}(\bar{x}) \quad < \left| \frac{M(\bar{x}-x_0)\ldots(\bar{x}-x_n)}{(n+1)!} \right|$$

Using the table of example 2 compare the computed error with max error given by the above formula. Remember that $x \in [x_0, x_n]$.

b. Discuss why a computed error using the computer may exceed the maximum predicted error.

**

6. Modify the program INTERPOLATING POLYNOMIAL to include a provision for evaluating the interpolating polynomial for values of x entered by the user. Use appropriate test data to verify your modification.

***

7. a. For each of the polynomials below compute sufficient points on the graph and then use program INTERPOLATING POLYNOMIAL to determine an interpolating polynomial of the same degree as the given polynomial.

$$\text{i)} \quad f(x)=3x^2+2x-1$$

$$\text{ii)} \quad f(x)=x^3-4x^2+2x+4$$

$$\text{iii)} \quad f(x)=x^4+x^2-1$$

b. Make a conjecture concerning the interpolating polynomial of degree n for a polynomial of degree n.

c. Prove the conjecture of part b.

8. Besides the Lagrange form for interpolating polynomials, there are many other forms. (For a discussion of interpolating polynomials see a text on numerical analysis.) However, all interpolating polynomial forms have the following two properties:

a. Given n+1 points on the curve the interpolating polynomial is of degree n.

b. If $(x_i, f(x_i))$ is one of the given points on the curve and $p_n(x)$ is the interpolating polynomial, then $p_n(x_i)=f(x_i)$.

Prove that any two interpolating polynomials formed from the same set of points are identical.   <u>Hint</u>: How many roots does a polynomial of degree n have?

9. Simpson's Rule results from approximating the function being integrated by a polynomial of degree 2 or less on two successive subintervals and then integrating this polynomial on   these subintervals.   As   a   variation   of   Simpson's   Rule   we   could approximate the function being integrated by an nth degree polynomial on n successive subintervals and then integrate this polynomial.   Hopefully, this would lead to more accurate   results. Using this idea approximate

$$\int_0^1 xe^x \, dx$$

by using a polynomial of degree 8.

<u>Hint</u>: Consider the following steps.

   a.   Divide [0,1] into eight subintervals of equal length with endpoints x_0, x_1,..., x_8.

   b.   For each x_i determine the point $(x_i, x_i \exp(x_i))$

   c.   Use the nine points of step b to   determine   an interpolating polynomial of degree eight.

   d.   Integrate the polynomial of step c.

CHALLENGE ACTIVITY

The most direct approach to determining the interpolating polynomial of degree n given the $n+1$ points $(x_0, f(x_0)), \ldots, (x_n, f(x_n))$ is to assume the interpolating polynomial has the form

$$p_n(x) = A_n x^n + A_{n-1} x^{n-1} + \ldots + A_1 x + A_0$$

and use substitution making use of the fact that for an interpolating polynomial $p_n(x_i) = f(x_i)$. We then obtain the following system of equations in which $A_0, \ldots, A_n$ are the unknowns.

$$f(x_0) = A_n x_0^n + A_{n-1} x_0^{n-1} + \ldots + A_1 x_0 + A_0$$

$$f(x_1) = A_n x_1^n + A_{n-1} x_1^{n-1} + \ldots + A_1 x_1 + A_0$$

$$\vdots$$

$$f(x_n) = A_n x_n^n + A_{n-1} x_n^{n-1} + \ldots + A_1 x_n + A_0$$

Write a program to solve this system of equations and thus obtain the interpolating polynomial. You may use any method, such as Cramer's Rule or Gauss Elimination, to solve the system, but your program must be designed to handle any number of points. As test data use the points $(-1,6)$, $(2,27)$, $(1,4)$ and $(0,1)$ which determine the interpolating polynomial

$$p_3(x) = 2x^2 + 4x^2 - 3x + 1$$

When using Cramer's Rule or Gauss Elimination simplify the program by assuming that the matrix will be entered with no zero values on the main diagonal. Other methods which you may want to try are Jacobi iteration or Gauss-Seidel iteration.

CHAPTER 13
Improper Integrals

The average velocity of a hydrogen molecule at room temperature can be approximated by the integral

$$0.006 \int_{-\infty}^{\infty} x^3 \exp(-x^2/50) \, dx$$

However, this integral is unlike those encountered in previous chapters in that the limits of integration are unbounded or infinite. Integrals of this type along with those for which the function being integrated has an infinite discontinuity are referred to as improper integrals. In this chapter we will study both types of improper integrals as well as several of their applications.

13.1 INFINITE LIMITS

Programming Problem: Estimate the improper integral

$$\int_{a}^{\infty} f(x) \, dx$$

Output: A table listing the value used as the upper limit of integration and the approximation of the corresponding definite integral. Also, appropriate messages which summarize the results.

Input: The lower limit of integration and the function being integrated.

Strategy: The improper integral

$$\int_a^\infty f(x)\ dx$$

is evaluated by using

$$\lim_{b\to\infty}\int_a^b f(x)\ dx$$

Thus, we should be able to estimate the integral by a process similar to that used in section 6.1 for infinite limits--namely, approximate

$$\int_a^b f(x)\ dx$$

for increasingly larger values of b until two successive approximations are within the desired precision. We will use Simpson's Rule to approximate the integrals, starting with an upper limit of 25 and doubling the upper limit for each approximation. However, we must also consider the possibility that the improper integral does not converge and provide a way to terminate the program if this appears to be the case.

Since the improper integral can diverge by either becoming unbounded or by remaining bounded but not approaching a limit, two different tests for divergence should be incorporated into the program. Assuming that the integral diverges if the absolute value of an approximation exceeds 10^6 provides an easy test for integrals which become unbounded. However, the case where the integral remains bounded yet does not have a limit is much more difficult since successive approximations may be quite small in magnitude yet never be within the desired precision of each other. We will assume that this type of divergence is occurring if the integral has not converged by the time the upper limit of integration is 200. It should be noted, however, that this may cause us to conclude that some improper integrals diverge when they actually converge.

The selection of 10^6 as the criteria for the integral becoming unbounded and 200 as the bound on the upper limit of integration by which we must have convergence are arbitrary and can be replaced by other values. To facilitate the change of these threshold values we will initialize the variable D to the value for the criteria used in deciding if the integral becomes unbounded and the variable E to values of the highest upper limits of integration which will be allowed. Thus, it will be possible to change the threshold values by changing the values in the assignment statements.

Summarizing, we will use the following steps in deciding if the integral

$$\int_a^\infty f(x)\ dx$$

converges.

1. Input the lower limit of integration.

2. Initialize the threshold values and the upper limit of integration.

3. Use Simpson's Rule to approximate the integral.

4. Test to see if the integral is unbounded. If so, print an appropriate message and end.

5. Test to see if the improper integral has converged. If so, print an appropriate message and end.

6. Test to see if the upper limit of integration is equal to E. If so, print an appropriate message and end.

7. Double the upper limit of integration and return to step 3.

Although this procedure will work, we will save a large amount of computer time if we perform the test for an unbounded integral after each approximation within the Simpson's Rule process rather than after we have approximated the integral for a given upper bound. This will save time because, if the integral is unbounded, then all the approximations in the Simpson's Rule process will be quite large, and it will require a large number of subdivisions before two successive approximations are within the desired precision .0001.

Program:

```
100 REM ** INFINITY **
110 REM ** FUNCTION **
120 DEF FNF(X)=.....
130 PRINT "ENTER THE LOWER LIMIT OF INTEGRATION."
140 INPUT A
150 REM ** INITIALIZING THRESHOLD VALUES **
160 LET D=1000000
170 LET E=200
180 REM ** INITIALIZING UPPER LIMIT **
190 REM ** OF INTEGRATION **
200 LET B=25
210 PRINT
220 REM ** HEADINGS **
230 PRINT "UPPER LIMIT";TAB(17);"APP.  OF INTEGRAL"
240 PRINT
250 REM ** VARIABLE USED TO SAVE THE LAST **
260 REM ** APPROXIMATION OF THE INTEGRAL **
270 LET L=0
280 REM ** SIMPSON'S RULE **
290 LET S2=0
300 LET N=20
310 LET DX=(B-A)/N
320 LET S1=FNF(A)+FNF(B)
330 FOR I=1 TO (N-1) STEP 2
340 LET X=A+I*DX
350 LET S1=S1+4*FNF(X)
360 NEXT I
370 FOR I=2 TO (N-2) STEP 2
380 LET X=A+I*DX
390 LET S1=S1+2*FNF(X)
400 NEXT I
410 LET S1=S1*DX/3
420 GOTO 520
430 IF ABS(S1-S2)<.0001 THEN 470
440 LET N=2*N
450 LET S2=S1
460 GOTO 310
470 PRINT B;TAB(17);S1
480 GOTO 600
490 REM ** VALUE OF S1 IS NEW APPROXIMATION **
500 REM ** OF THE INTEGRAL **
510 REM ** TESTING FOR UNBOUNDED INTEGRAL **
520 IF S1<D THEN 430
530 PRINT
540 PRINT "FOR AN UPPER LIMIT OF INTEGRATION"
550 PRINT "EQUAL TO ";B;" THE INTEGRAL IS"
560 PRINT "APPROXIMATELY ";S1;" AND "
570 PRINT "APPEARS TO BE UNBOUNDED."
```

```
580 GOTO 730
590 REM ** TESTING FOR CONVERGENCE **
600 IF ABS (S1-L)>= .0001 THEN 660
610 PRINT
620 PRINT "THE INTEGRAL APPEARS TO CONVERGE"
630 PRINT "TO A VALUE OF ";S1
640 GOTO 730
650 REM ** TESTING FOR UPPER LIMIT EQUAL TO E **
660 IF B>=E THEN 700
670 LET L=S1
680 LET B=2*B
690 GOTO 290
700 PRINT
710 PRINT "THE INTEGRAL HAS NOT CONVERGED BY THE"
720 PRINT "TIME THE UPPER LIMIT REACHED ";E;"."
730 END
```

Test Data:

a. $\displaystyle\int_0^\infty e^{-x}\ dx = \lim_{b\to\infty} \int_0^b e^{-x}\ dx =$

$\displaystyle\lim_{b\to\infty}\ -e^{-x}\ \Big|_0^b = \lim_{b\to\infty}\ -e^{-b}+1 = 1$

```
120 DEF FNF(X)=EXP(-X)
```

```
RUN
```
ENTER THE LOWER LIMIT OF INTEGRATION.
```
?0
```

UPPER LIMIT	APP. OF INTEGRAL
25	1.0000033
50	1.0000033

THE INTEGRAL APPEARS TO CONVERGE
TO A VALUE OF 1.0000033

b. $\displaystyle\int_0^\infty e^x\ dx = \lim_{b\to\infty} \int_0^b e^x\ dx =$

$$\lim_{b\to\infty} e^x \Big|_0^b = \lim_{b\to\infty} e^b - 1 = \infty$$

```
120 DEF FNF(X)=EXP(X)
```

RUN
ENTER THE LOWER LIMIT OF INTEGRATION.
?0

UPPER LIMIT   APP. OF INTEGRAL

**FOR AN UPPER LIMIT OF INTEGRATION
EQUAL TO 25 THE INTEGRAL IS
APPROXIMATELY 7.28255601E+10 AND
APPEARS TO BE UNBOUNDED.**

c. $\displaystyle\int_0^\infty \sin x \, dx = \lim_{b\to\infty} -\cos x \Big|_0^b = \lim_{b\to\infty} -\cos b + 1$

which diverges.

```
120 DEF FNF(X)=SIN(X)
```

RUN
ENTER THE LOWER LIMIT OF INTEGRATION.
?0

UPPER LIMIT	APP. OF INTEGRAL
25	8.79766091E-03
50	.0350358578
100	.137681595
200	.512814034

**THE INTEGRAL HAS NOT CONVERGED BY THE
TIME THE UPPER LIMIT REACHED 200.**

Although the approximations for the integral remain bounded, they do not appear to approach a limit. Thus, we should conclude that the integral does not converge.

Example 1: For the figure of revolution obtained by revolving the curve xy=1, x≥1, about the x-axis:

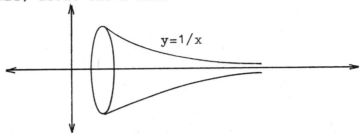

a.   Estimate the area under the curve xy=1, x≥1, and the x-axis.

b.   Estimate the volume enclosed by the figure of revolution.

c.   Estimate the surface area of the figure of revolution.

a. The area is given by the integral $\displaystyle\int_1^\infty dx/x$


```
    120 DEF FNF(X)=1/X

    RUN
    ENTER THE LOWER LIMIT OF INTEGRATION.
    ?1
```

UPPER LIMIT	APP. OF INTEGRAL
25	3.21887687
50	3.91202413
100	4.60517137
200	5.29831856

THE INTEGRAL HAS NOT CONVERGED BY THE TIME THE UPPER LIMIT REACHED 200.

Since the approximations are getting larger, it is likely that the area is actually unbounded even though the output yields finite values.

b. The volume is given by the integral $\pi \int_1^\infty dx/x^2$

Using the program INFINITY to approximate $\int_1^\infty dx/x^2$

we can then approximate the volume by multiplying by π.

```
120 DEF FNF(X)=1/(X*X)

RUN
```
ENTER THE LOWER LIMIT OF INTEGRATION.
?1

UPPER LIMIT	APP. OF INTEGRAL
25	.960004139
50	.980004493
100	.990004671
200	.995004765

THE INTEGRAL HAS NOT CONVERGED BY THE
TIME THE UPPER LIMIT REACHED 200.

Even though we have a message that the integral appears to diverge, it does appear that the approximations may be converging toward a limit. To investigate further we can increase the threshold values for the upper limit to 12800 and rerun the program.

```
170 LET E=12800

RUN
ENTER THE LOWER LIMIT OF INTEGRATION.
?1

UPPER LIMIT        APP. OF INTEGRAL

25                 .960004139
50                 .980004493
100                .990004671
200                .995004765
400                .99750482
800                .998754826
1600               .999379849
3200               .999692335
6400               .999848573
12800              .999925877

THE INTEGRAL APPEARS TO CONVERGE
TO A VALUE OF .999925877
```

Considering round-off error we might conclude that the actual limit is 1 which is the correct value of the integral.

c.   The surface area of a curve revolved about the x-axis is given by

$$2\pi \int f(x) \sqrt{1 + (f'(x))^2} \ dx$$

Since $f(x) = 1/x$, $f'(x) = -1/x^2$ and the surface area is given by

$$2\pi \int_{1}^{\infty} 1/x \sqrt{1 + (1/x^4)} \ dx$$

As in part b we can approximate the surface area by multiplying by 2π.

```
120 DEF FNF(X)=SQR(1+1/(X*X*X*X))/X

RUN
```
ENTER THE LOWER LIMIT OF INTEGRATION.
```
?1
```

UPPER LIMIT	APP. OF INTEGRAL
25	3.33186973
50	4.02501728
100	4.71816446
200	5.41131173

THE INTEGRAL HAS NOT CONVERGED BY THE TIME THE UPPER LIMIT REACHED 200.

For the same reason as used in part a the integral is most likely unbounded.

This example illustrates that an infinite surface can enclose a finite volume. Also, it demonstrates the necessity of interpreting carefully output produced by the computer.

Ch. 13   IMPROPER INTEGRALS

Exercise Set 13.1

In exercises 1-6 use program INFINITY to estimate the given improper integral.

1. $\displaystyle\int_{3}^{\infty} dx/\sqrt{x+1}$

2. $\displaystyle\int_{0}^{\infty} (\sin x)/(x^2+1)\ dx$

3. $\displaystyle\int_{0}^{\infty} x/\sqrt{x^4+x^2+3}\ dx$

4. $\displaystyle\int_{1}^{\infty} \sin x/x^2\ dx$

5. $\displaystyle\int_{1}^{\infty} ((\ln(1+x))/x^3)^2\ dx$

6. $\displaystyle\int_{2}^{\infty} (x^2-1)/\sqrt{x^{18}+16}\ dx$

7. For the figure of revolution obtained by revolving the curve $y=(\ln x)/x$, $x\geq1$, about the x-axis, estimate:

a. the area under the curve and above the x-axis.

b. the volume bounded by the figure of revolution.

c. the surface area of the figure of revolution.

Hint: See example 1.

8. Same as exercise 7 except for the curve $y=\exp(-x^2)$, $x\geq0$.

*
9. The force due to gravity is given by the equation $f(x)=k/r^2$
where r is the distance between the two objects. Thus, the work,
neglecting all other forces, in moving an object against gravity
is

$$\int_{r1}^{r2} k/r^2 \, dr$$

For example, the work needed to move a 5 ton rocket 1000 miles
above the earth would be

$$\int_{4000}^{5000} \frac{80000000}{r^2} \, dr$$

The radius of the earth is 4000 miles, and, since weight is a
force, we could determine k by solving

$$5 = \frac{k}{(4000)^2} \quad \text{force at the surface of the earth}$$

Use program INFINITY to estimate how much work would be required
to send the rocket an infinite distance away from the earth.

Hint: Bring k outside the integral and change the limits of
integration used in the program.

10. Modify program INFINITY to estimate the improper integral

$$\int_{-\infty}^{b} f(x) \, dx$$

Use as test data

$$\int_{-\infty}^{0} e^x \, dx = 1 \quad \text{and} \quad \int_{-\infty}^{0} e^{-x} \, dx$$

which is unbounded.

11.   Use the program from exercise 10 to estimate

$$\int_{-\infty}^{0} e^{x} \cos x \, dx$$

12.   Use the program written in exercise 10, the program INFINITY, and the fact that

$$\int_{a}^{b} f(x) \, dx = \int_{a}^{c} f(x) \, dx + \int_{c}^{b} f(x) \, dx$$

to estimate the improper integral $\int_{-\infty}^{\infty} \exp(-x^2) \, dx$

**
13.   Modify an existing program or write a program which will estimate the improper integral

$$\int_{-\infty}^{\infty} f(x) \, dx \qquad by \qquad \lim_{a \to \infty} \int_{-a}^{a} f(x) \, dx$$

a. Use this program to estimate $\int_{-\infty}^{\infty} x^3 \, dx$ and $\int_{-\infty}^{\infty} x^2 \, dx$

b.   Evaluate these integrals to determine the exact value.

c.   Are there any discrepancies between parts a and b?

d.   Explain your answer to part c.

***
14.   The integral $\int_{0}^{\infty} x^{n-1} e^{-x} \, dx$

occurs frequently in the study of differential equations and probability functions. Known as the gamma function, this integral is denoted by $\Gamma(n)$.

a.   Using program INFINITY estimate $\Gamma(2)$, $\Gamma(3)$, $\Gamma(4)$, $\Gamma(5)$.

b.   Using the results of part a and further evaluations of $\Gamma(n)$ for different n, determine an expression for $\Gamma(n+1)$ which does not involve an integral.

c.   Another expression for $\Gamma(n+1)$ is

$$\Gamma(n+1) = \sqrt{2\pi n}\ n^n\ e^{-n}\ \exp(\theta/12(n+1))\ ;\ 0<\theta<1$$

However, this is not the intended answer to part b. If n is large, then $\exp(\theta/12(n+1))$ is very close to 1 and

$$\Gamma(n+1) \approx \sqrt{2\pi n}\ n^n\ e^{-n}$$

Verify this formula for n=68 by estimating $\Gamma(n)$ using the expression in part b and $\sqrt{2\pi(68)}\ 68^{68}\ e^{-68}$ with a calculator. The expression $\sqrt{2\pi n}\ n^n\ e^{-n}$ is called Stirling's factorial approximation.

d.   Use program INFINITY to estimate $\Gamma'(1) = \displaystyle\int_0^\infty e^{-x} \ln x\ dx$

The answer is the negative of Euler's constant, $\gamma = .577215....$ An interesting fact about γ is that it is not known whether it is a rational or an irrational number.

e.   Determine a formula for $\displaystyle\int_0^\infty x^n e^{-x}\ dx$

without using the integral symbol.

The gamma function in part e is sometimes referred to as Euler's gamma function since this integral was his solution to the problem of finding a function whose values for any positive integer n is n!

13.2 INFINITE DISCONTINUITIES

In section 13.1 we investigated integrals in which one of the limits of integration was infinity. Another type of improper integral occurs when the function being integrated becomes infinite at one of the limits of integration. For example, consider the following integral:

$$\int_{1}^{3} dx/(x-1)^{1/4}$$

At the lower limit of integration, 1, the function being integrated, $f(x)=1/(x-1)^{1/4}$, becomes infinite. In a manner similar to what is done for improper integrals with infinite limits of integration, the value of the integral is defined to be the limit

$$\lim_{c \to 1^{+}} \int_{c}^{3} dx/(x-1)^{1/4}$$

Or, in general, we have the following definition.

If $\lim_{x \to a^{+}} f(x) = \pm\infty$, then $\int_{a}^{b} f(x) \, dx = \lim_{c \to a^{+}} \int_{c}^{b} f(x) \, dx.$

The most direct approach to approximate this type of integral using a computer is:

1.   Select a value for the lower limit of integration "close" to but greater than "a".

2.   Approximate the resulting definite integral to the desired precision.

3.   Compare the approximations obtained in step 2 with previous approximations.

4.   If the last two approximations are within the desired precision, use it to estimate the improper integral. If not, choose a value for the lower limit "closer" to "a" and return to step 2.

Unfortunately, this approach has a serious drawback which makes it almost impossible to use on a computer. In order to have confidence in our estimation of the improper integral, we should choose values quite close to "a" for the lower limit of integration. Yet the number of subintervals required to approximate the definite integral to a desired precision grows rapidly as we choose values closer to "a".

To illustrate, consider the integral from above. Suppose we wish to estimate

$$\int_{1}^{3} dx/(x-1)^{1/4}$$

to three decimal places. Choosing 1.001 as the lower limit of integration, we approximate the definite integral using the program SIMPSON'S RULE from chapter 11.

```
120 DEF FNF(X)=1/(X-1)^.25

RUN
ENTER THE LOWER LIMIT OF INTEGRATION.
?1.001
ENTER THE UPPER LIMIT OF INTEGRATION.
?3

  N        APP. SIMP. RULE
 10        2.42740976
 20        2.31588321
 40        2.26709448
 80        2.24671606
160        2.23876842
320        2.23597394
640        2.23513375

THE APPROXIMATION OF THE INTEGRAL
FROM 1.001 TO 3 IS 2.23513375
```

After 640 subdivisions, the last two approximations have an error less than .001. Thus, we would approximate the value of the definite integral to be 2.235 and feel confident that it is correct to three decimal places. However, at this point we have no way to decide if this would be a good estimation of the improper integral. Thus, we should choose a value closer to 1 as the lower limit of integration and approximate a second integral.

Choosing 1.00001, we obtain the following results for approximating

$$\int_{1.00001}^{3} dx/(x-1)^{1/4}$$

```
120 DEF FNF(X)=1/(X-1)^.25

RUN
```
ENTER THE LOWER LIMIT OF INTEGRATION.
```
?1.00001
```
ENTER THE UPPER LIMIT OF INTEGRATION.
```
?3
```

N	APP. SIMP. RULE
10	3.23993592
20	2.72337031
40	2.47229311
80	2.35103772
160	2.29295519
320	2.26542531
640	2.25255669
1280	2.24665257
2560	2.24401249
5120	2.24287416

THE APPROXIMATION OF THE INTEGRAL
FROM 1.00001 TO 3 IS 2.24287416

For this integral it takes 5120 subintervals before we have an approximation which is accurate to roughly three decimal places, the difference between the last two approximations being about .00114. Yet we still do not know if we have a good approximation of the improper integral since the approximations for the two integrals

$$\int_{1.001}^{3} dx/(x-1)^{1/4} \qquad \text{and} \qquad \int_{1.00001}^{3} dx/(x-1)^{1/4}$$

are not within 0.001 of each other. Thus, we should approximate the integral again with a lower limit of integration closer to 1 than 1.001 and 1.00001.

Instead of running the program again to approximate the next integral

$$\int_{1.000001}^{3} dx/(x-1)^{1/4}$$

we will try to determine the number of subintervals required to have an approximation accurate to .001 The discretization error for Simpson's Rule is bounded by

$$E \le \frac{K(b-a)^5}{180N^4} \text{ where } K=\max \left| f^{(4)}(x) \right| \text{ for all } x \in [a,b]$$

(See section 11.2 and exercise 10, section 11.2)

For

$$\int_{1.000001}^{3} dx/(x-1)^{1/4}$$

b=3, a=1.000001 and $f(x)=1/(x-1)^{1/4}$.

Thus, $f^{(4)}(x) = \dfrac{-585}{256} \dfrac{1}{(x-1)^{17/4}}$

which is a decreasing function on [a,b]

and

$$K = \max \frac{585}{256(x-1)^{17/4}} \quad , \quad x \in (1.000001, 3)$$

$$= \frac{585}{256(.000001)^{17/4}}$$

since the maximum will occur at the left-hand endpoint of [1.000001,3].

Therefore, if $E_n < .001$, we must have

$$\frac{\dfrac{585}{256(.000001)^{17/4}}(3-1.000001)^5}{180N^4} < .001$$

$$\frac{.40624898}{0.001(.000001)^{17/4}} < N^4$$

$$\frac{406.24898}{3.1622776 \times 10^{-26}} < N^4$$

$$1.284672 \times 10^{28} < N^4$$

$$10,646,284.72 < N$$

Even if the bound on N is off by a factor of 1000, it would still take about 11,000 subintervals and several hours of computer time!!

Due to this rapid increase in the number of subintervals, the time to approximate successive integrals becomes extremely long, thus reducing the usefulness of this approach in estimating improper integrals. Unfortunately, there are no other simple methods for estimating improper integrals.

Programming Problem: Estimate the value of an improper integral where the function being integrated becomes unbounded or has an infinite discontinuity at the lower limit of integration.

Output: A table listing the lower limit of integration used and an approximation of the corresponding definite integral for three different values of the lower limit.

Input: The function being integrated and the limits of integration.

Strategy: From the above discussion we know that in trying to estimate the improper integral

$$\int_a^b f(x)\ dx \quad \text{where} \quad \lim_{x \to a^+} f(x) = \pm\infty$$

two conflicting conditions exist.

1. As we choose values close to "a" an increasingly larger number of subdivisions, and, hence, computer time, is required to approximate the corresponding definite integral.

2. To obtain a good estimation of the improper integral values close to "a" must be used.

For our program a compromise between these two conditions will be used. We will select values somewhat close to "a" and limit the number of subdivisions used in the approximation of the definite integral. The restrictions on both conditions are necessary. If we only limit the number of subdivisions, then the approximation for the definite integrals with lower limits of integration chosen closer and closer to "a" would become increasingly inaccurate. On the other hand, if we restricted ourselves to choosing only values near "a" which do not require an excessive number of subdivisions to approximate the corresponding definite integral, the results would probably not be very close to the value of the improper integral.

Specifically, we will use Simpson's Rule with 640 subdivisions to approximate three definite integrals where the lower limits of integration are a+ .01, a+ .001, a+ .0001 These values were chosen arbitrarily as a compromise between accuracy and computer time. If the three approximations appear to "cluster" about a particular number, we will assume that the improper integral converges and estimate its value as the last approximation. Otherwise, we will assume that the improper integral diverges.

Program:

```
100 REM ** LOWER LIMIT **
110 REM ** FUNCTION TO BE INTEGRATED **
120 DEF FNF(X)=.....
130 REM ** INPUT **
140 PRINT "ENTER THE LOWER LIMIT OF INTEGRATION."
150 INPUT A
160 PRINT
170 PRINT "ENTER THE UPPER LIMIT OF INTEGRATION."
180 INPUT B
190 PRINT
200 REM ** MAIN PROGRAM **
210 REM ** FINDING THREE LOWER LIMITS **
220 LET X1=A+.01
230 LET X2=A+.001
240 LET X3=A+.0001
250 REM ** NUMBER OF SUBDIVISIONS **
260 LET N=640
270 REM ** SIMPSON'S RULE **
280 LET S1=FNF(X1)+FNF(B)
290 LET S2=FNF(X2)+FNF(B)
300 LET S3=FNF(X3)+FNF(B)
310 LET D1=(B-X1)/N
320 LET D2=(B-X2)/N
330 LET D3=(B-X3)/N
340 FOR I=1 TO (N-1) STEP 2
350 LET P1=X1+I*D1
360 LET P2=X2+I*D2
370 LET P3=X3+I*D3
380 LET S1=S1+4*FNF(P1)
390 LET S2=S2+4*FNF(P2)
400 LET S3=S3+4*FNF(P3)
410 NEXT I
420 FOR I=2 TO (N-2) STEP 2
430 LET P1=X1+I*D1
440 LET P2=X2+I*D2
450 LET P3=X3+I*D3
460 LET S1=S1+2*FNF(P1)
470 LET S2=S2+2*FNF(P2)
480 LET S3=S3+2*FNF(P3)
490 NEXT I
500 LET S1=S1*D1/3
510 LET S2=S2*D2/3
520 LET S2=S2*D3/3
530 REM ** OUTPUT **
540 PRINT "LOWER LIMIT";TAB(15);"APP.  OF INTEGRAL"
550 PRINT X1;TAB(15);S1
560 PRINT X2;TAB(15);S2
570 PRINT X3;TAB(15);S3
```

580 END

Test Data:

a. $\displaystyle\int_1^3 dx/(x-1)^{1/4} = \lim_{c\to 1}\int_c^3 dx/(x-1)^{1/4}$

$$= \lim_{c\to 1} 4/3\ (x-1)^{3/4}\ \Big|_c^3$$

$$= \lim_{c\to 1} [(4/3)(2)^{3/4} - (4/3)(c-1)^{3/4}]$$

$$= (4/3)(2)^{3/4}$$

$$\approx 2.242390441$$

120 DEF FNF(X)=1/(X-1)^.25

RUN
ENTER THE LOWER LIMIT OF INTEGRATION.
?1

ENTER THE UPPER LIMIT OF INTEGRATION.
?3

LOWER LIMIT	APP. OF INTEGRAL
1.01	2.20022775
1.001	2.23513375
1.0001	2.24399257

Since the three approximations appear to "cluster" about 2.2, we will assume that the improper integral

$$\int_1^3 dx/(x-1)^{1/4}$$

converges and estimate its value by 2.243992559. Note that our estimation is accurate to two decimal places.

b. $\displaystyle\int_1^2 dx/(x-1)^3 = \lim_{c \to 1^+} \int_c^2 dx/(x-1)^3$

$$= \lim_{c \to 1^+} (-1/2)(x-1)^2 \Big|_c^2$$

$$= \lim_{c \to 1^+} (-1/2) + 1/(2(c-1)^2)$$

$$= \infty$$

Thus, the integral diverges.

```
120 DEF FNF(X)=1/((X-1)*(X-1)*(X-1))

RUN
ENTER THE LOWER LIMIT OF INTEGRATION.
?1

ENTER THE UPPER LIMIT OF INTEGRATION.
?2

LOWER LIMIT    APP. OF INTEGRAL
1.01           5001.21344
1.001          681337.018
1.0001         521301886
```

Since the approximations of the integral are increasing we conclude that it diverges.

Example 1: Approximate the area bounded by the curve $y=1/(x^2 + 2\sqrt{x})$, $x=0$, $x=1$, and $y=0$.

The area bounded by the given curve is the improper integral

$$\int_0^1 dx/(x^2 + 2\sqrt{x})$$

```
120 DEF FNF(X)=1/(X*X+2*SQR(X))

RUN
ENTER THE LOWER LIMIT OF INTEGRATION.
?0

ENTER THE UPPER LIMIT OF INTEGRATION.
?1

LOWER LIMIT    APP. OF INTEGRAL
1.01           .801657039
1.001          .870290568
1.0001         .901040585
```

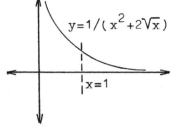

The area is approximately .9

Example 2: If the velocity at any time t of a particle is given by v(t), then the total distance travelled by the particle between times t_1 and t_2 is given by

$$\int_{t_1}^{t_2} v(t)\ dt$$

For the velocity function

$$v(t)=(t-1)e^x/(t^3 - 4t^2)$$

estimate the total distance travelled between times t=4 and t=20.

Since $\left| e^t (t-1)/(t^3 - 4t^2) \right| = e^t (t-1)/(t^3 - 4t^2)$, the total distance is given by

$$\int_{4}^{20} e^t (t-1)/(t^3 - 4t^2)\ dt$$

which can be estimated using program LOWER LIMIT.

```
120 DEF FNF(X)=(X-1)*EXP(X)/(X*X*X-4*X*X)

RUN
ENTER THE LOWER LIMIT OF INTEGRATION.
?4

ENTER THE UPPER LIMIT OF INTEGRATION.
?20

LOWER LIMIT     APP. OF INTEGRAL
1.01            1633307.2
1.001           1633389.34
1.0001          1634157.79
```

Although the approximations are quite large, they do appear to "cluster" about 1634157. Thus, we will estimate the total distance travelled by this value.

Exercise Set 13.2

In exercises 1-5 use program LOWER LIMIT to estimate the value of each integral.

1. $\displaystyle\int_{2}^{3} dx/(x-2)^{1/5}$

2. $\displaystyle\int_{2}^{5} dx/x^2(x^3 - 8)^{2/3}$

3. $\displaystyle\int_{0}^{\pi/2} e^{-x}(\cos x)/x \ dx$

4. $\displaystyle\int_{0}^{4} dx/x^x$

5. $\displaystyle\int_{0}^{\pi/2} \ln \sin x \ dx$

In exercises 6-8 use program LOWER LIMIT to estimate the area bounded by the indicated curves.

6. $y^2 = x^2/(x-1)$, $x=1$, $x=8$, $y=0$

7. $y=\exp(-x^2/2/x$, $x=0$, $x=5$, $y=0$

8. $y=(\cos x)/x$, $x=0$, $x=\pi/3$, $y=0$

9. Estimate the total distance travelled by a particle between time $t=5$ and $t=7$ if its velocity at any time t is given by $v(t)=1/\sqrt{(t-5)(t-8)^6}$.

*
10.   Estimate using program LOWER LIMIT the volume   enclosed   when the   plane   region   bounded   by x=1, y=0, x=4, and y=$(x^2+2)/(\sin(x-1))$ is rotated about the x-axis.

11.   Estimate the surface area if the curve y=$(x^2+2)/(\sin(x-1))$ between x=1 and x=4 is revolved about the x-axis.

**
12.   Modify program LOWER LIMIT to estimate the improper integral

$$\int_a^b f(x)\ dx$$

where $\lim_{x \to b} f(x) = \pm\infty$ and f(a) is defined.   As test data use

$$\int_0^1 dx/(1-x) \ = 3/2 \qquad \text{and} \qquad \int_0^1 dx/(1-x)^3$$

which is unbounded.

13.   Use   your   program written in exercise 12 to estimate each of the following integrals.

a. $\int_0^1 dx/[(1-x^2)(1-.5x^2)]$ (An elliptic integral)

b. $\int_0^2 dx/[(x+1)\ \sqrt{4-x^2}]$

c. $\int_0^5 [x^2/(5-x)^2]\ dx$

14. Use program LOWER LIMIT, the program written in exercise 12, and the fact that

$$\int_a^b f(x)\ dx= \int_a^c f(x)\ dx\ +\ \int_c^b f(x)\ dx$$

to estimate each of the following integrals.

a. $\int_1^2 dx/(x^2-3x+2)$

b. $\int_0^4 dx/(x-1)^5$

c. $\int_0^{\pi/2} (x^2+3)/(\tan x)\ dx$

CHALLENGE ACTIVITY

If f(x) is a probability density function associated with an experiment, then the probability that a randomly selected outcome is bounded between a and b is given by

$$\int_a^b f(x) \; dx$$

Also associated with a probability density function is its expected value or mean, X, and its variance, σ^2, which are defined by the improper integrals below.

$$X = \int_{-\infty}^{\infty} xf(x) \; dx \qquad \sigma^2 = \int_{-\infty}^{\infty} (x-\overline{X})^2 \; f(x) \; dx$$

a.   Write a program, which for a given probability density function f(x) will estimate both its mean and variance. Test your program using the unit normal probability density function

$$f(x)=(1/ \sqrt{2\pi})e^{-x*x/2}$$

which has a mean 0 and a variance of 1.

b.   In the introduction to this unit we gave an integral which would estimate the average velocity of a hydrogen molecule at room temperature.   This integral resulted from the above formula for the mean since the velocity of gas molecules have a Maxwell probability density function with parameter $\alpha = (2kt/m)^{1/2}$.

Maxwell probability density function:
$$f(x)=(4/\sqrt{\pi})(1/\alpha^3)x^2\exp(-x^2/\alpha^2)$$

T: absolute temperature (degrees Kelvin)=300 K for room temperature

m: mass of molecule=$(1.67)10^{-27}$ for hydrogen

k: Boltzmann's Constant=$(1.38)10^{-28}$

Using the program from part a estimate the average velocity and variance for hydrogen molecules at room temperature.

CHAPTER 14
Sequences And Series

In many instances in pure and applied mathematics results are not obtained through a single calculation but rather through an iterative process such as that used in approximating roots by Newton's Method. Such iterative processes generally produce what is called an infinite sequence or in some instances an infinite series. In this chapter we will investigate convergence and other related properties associated with sequences and series.

14.1 SEQUENCES

Programming Problem: Approximate the limit of an infinite sequence.

Output: A table listing the value of N and the corresponding N-th term.

Input: Several BASIC statements defining the n-th term, the index of the first and last term to be approximated and the increment of the index between successive terms being approximated.

Strategy: Since the limit of the infinite sequence {an} is found by taking the

$$\lim n \text{->} \infty \; a_n$$

we will approximate the limit in a manner similar to that used for infinite limits in Chapter 6. Namely, we will generate a table giving the values of terms in the sequence for increasingly larger values of n. If the numbers appear to be approaching a fixed value, we will use it as the approximation for the limit. If not, we will assume the sequence diverges.

Unfortunately, for many sequences the expression for the n-th term may become rather complicated. In these instances it will not be possible to express the n-th term by a single DEF statement, but it will require several BASIC statements which we will include in the program between lines 270 and 300.

Program:

```
100 REM ** SEQUENCE **
110 REM ** INPUT **
120 PRINT "ENTER THE INDEX OF THE FIRST"
130 PRINT " TERM TO BE PRINTED."
140 INPUT F
150 PRINT
160 PRINT "ENTER THE INDEX OF THE HIGHEST"
170 PRINT "TERM TO BE PRINTED."
180 INPUT H
190 PRINT
200 PRINT "ENTER THE INCREMENT BETWEEN TERMS."
210 INPUT S
220 PRINT
230 REM ** HEADINGS **
240 PRINT " N";TAB(10);"N-TH TERM"
250 REM ** MAIN PROGRAM **
260 FOR N=F TO H STEP S
270 REM ** CALCULATING N-TH TERM **
271 REM ** LINES 271 THROUGH 299 **
298 REM ** ARE USED TO DESCRIBE **
299 REM ** THE N-TH TERM **
300 REM ** OUTPUT **
310 PRINT N;TAB(10);A
320 NEXT N
330 END
```

Test Data: The sequence with an = $(7n-1)/(2n+1)$ has a limit of 3.5

```
271 LET A=(7*N-1)/(2*N+1)

RUN
ENTER THE INDEX OF THE FIRST
TERM TO BE PRINTED.
?1

ENTER THE INDEX OF THE HIGHEST
TERM TO BE PRINTED.
?36001

ENTER THE INCREMENT BETWEEN TERMS.
?2000
```

N	N-TH TERM
1	2
2001	3.49887584
4001	3.49943771
6001	3.49962509
8001	3.4997188
10001	3.49977503
12001	3.49981253
14001	3.4998393
16001	3.49985939
18001	3.49987501
20001	3.49988751
22001	3.49989774
24001	3.49990625
26001	3.49991347
28001	3.49991965
30001	3.499925
32001	3.49992969
34001	3.49993383
36001	3.4999375

Considering round-off error we would probably conclude that the limit is 3.5 which is the expected answer. However, notice how slowly we are approaching the limit--two-place decimal accuracy with N=10001 but only four-place accuracy with N=36001.

Example 1: Show that a bounded sequence does not have to have a limit by investigating the sequence {an} where an = $(-1)^n(1 + 1/n)^2$.

Since $(1 + 1/n)^2 \leq 4$ for all n we can conclude that $-4 \leq a_n \leq 4$. Also, since the sequence has a factor of -1 raised to the n-th power, the increment has to be odd so that both odd and even terms are computed.

```
271 LET A=((-1)^N)*(1+1/N)*(1+1/N)

RUN
ENTER THE INDEX OF THE FIRST
TERM TO BE PRINTED.
?1

ENTER THE INDEX OF THE HIGHEST
TERM TO BE PRINTED.
?36001

ENTER THE INCREMENT BETWEEN TERMS.
?2999
```

N	N-TH TERM
1	-4
3000	1.00066678
5999	-1.00033342
8998	1.00022228
11997	-1.00016672
14996	1.00013337
17995	-1.00011114
20994	1.00009527
23993	-1.00008336
26992	1.0000741
29991	-1.00006669
32990	1.00006063
35989	-1.00005557

From the output the terms of the sequence appear to be oscillating between values close to 1 and -1, and, thus, there is no limit. However, the output does indicate that there may be two different subsequences, one converging to 1 and the other to -1. This illustrates a general property of bounded sequences which is stated in the Bolzano-Weierstrass Theorem: If {a_n} is a bounded sequence, then {a_n} has a convergent subsequence.

To further illustrate the Bolzano-Weierstrass Theorem we rerun the program once for even terms and once for odd terms.

```
RUN
ENTER THE INDEX OF THE FIRST
TERM TO BE PRINTED.
?1

ENTER THE INDEX OF THE HIGHEST
TERM TO BE PRINTED.
?36001

ENTER THE INCREMENT BETWEEN TERMS.
?3000

   N        N-TH TERM
   1        -4
   3001     -1.00066656
   6001     -1.00033331
   9001     -1.00022221
   12001    -1.00016666
   15001    -1.00013333
   18001    -1.00011111
   21001    -1.00009524
   24001    -1.00008333
   27001    -1.00007407
   30001    -1.00006667
   33001    -1.00006061
   36001    -1.00005556


RUN
ENTER THE INDEX OF THE FIRST
TERM TO BE PRINTED.
?2

ENTER THE INDEX OF THE HIGHEST
TERM TO BE PRINTED.
?36002

ENTER THE INCREMENT BETWEEN TERMS.
?3000

   N        N-TH TERM
   2        2.25
   3002     1.00066633
   6002     1.00033325
   9002     1.00022219
   12002    1.00016665
   15002    1.00013332
   18002    1.00011111
   21002    1.00009523
```

```
24002     1.00008333
27002     1.00007407
30002     1.00006666
33002     1.0000606
36002     1.00005555
```

Notice that the odd terms appear to converge to -1 while the even terms appear to converge to 1.

Example 2: Investigate the sequence with

$$an = \frac{(n+2)!}{n!(4n^2+7n+8)}$$

In chapter 4 the difficulties involved in dealing with factorials on a computer were detailed. As shown there the value of factorials grow large rapidly. Thus, we must be careful in choosing the index for the highest term of our sequence. To actually compute the factorial we can use a FOR-NEXT loop as illustrated below.

```
271 LET N1=1                  initializing variable for n!
272 LET N2=1                  initializing variable for (n+2)!
273 FOR I=1 TO N
274 LET N1=N1*I               computing n!
275 NEXT I
276 LET N2=N1*(N+1)*(N+2)     multiplying to obtain (n+2)!
277 LET A=N2/(N1*(4*N*N+7*N+8))
```

```
RUN
ENTER THE INDEX OF THE FIRST
TERM TO BE PRINTED.
?1

ENTER THE INDEX OF THE HIGHEST
TERM TO BE PRINTED.
?31

ENTER THE INCREMENT BETWEEN TERMS.
?3
```

313

N	N-TH TERM
1	.315789474
4	.3
7	.28458498
10	.276150628
13	.270967742
16	.267482518
19	.264984227
22	.263107722
25	.26164741
28	.260479042
31	.259523224

From the output it is difficult to make a conjecture about the limit of the sequence. However, if we increase the number of terms to be printed we will create an overflow error due to the factorials. Fortunately, a little work beforehand can eliminate the factorials and allow more terms to be printed. For example, we can rewrite the general term of this sequence as

$$\frac{(n+2)!}{n!(4n^2+7n+8)} = \frac{n!(n+1)(n+2)}{n!(4n^2+7n+8)}$$

$$= \frac{(n+1)(n+2)}{4n^2+7n+8}$$

$$= \frac{n^2+3n+2}{4n^2+7n+8}$$

which can then be used in computing the limit.

```
271 LET A=(N*N+3*N+2)/(4*N*N+7*N+8)

RUN
ENTER THE INDEX OF THE FIRST
TERM TO BE PRINTED.
?1

ENTER THE INDEX OF THE HIGHEST
TERM TO BE PRINTED.
?3601

ENTER THE INCREMENT BETWEEN TERMS.
?200
```

N	N-TH TERM
1	.315789474
201	.251541231
401	.250775906
601	.250518454
801	.250389286
1001	.250311642
1201	.250259821
1401	.250222776
1601	.250194977
1801	.250173346
2001	.250141868
2201	.250141868
2401	.250130059
2601	.250120065
2801	.250111498
3001	.250104071
3201	.250097572
3401	.250091837
3601	.250086739

Example 3: Investigate the sequence defined recursively by

$$a_n = a_{n-1} - a_{n-1}^2 \quad , \quad a_0 = .5$$

To illustrate how to generate terms from a recursive definition we will compute the first three terms of this sequence.

$$a_0 = .5$$

$$
\begin{aligned}
a_1 &= a_{1-1} - a_{1-1}^2 \\
&= a_0 - a_0^2 \\
&= .5 - .5^2 \\
&= .25
\end{aligned}
$$

$$
\begin{aligned}
a_2 &= a_{2-1} - a_{2-1}^2 \\
&= a_1 - a_1^2 \\
&= .25 - .25^2 \\
&= .25 - .0625 \\
&= .1875
\end{aligned}
$$

Now in dealing with a sequence defined recursively several changes must be made from previous examples. First is the fact that not all of the terms are defined by the same formula but are given as specific values. In this example the zero term, a_0, is .5 Thus, special statements must be used when determining the values of the terms. Also, each term of the sequence must be computed in order, which means the increment between terms must be 1. Therefore, we will add the following lines to our program and run the program for the recursive sequence defined above.

```
275 IF N<>F THEN 290        testing for first term
280 LET A=.5
285 GOTO 310
290 LET A=A-A*A
```

```
RUN
ENTER THE INDEX OF THE FIRST
TERM TO BE PRINTED.
?1

ENTER THE INDEX OF THE HIGHEST
TERM TO BE PRINTED.
?20

ENTER THE INCREMENT BETWEEN TERMS.
?1
```

N	N-TH TERM
1	.5
2	.25
3	.1875
4	.15234375
5	.129135132
6	.11245925
7	.0998121668
8	.0898496981
9	.0817767298
10	.0750892963
11	.0694508939
12	.0646274672
13	.0604507577
14	.0567964636
15	.0535706253
16	.0507008134
17	.0481302409
18	.0458137208
19	.0437148238
20	.041803838

Since the terms appear to be steadily decreasing and positive, the most likely conjecture for the limit is zero. Yet, how much faith can we have in our conjecture after only 20 terms? If we re-run the program with a larger value for the highest index we would gain more information about the limit, but we would also have to print out a large number of terms since the step between terms must be 1. Therefore, it would be helpful to modify the program so that each term is computed but only terms at selected intervals are printed. This can be accomplished by making two changes in the program. First, change line 260 to have a step size of 1 since we need to compute each term.

```
260 FOR N=F TO H
```

Secondly, we need a test which will skip over the print statement of line 310 if the term is not to be printed. We accomplish this by adding

```
299 IF ((N-F)/S)<>INT((N-F)/S) THEN 320
```

which will result in only the terms at the specified increment being printed.

```
260 FOR N=F TO H
299 IF ((N-F)/S)<>INT((N-F)/S) THEN 320
```

```
RUN
ENTER THE INDEX OF THE FIRST
TERM TO BE PRINTED.
?1

ENTER THE INDEX OF THE HIGHEST
TERM TO BE PRINTED.
?3601

ENTER THE INCREMENT BETWEEN TERMS.
?300

   N        N-TH TERM
   1        .5
   301      3.25208728E-03
   601      1.64425644E-03
   901      1.10061936E-03
   1201     8.27222216E-04
   1501     6.62652351E-04
   1801     5.52709593E-04
   2101     4.74063527E-04
   2401     4.15014394E-04
   2701     3.69048303E-04
   3001     3.3225048E-04
   3301     3.02126443E-04
   3601     2.77011434E-04
```

It appears that our conjecture that the limit is 0 is correct.

Example 4: Investigate the limit of the sequence defined by

$$a_1 = .5$$

$$a_2 = f(a_1) = f(.5)$$

$$a_3 = f(a_2) = f(f(a_1))$$

$$a_4 = f(a_3) = f(f(f(a_1)))$$

$$a_n = f(a_{n-1}) = \underbrace{(f \circ f \circ \ldots \ldots \circ f)}_{n-1 \text{ times}}(a_1)$$

where $f(x) = \dfrac{e^x (x^2 - 2x + 2)}{3}$

318

```
271 IF N<>1 THEN 274
272 LET A=.5
273 GOTO 310
274 DEF FNF(X)=EXP(X)*(X*X-2*X+2)/3
275 LET A=FNF(FNF(A))                    computing two terms
```

RUN
**ENTER THE INDEX OF THE FIRST
TERM TO BE PRINTED.**
?1

**ENTER THE INDEX OF THE HIGHEST
TERM TO BE PRINTED.**
?25

ENTER THE INCREMENT BETWEEN TERMS.
?2

N	N-TH TERM
1	.5
3	.727483268
5	.746331653
7	.749107809
9	.74953978
11	.749607549
13	.749618195
15	.749619867
17	.74962013
19	.749620171
21	.749620178
23	.749620179
25	.749620179

It appears that for for n larger than 22, a_n = .749620179.
However, $a_n = f(a_{n-1})$, which implies that
f(.749620179)=.749620179

A value "a" such that f(a)=a is called a "fixed point" for
the function f. This example has illustrated a general theorem
about fixed points which states:

> If f is continuous and the sequence x, f(x),
> f(f(x)),... converges to "a", then "a" is a
> fixed point for f.

Exercise Set 14.1

In exercises 1-10 use program SEQUENCE to approximate the limit of the given sequence. If possible, determine the actual value of the limit and compare it with the approximation.

1. $a_n = n/(4n+1)$

2. $a_n = \cos(n + \pi/2)$

3. $a_n = \sqrt{n^2+1} - \sqrt{n-1}$

4. $a_n = n^{(-1)^n}$

5. $a_n = n\sqrt{5^n/(3n+1)}$

6. $a_n = n\sqrt{n}$

7. $a_n = (-2)^{-n} + (-1)^n/n$

8. $a_n = 6^n/n!$

9. $a_n = n!/n^n$

10. $a_n = 1/n^2 + 1/(n+1)^2 + \ldots + 1/(2n)^2$

In exercises 11-15 use program SEQUENCE to approximate the limit of the sequence defined recursively.

11. $a_1 = 1/2$, $a_{n+1} = (1/2)^{a_n}$

12. $a_{n+1} = 1/(3+a_n)$, $a_n = 2$

13. $a_1 = \sqrt{2}$, $a_{n+1} = \sqrt{2+a_n}$

14. $a_1 = 0$, $a_{2n} = (a_{2n-1})/2$, $a_{2n+1} = 1/2 + a_{2n}$

320

15. $a_1 = 2$, $a_{n+1} = a_n + \sin a_n$

*
16.   Let $f(x) = e^x - 1$.   Approximate the limit of the sequence defined by $a_n = n f(1/n)$.

17.   Approximate a fixed point of $f(x) = (x-1)e^{x-1}$.   <u>Hint</u>: See example 4.

18.   Let $a_n = 1 + 1/2 + 1/3 + \ldots + 1/n - \ln n$. Approximate the limit of this sequence. The limit of this sequence is known as Euler's number.

19. Use program SEQUENCE to illustrate the validity of Cauchy's Limit Theorem using the sequence $a_n = 1 + 1/n$ which has a limit of 1

> <u>Cauchy's Limit Theorem</u>: If the sequence $\{a_n\}$ converges to L, then the sequence S_n defined by $S_n = (a_1 + \ldots + a_n)/n$ also converges to L.

**
20.   Let $a_n = (2-n)/(n \, 4^n)$.   Define sequences A_n and B_n by $A_n = a_{n+1}/a_n$ and $B_n = \sqrt[n]{a_n}$. Using program SEQUENCE show that A_n and B_n have the same limit value. This illustrates the equivalency of the root and ratio tests for infinite series.

21.   The   Fibonacci   sequence   is   given   by $a_1 = 1$, $a_2 = 1$, $a_n = a_{n-1} + a_{n-2}$.

a. Give the first six terms of this sequence.

b. Approximate the limit of the sequence $A_n = a_{n+1}/a_n$ where a_{n+1} and a_n are the nth and (n+1)st terms of the Fibonacci sequence.

c. Show that the approximation obtained in part b is approximately $(1/2)(1 + \sqrt{5})$.

d.   Approximate the limit of the sequence defined by $a_1 = 1$, $a_{n+1} = \sqrt{a_n} + 1$.

e. Show that the limit of part d is approximately $(1/2)(1 + \sqrt{5})$.

f. Are the sequences of parts b and e the same?

*** <u>Exercises 22 and 23 are very difficult</u>.
22.   Let   an $= ^n\sqrt{x^n + y^n}$. By running program SEQUENCE for different values of x and y make a conjecture about the limit of this sequence. Prove your conjecture if possible.

23.   Let $a_{n+1} = (1/2)(a_n + x/a_n$) with $x > 0$ and $a_1 = 1$. By running program SEQUENCE for different values of x make a conjecture   about   the   limit   of   this   sequence.   Prove   your conjecture if possible.

14.2 INFINITE SERIES

Programming Problem: Approximate the limit of an infinite series.

Output: A table listing the value of n and the corresponding nth partial sum.

Input: Several BASIC statements defining the general term of the series, the initial index of the series, the final index of the series to be used in the approximation, and the number of terms to be summed between printing values.

Strategy: By definition the value of an infinite series

$$\sum_{i=1}^{\infty} a_i$$

is the limit of the corresponding sequence of partial sums. That is,

$$\sum_{i=1}^{\infty} a_i = \lim_{n \to \infty} S_n$$

where $S_n = a_1 + \ldots + a_{n-1} + a_n$.

Thus, the sum of the infinite series reduces to finding the limit of an infinite sequence. If we use the recursive definition for S_n, $S_n = S_{n-1} + a_n$, then we can approximate the infinite series by a program similar to that developed in example 3 of the previous section.

323

Program:

```
100 REM ** SERIES **
110 REM ** INPUT **
120 PRINT "ENTER THE INDEX OF THE FIRST"
130 PRINT "TERM TO BE PRINTED."
140 INPUT F
150 PRINT
160 PRINT "ENTER THE INDEX OF THE HIGHEST"
170 PRINT "TERM TO BE PRINTED."
180 INPUT H
190 PRINT
200 PRINT "ENTER THE INCREMENT BETWEEN TERMS."
210 INPUT S
220 PRINT
230 REM ** HEADINGS **
240 PRINT " N";TAB(10);" N-TH PARTIAL SUM"
250 REM ** MAIN PROGRAM **
260 REM ** INITIALIZING THE SUM **
270 LET SUM=0
280 FOR N=F TO H
290 REM ** CALCULATING N-TH PARTIAL SUM **
291 REM ** LINES 291 THROUGH 309 **
300 REM ** ARE USED TO DESCRIBE THE **
309 REM ** N-TH TERM OF THE SERIES.  **
310 LET SUM=SUM+A
320 IF ((N-F)/S)<>INT((N-F)/S) THEN 350
330 REM ** OUTPUT **
340 PRINT N;TAB(10;SUM
350 NEXT N
360 END
```

Test Data: The series

$$\sum_{n=1}^{\infty} 3(1/2)^{n-1}$$

is a geometric series with first term 3 and ratio 1/2. Thus, the sum is $3/(1-(1/2)) = 6$.

```
291 LET A=3*(.5)^(N-1)
```

```
RUN
ENTER THE INDEX OF THE FIRST
TERM TO BE PRINTED.
?1

ENTER THE INDEX OF THE HIGHEST
TERM TO BE PRINTED.
?43

ENTER THE INCREMENT BETWEEN TERMS.
?3

 N          N-TH PARTIAL SUM
 1          3
 4          5.625
 7          5.953125
 10         5.99414063
 13         5.99926758
 16         5.99990845
 19         5.99998856
 22         5.99999857
 25         5.99999982
 28         5.99999998
 31         6
 34         6
 37         6
 40         6
 43         6
```

Example 1: Run program SERIES for the harmonic series

$$\sum_{n=1}^{\infty} 1/n$$

```
291 LET A=1/N

RUN
ENTER THE INDEX OF THE FIRST
TERM TO BE PRINTED.
?1

ENTER THE INDEX OF THE HIGHEST
TERM TO BE PRINTED.
?3601
```

```
ENTER THE INCREMENT BETWEEN TERMS.
?200

  N          N-TH PARTIAL SUM
  1          1
  201        5.88300607
  401        6.57242345
  601        6.9766423
  801        7.2637007
  1001       7.48646987
  1201       7.66854176
  1401       7.82251408
  1601       7.95591166
  1801       8.07359062
  2001       8.17886786
  2201       8.27410992
  2401       8.3610645
  2601       8.44105913
  2801       8.5151259
  3001       8.5840831
  3201       8.64859038
  3401       8.70918744
  3601       8.76632138
```

From the output, the sequence of partial sums is increasing,
and we should conclude that the series diverges. This is indeed
the case for the harmonic series. However, if we do not know
about the convergence of a series, we should be cautious in
interpreting results. Although the sequence of partial sums is
increasing as in this example, the series may have a limit as N
goes to infinity.

Example 2: Investigate the alternating series

$$\sum_{n=1}^{\infty} (-1)^{n-1} \sin(1/\sqrt{n})/(2n-1)$$

```
291 LET A=(-1)^(N-1)*SIN(1/SQR(N))/(2*N-1)
```

```
RUN
ENTER THE INDEX OF THE FIRST
TERM TO BE PRINTED.
?1

ENTER THE INDEX OF THE HIGHEST
TERM TO BE PRINTED.
?1415

ENTER THE INCREMENT BETWEEN TERMS.
?101
```

N	N-TH PARTIAL SUM
1	.841470985
101	.693058683
203	.693386627
304	.693253268
405	.693331011
506	.693278424
607	.693317075
708	.693287106
809	.693311227
910	.693291265
1011	.693308142
1112	.693293629
1213	.693306284
1314	.693295121
1415	.693305064

The series appears to converge to about .693305. However, a measure of the accuracy for the approximation can be obtained using the formula

$$\left| L - \sum_{n=1}^{N} a_n \right| < \left| a_{n+1} \right|$$

which gives the bound in the error of approximating an alternating series by the first n terms.

Now, $a_{1416} = (\sin (1/\sqrt{1416}))/(2(1415)-1)$

$$\approx .000009386$$

Thus, .693305 is accurate to within ±.000009386 of the actual value of the series neglecting round-off error.

Example 3: Investigate the alternating series

$$\sum_{i=1}^{\infty} (-1)^{n+1} (\sqrt{n+1} - \sqrt{n})$$

First, we will test for absolute convergence.

```
291 LET A=SQR(N+1)-SQR(N)

RUN
ENTER THE INDEX OF THE FIRST
TERM TO BE PRINTED.
?1

ENTER THE INDEX OF THE HIGHEST
TERM TO BE PRINTED.
?3601

ENTER THE INCREMENT BETWEEN TERMS.
?200
```

N	N-TH PARTIAL SUM
1	.414213562
201	13.2126704
401	19.0499377
601	23.5356883
801	27.3196045
1001	30.6543836
1201	33.6698717
1401	36.4432905
1601	39.0249922
1801	41.4499706
2001	43.7437147
2201	45.9254729
2401	48.0102031
2601	50.0098031
2801	51.9339212
3001	53.7905102
3201	55.5862175
3401	57.3266664
3601	59.0166645

The series is not absolutely convergent.

Testing for conditional convergence we obtain the following.

    291 LET A=(-1)^(N+1)*(SQR(N+1)-SQR(N))

    RUN
    ENTER THE INDEX OF THE FIRST
    TERM TO BE PRINTED.
    ?1

    ENTER THE INDEX OF THE HIGHEST
    TERM TO BE PRINTED.
    ?3601

    ENTER THE INCREMENT BETWEEN TERMS.
    ?200

N	N-TH PARTIAL SUM
1	.414213562
201	.257382208
401	.252261205
601	.249981529
801	.248620016
1001	.247689909
1201	.247002991
1401	.246468866
1601	.24603824
1801	.245681276
2001	.245379339
2201	.245119663
2401	.244893206
2601	.244693354
2801	.244515072
3001	.244354833
3201	.244210038
3401	.244078236
3601	.243957657

Although the sequence of partial sums does not appear to cluster about a value, it is decreasing and always positive. Thus, we have a decreasing sequence which is bounded below by at least zero, and we should conclude that it converges. Therefore, the series is conditionally convergent.

Example 4: Use the power series for e^x to approximate the value of e^2.

The power series for e^x is

$$e^x = \sum_{n=0}^{\infty} x^n / n!$$

Thus, $e^2 = \sum_{n=0}^{\infty} 2^n / n!$

```
291 LET NF=1
292 FOR I=1 TO N
293 LET NF=NF*I
294 NEXT I
295 LET A=2^N/NF

RUN
ENTER THE INDEX OF THE FIRST
TERM TO BE PRINTED.
?0

ENTER THE INDEX OF THE HIGHEST
TERM TO BE PRINTED.
?21

ENTER THE INCREMENT BETWEEN TERMS.
?3
```

N	N-TH PARTIAL SUM
0	1
3	6.33333334
6	7.35555556
9	7.38871252
12	7.38905457
15	7.3890561
18	7.3890561
21	7.3890561

Thus, $e^2 \approx 7.3890561$

Exercise Set 14.2

In exercises 1-10 use program SERIES to approximate each series.

1. $\displaystyle\sum_{n=0}^{\infty} 4/5^n$

6. $\displaystyle\sum_{n=2}^{\infty} 1/(n(\ln n))$

2. $\displaystyle\sum_{n=1}^{\infty} (\sqrt{n+1} - \sqrt{n})/(\sqrt{n}\sqrt{n+1})$

7. $\displaystyle\sum_{n=1}^{\infty} (2^n + 3^n)/6^n$

3. $\displaystyle\sum_{n=1}^{\infty} \cos 1/n$

8. $\displaystyle\sum_{n=4}^{\infty} 1/(n (\ln n) (\ln (\ln n))^2)$

4. $\displaystyle\sum_{n=1}^{\infty} (n!/n^n)^n$

9. $\displaystyle\sum_{n=1}^{\infty} n^2 e^{-n*n*n}$

5. $\displaystyle\sum_{n=2}^{\infty} 1/(4^{n-1}-1)$

10. $\displaystyle\sum_{n=1}^{\infty} (\ln n)^n/n^2$

In exercises 11-13 use program SERIES and the formula for the approximation of a convergent alternating series to approximate the series to an accuracy of at least .0001

11. $\displaystyle\sum_{n=1}^{\infty} (-1)^n/(2^n(n+1))$

12. $\displaystyle\sum_{n=1}^{\infty} (-1)^n n^2/2^n$

13. $\displaystyle\sum_{n=1}^{\infty} (-1)^n \tan^{-1} n/(n^2+1)$

In exercises 14-16 use program SERIES to decide if the series is absolutely or conditionally convergent.

14. $\displaystyle\sum_{n=2}^{\infty} (-1)^n / n \sqrt{\ln n}$

15. $\displaystyle\sum_{n=1}^{\infty} (-1)^n 2^{n+1} / (n+6^n)$

16. $\displaystyle\sum_{n=1}^{\infty} (-1)^n / ((n+2)e^n)$

*

In exercises 17-20 use the power series given to find an approximation for the indicated value.

17. $\sin 4;$ $\displaystyle \sin x = \sum_{n=1}^{\infty} (-1)^{n-1} x^{2n-1} / (2n-1)!$

18. $\cos -2;$ $\displaystyle \cos x = \sum_{n=1}^{\infty} (-1)^{n-1} x^{2n-1} / (2n-2)!$

19. $\ln 2;$ $\displaystyle \ln (1+x) = \sum_{n=1}^{\infty} (-1)^{n-1} x^n / n$

20. $\tan 1;$ $\displaystyle \tan x = \sum_{n=1}^{\infty} (-1)^{n-1} x^{2n-1} / (2n-1)$

***
21.   If the series

$$\sum_{n=1}^{\infty} an$$

diverges and S_n is the nth partial sum, investigate each of the series below with an = 1/n.

a. $\sum_{n=1}^{\infty} an/(1+an)$

b. $\sum_{n=1}^{\infty} an/(1+n*an)$

c. $\sum_{n=1}^{\infty} an/(1+n^2*an)$

d. $\sum_{n=1}^{\infty} an/S_n$

e. $\sum_{n=1}^{\infty} an/S_n^2$

CHALLENGE ACTIVITY

Write a program which will approximate the interval of convergence of a power series in x or x-a. As test data use:

1. $\sum\limits_{n=1}^{\infty} 4^n x^n /n^2$ which converges for $-1/4 \le x \le 1/4$.

2. $\sum\limits_{n=1}^{\infty} (-1)^n (x-1)^n /n$ which converges for $0 < x \le 2$.

Be sure to investigate the endpoints.

CHAPTER 15
Analytic Geometry

The second degree equation $Ax^2 + Bxy + Cy^2 + Dx + Ey + F = 0$ has a graph which is one of the conic sections: a circle, an ellipse, a parabola, a hyperbola, or a degenerate conic section. By appropriate substitutions the general second degree equation can be reduced to the equation for the standard form of one of the conics. In this chapter we will develop a program which will identify the type of conic represented by a second degree equation and transform that equation to the standard form of a conic. In addition, we will write a program which can be used to obtain the graph of the conic section or of any function.

15.1 THE SECOND DEGREE EQUATION

Programming Problem: For the second degree equation

$$Ax^2 + Bxy + Cy^2 + Dx + Ey + F = 0$$

determine:

a. the type of conic represented

b. the angle of rotation

c. the translation

d. the equation in standard form after the effect of the rotation is taken into account.

Output: The output will consist of the following messages for an ellipse or hyperbola, the circle being considered a special case of the ellipse.

    TYPE:
    ANGLE OF ROTATION:
    CENTER:
    DENOMINATOR X^2:
    DENOMINATOR Y^2:
    EQUATION:

For a parabola the output will contain the information indicated
below.

    TYPE:
    ANGLE OF ROTATION:
    CENTER:
    EQUATION:

Input: The coefficients A, B, C, D, E, and F.

Strategy: The type of conic section represented by

$$Ax^2 + Bxy + Cy^2 + Dx + Ey + F = 0$$

can be determined by computing the value of $B^2 - 4AC$ and using the
following. The conic section is:

    a.   a parallel lines or imaginary if
$B^2 - 4AC = 0$.

    b.   an ellipse, a circle, a single point, or imaginary if
$B^2 - 4AC < 0$.

    c.   a hyperbola or two intersecting lines if $B^2 - 4AC > 0$.

Furthermore, we know that by choosing θ such that cot $2\theta = (A-C)/B$,
a substitution of

$$x = x'\cos\theta - y'\sin\theta \quad \text{and} \quad y = x'\sin\theta + y'\cos\theta$$

will result in the elimination of the xy-term and the equation
being rewritten as

$$A'(x')^2 + C'(y')^2 + D'x' + E'y' + F = 0$$

where, without derivation, the resulting expressions for A',C',D', and E' are:

$$A' = A\cos^2\theta + B\sin\theta\,\cos\theta + C\sin^2\theta$$

$$C' = A\sin^2\theta - B\sin\theta\,\cos\theta + C\cos^2\theta$$

$$D' = D\cos\theta + E\sin\theta$$

$$E' = -D\sin\theta + E\cos\theta$$

With the xy-term eliminated and by completing the square on x' and y' we can rewrite the equation in one of the following three forms.

1. Ellipse, circle, or hyperbola

$$\frac{(x'-(-D'/2A'))^2}{H/A'} + \frac{(y'-(-E'/2C'))^2}{H/C'} = 1$$

$$H = -F + E'^2/4C' + D'^2/4A'$$

2. Parabola in y'^2

$$(y'-(-E'/2C'))^2 = (-D'/C')(x'-(-F/D' + E'^2/4D'C'))$$

3. Parabola in x'^2

$$(x'-(-D'/2A'))^2 = (-E'/A')(y'-(-F/E' + D'^2/4E'A'))$$

Finally, in writing the program we will round the values to two decimal places for the angle of rotation, A',C',D', and E' to minimize the effect of round-off error and provide an output which is easier to read. Since all of the values may not be positive, it will be necessary to use the BASIC statement

```
X=SGN(X)*(INT(ABS(X) * 100+.5))/100
```

for rounding. Also, BASIC provides no built-in function for finding the Arccotangent of an angle. However, the statement

$$Z=-ATN(X)+1.57079633$$

which is known as a derived function (see Appendix C), does return the arccot of the value x and will be used in the program.

Program:

```
100 REM ** SECOND DEGREE **
110 REM ** INPUT **
120 PRINT "AX^2+BXY+CY^2+DX+EY+F=0"
130 PRINT
140 PRINT "ENTER THE COEFFICIENTS :A--B--C--D--E--F"
150 INPUT A,B,C,D,E,F
160 PRINT
170 REM ** DETERMINING TYPE **
180 LET G=B*B-4*A*C
190 IF G=0 THEN 250
200 IF G<0 THEN 230
210 PRINT "TYPE: HYPERBOLA"
220 GO TO 270
230 PRINT "TYPE: ELLIPSE"
240 GO TO 270
250 PRINT "TYPE: PARABOLA"
260 REM ** DETERMINING ANGLE OF ROTATION **
270 IF B=0 THEN 320
280 LET Z=(-ATN((A-C)/B)+1.57079633)/2
290 LET Y=45*Z/ATN(1)                       changing to degrees
300 LET Y=(INT(Y * 100+.5))/100             rounding, y always +
310 GO TO 340
320 LET Z=0
330 LET Y=0
340 PRINT "ANGLE OF ROTATION: ";Y;" DEGREES"
350 REM ** FINDING COEFFICIENTS AFTER ROTATION **
360 LET AP=A*COS(Z)*COS(Z)+B*COS(Z)*SIN(Z)+C*SIN(Z)*SIN(Z)
370 LET CP=A*SIN(Z)*SIN(Z)-B*COS(Z)*SIN(Z)+C*COS(Z)*COS(Z)
380 LET DP=D*COS(Z)+E*SIN(Z)
390 LET EP=-D*SIN(Z)+E*COS(Z)
400 REM ** ROUNDING COEFFICIENTS **
410 LET AP=SGN(AP)*INT(ABS(AP) * 100+.5)/100
420 LET CP=SGN(CP)*(INT(ABS(CP) * 100+.5)/100
430 LET DP=SGN(DP)*(INT(ABS(DP) * 100+.5)/100
440 LET EP=SGN(EP)*(INT(ABS(EP) * 100+.5)/100
450 REM ** TESTING TO DECIDE IF PARABOLA **
460 IF G=0 THEN 620
470 REM ** WORKING WITH ELLIPSE OR HYPERBOLA **
```

```
480 REM ** FINDING CENTER **
490 LET X=-DP/(2*AP)
500 LET Y=-EP/(2*CP)
510 PRINT "CENTER: (";X;",";Y;")"
520 REM ** FINDING DENOMINATORS **
530 LET H=-F+(EP*EP/(4*CP))+(DP*DP/(4*AP))
540 LET H=SGN(H)*INT(ABS(H) * 100+.5)/100
550 PRINT "DENOMINATOR X^2: ";H;"/";AP;"=";H/AP
560 PRINT "DENOMINATOR Y^2: ";H;"/";CP;"=";H/CP
570 REM ** STANDARD FORM **
580 PRINT "EQUATION:";
590 PRINT "(X-(";X;"))^2/";H/AP;" + (Y-(";Y;"))^2/";H/CP;" =1"
600 GO TO 740
610 REM ** WORKING WITH PARABOLA **
620 IF CP=0 THEN 700
630 REM ** NO SQUARED X TERM **
640 LET X=-F/DP+(EP*EP/(4*DP*CP))
650 LET Y=-EP/(2*CP)
660 PRINT "CENTER: (";X;",";Y;")"
670 PRINT "EQUATION: (Y-(";Y;"))^2=";-DP/CP;"(X-(";X;"))"
680 GO TO 740
690 REM ** NO SQUARED Y TERM **
700 LET X=-DP/(2*AP)
710 LET Y=-F/EP+(DP*DP/(4*EP*AP))
720 PRINT "CENTER: (";X;",";Y;")"
730 PRINT "EQUATION: (X-(";X;"))^2=";-EP/AP;"(Y-(";Y;"))"
740 END
```

Test Data: The equation xy=1 or xy-1=0 is a hyperbola rotated 45 with final equation

$$(x'^2/2)-(y'^2/2)=1$$

$B^2-4AC=1$ and $\cot 2\theta =(0-0)/1$

Hyperbola

$\cot 2\theta = 0$
$2\theta = \text{Arccot } 0$
$2\theta = 90$
$\theta = 45$

x=x'cos 45 + y'sin 45   y=x'sin 45 - y'cos 45

$$1 = xy = \left(\frac{\sqrt{2}}{2} x' + \frac{\sqrt{2}}{2} y'\right) \left(\frac{\sqrt{2}}{2} x' - \frac{\sqrt{2}}{2} y'\right)$$

$$1 = \frac{(x')^2}{2} - \frac{(y')^2}{2}$$

```
RUN
AX^2+BXY+CY^2+DX+EY+F=0

ENTER THE COEFFICIENTS:A--B--C--D--E--F
?0,1,0,0,0,-1

TYPE: HYPERBOLA
ANGLE OF ROTATION: 45 DEGREES
CENTER: (0,0)
DENOMINATOR X^2: 1/.5=2
DENOMINATOR Y^2: 1/-.5=-2
EQUATION: (X-(0))^2/2 + (Y-(0))^2/-2 =1
```

Example 1: Use program SECOND DEGREE to aid in graphing

$16x^2-24xy+9y^2+100x-200y+100=0$.

```
RUN
AX^2+BXY+CY^2+DX+EY+F=0

ENTER THE COEFFICIENTS:A--B--C--D--E--F
?16,-24,9,100,-200,100

TYPE: PARABOLA
ANGLE OF ROTATION: 53.13 DEGREES
CENTER: (-3,4)
EQUATION: (Y-(4))^2=4(X-(-3))
```

The equation $(y-4)^2=4(x-(-3))$ is that of a parabola with vertex at $(-3,4)$ in the rotated coordinate system.

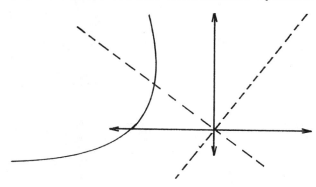

Example 2: Use program SECOND DEGREE to show that

$$x^2 + xy + y^2 + x + y = 1$$

is an ellipse. Also, determine the center and the length of the
semi-major and semi-minor axes and use this information to sketch
the graph.

RUN
AX^2+BXY+CY^2+DX+EY+F=0

ENTER THE COEFFICIENTS:A--B--C--D--E--F
?1,1,1,1,1,-1

TYPE: ELLIPSE
ANGLE OF ROTATION: 45 DEGREES
CENTER: (-.47,0)
DENOMINATOR X^2: 1.33/1.5=.886666667
DENOMINATOR Y^2: 1.33/.5=2.66
EQUATION: (X-(-.47))^2/.886666667 + (Y-(0))^2/2.66 =1

From the output we see that the equation is an ellipse with center
at (-.47,0). The square of the lengths of the semi-major and
semi-minor axes are the denominators of the x-squared and
y-squared terms with the largest being associated with the
semi-major axis. Thus, the length of the semi-major axis is $\sqrt{2.66}$
= 1.63 and the length of the semi-minor axis is $\sqrt{.8867}$ = .9416

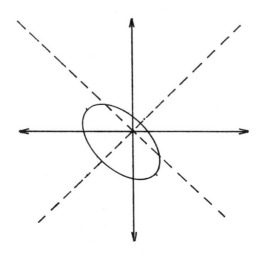

Exercise Set 15.1

In exercises 1-11 use program SECOND DEGREE to aid in sketching the graph of each second degree equation.

1. $x^2 + 8xy + 7y^2 - 1 = 0$

2. $x^2 - 2xy + 4y^2 - 4x = 0$

3. $2x^2 + 9xy + 14y^2 - 5 = 0$

4. $x^2 - 2xy + y^2 - 8x - 8y + 16 = 0$

5. $x^2 - xy + y^2 + x + y = 0$

6. $x^2 - 2xy + y - 2x - 2y + 1 = 0$

7. $x^2 - xy - y^2 + x + y = 0$

8. $5x^2 + 6xy + 5y^2 + 22x - 6y + 21 = 0$

9. $6x^2 - 5xy - 6y^2 + 78x + 52y + 26 = 0$

10. $11x^2 - 24xy + 4y^2 + 6x + 8y + 15 = 0$

11. $2x^2 - 3xy + 7y^2 + 8x - 4y - 10 = 0$

12.   a.   The BASIC statement

$$X = SGN(X)*(INT(ABS(X)*100+.5))/100$$

was used in program SECOND DEGREE to round values to two decimal places. To verify that this statement actually does round write a program to evaluate this statement and print the results for x=2.616 and -1.667.

b.   What changes should be made in the statement of part a to round to three decimal places? Verify your change by evaluating your new statement for x=2.6166 and -1.6647.

15.2 Graphing Functions and Relations

Note: The program in this section is written for use with a printer or an 80 column display. If neither is available, see Appendix E for a program which makes use of computer graphics capabilities. It may be possible to modify this program for the computer being used. If a printer is used, certain control characters, dependent upon the type of printer, may have to be sent to the printer before this program is executed.

Programming Problem: Approximate the graph of a function or relation.

Output: The graph of the function or relation.

Input: The function being graphed or the functions which determine the relation being graphed. Also, the range on the x and y-axes will be entered by an INPUT statement.

Strategy: Since the printer or an 80 column display will be used as the output device for our graph it will be necessary to print an entire line of the graph at one time. One method is to use the horizontal as the y-direction and the vertical as the x-direction and perform the following steps.

1.   Start with the smallest value of x.

2.   Determine all values of y which correspond to a given value of x.

3.   Print the corresponding points on the graph.

4.   Increment x to the next smallest value and return to step 2.

   To print a line of the graph we will use a character array, E\$, of dimension 70, where each cell will be associated with an endpoint of a subdivision of the y-axis. Thus, we will be subdividing the y-axis into 69 subdivisions (70 endpoints). If Y2 and Y1 represent the maximum and minimum values of y, respectively, then each subdivision will have length DY=(Y2-Y1)/69 and the correspondence between the array E\$ and the endpoints is as shown below.

E\$(1)   E\$(2)   E\$(3)   . . .   E\$(70)
$\updownarrow$ $\updownarrow$ $\updownarrow$ $\updownarrow$
Y1   Y1+DY   Y1+2*DY   Y1+69*DY=Y2

Now, for every value of y which corresponds to a given x, a "*" will be assigned to the cell of E$ which is associated with the endpoint closest to the y-value. If F represents the functional value, that is, the y value, then the appropriate cell of E$ is determined by computing to the nearest integer the number of times the subdivision length DY must be added to Y1 to obtain F and then adding 1 since the index of the array begins with 1. This can be accomplished by the BASIC statement

$$J=INT((F-Y1)/DY+.5)+1$$

Finally, provisions have been made in the program to include the x-axis and y-axis as part of the output. Also, we must remember to give each of the implicitly defined functions if a relation is being graphed. For this program each function should be entered somewhere between lines 441 and 469 followed by the two statements indicated below.

```
DEF  FNF(X)=.....
LET  J=INT((FNF(X)-Y1)/DY+.5)+1
LET  E$(J)="*"
```

Program:

```
100 REM ** GRAPH **
110 REM ** INPUT **
120 PRINT "ENTER THE RANGE OF THE Y-AXIS."
130 INPUT Y1,Y2
140 PRINT
150 PRINT "ENTER THE RANGE OF THE X-AXIS."
160 INPUT A,B
170 PRINT
180 PRINT
190 REM ** SETTING UP SCALE AND AXIS **
200 LET DY=(Y2-Y1)/69
210 LET DX=DY
220 REM ** FINDING LOCATION OF X-AXIS **
230 LET K=0
240 IF Y1>0 THEN 280                    testing for no x-axis
250 IF Y2<0 THEN 280
260 LET K=INT(ABS(Y1)/DY+.5)+1
270 REM ** HEADINGS **
280 PRINT TAB(10);"THE Y-AXIS GOES FROM ";Y1;
290 PRINT " TO ";Y2;" IN INCREMENTS OF ";DY
300 FOR I=1 TO 70
```

```
310 PRINT "-";
320 NEXT I
330 PRINT " X-VALUE"
340 PRINT
350 REM ** MAIN PROGRAM **
360 DIM E$(70)
370 LET X=A                              initializing x
380 REM ** INITIALIZING ARRAY FOR **
390 REM ** EACH ROW **
400 FOR I=1 TO 70
410 LET E$(I)=" "
420 NEXT I
430 LET E$(K)="."                        plotting x-axis
440 REM ** FUNCTIONS TO BE SKETCHED **
450 DEF FNF(X)= ** FIRST FUNCTION **
451 LET J=INT((FNF(X)-Y1)/DY+.5)+1   finding location in row
452 LET E$(J)="*"
460 DEF FNG(X)= ** SECOND FUNCTION **
461 LET J=INT((FNG(X)-Y1)/DY+.5)+1   finding location in row
462 LET E$(J))="*"                       plotting point
470 REM ** PRINTING LINE **
480 FOR I=1 TO 70
490 PRINT E$(I);
500 NEXT I
510 LET X1=SGN(X)*(INT(ABS(X)*10000+.5))/10000 rounding
520 PRINT " ";X1
530 LET X=X+DX
540 REM ** TESTING TO END PROGRAM **
550 IF X>B THEN 630
560 REM ** TESTING FOR Y-AXIS **
570 IF X< 0 THEN 400
580 IF (X-DX)> 0 THEN 400
590 FOR I=1 TO 70
600 LET E$(I)="."
610 NEXT I
620 GO TO 450
630 END
```

Test Data: The graph of

$$x^2 + y^2 = 1$$

is a circle of radius 1 and defines the implicit functions

$$y_1 = \sqrt{1-x^2} \quad \text{and} \quad y_2 = -\sqrt{1-x^2}$$

```
450 DEF FNF(X)=SQR(1-X*X)
460 DEF FNG(X)=-SQR(1-X*X)
```

```
RUN
ENTER THE RANGE OF THE Y-AXIS.
?-2,2

ENTER THE RANGE OF THE X-AXIS.
?-1,1
```

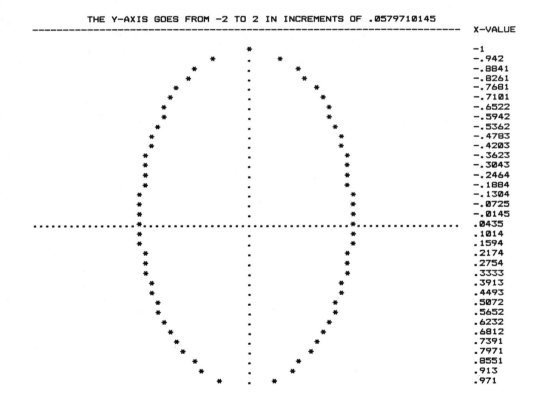

```
          THE Y-AXIS GOES FROM -2 TO 2 IN INCREMENTS OF .0579710145
------------------------------------------------------------------------  X-VALUE
                                                                          -1
                                                                          -.942
                                                                          -.8841
                                                                          -.8261
                                                                          -.7681
                                                                          -.7101
                                                                          -.6522
                                                                          -.5942
                                                                          -.5362
                                                                          -.4783
                                                                          -.4203
                                                                          -.3623
                                                                          -.3043
                                                                          -.2464
                                                                          -.1884
                                                                          -.1304
                                                                          -.0725
                                                                          -.0145
                                                                           .0435
                                                                           .1014
                                                                           .1594
                                                                           .2174
                                                                           .2754
                                                                           .3333
                                                                           .3913
                                                                           .4493
                                                                           .5072
                                                                           .5652
                                                                           .6232
                                                                           .6812
                                                                           .7391
                                                                           .7971
                                                                           .8551
                                                                           .913
                                                                           .971
```

The graph looks more like an ellipse than a circle!! This occurs because the distance between printed characters on a line is different than the distance the paper is advanced with a line feed. Thus, the scales on the x and y-axes are different. To correct the problem we simply need to adjust the scale on either axis. Since the graph from the test data should be a circle the horizontal and vertical diameters should be the same. By actual measurement the diameters are 150 mm and 90 mm.

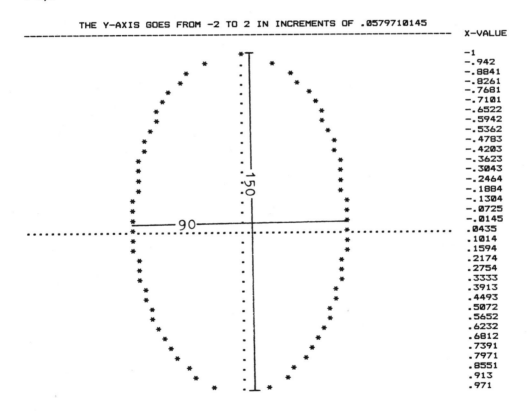

ENTER THE RANGE OF THE Y-AXIS.
?-2,2

ENTER THE RANGE OF THE X-AXIS.
?-1,1

Thus,

$$DX/DY = 150/90 \quad \text{or} \quad DX=DY(5/3).$$

This change of scale for the x-axis can be accomplished by modifying line 210 to

    210 LET DX=DY*5/3

Running the modified program for our test data we obtain the expected graph of a circle.

    450 DEF FNF(X)=SQR(1-X*X)
    460 DEF FNG(X)=-SQR(1-X*X

    RUN
    ENTER THE RANGE OF THE Y-AXIS.
    ?-2,2

    ENTER THE RANGE OF THE X-AXIS.
    ?-1,1

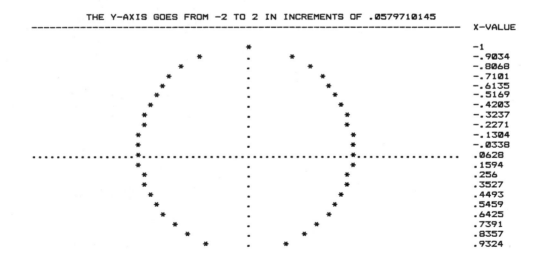

Example 1:

Graph y=cos 2x + 2sin 3x for $-\pi \leq x \leq \pi$

Since both cos 2x and sin 3x are bounded between -1 and 1, the range on the y-axis will be -3 to 3.

```
450 DEF FNF(X)=COS(2*X)+2*SIN(3*X)
460
461
462

RUN
```

ENTER THE RANGE OF THE Y-AXIS.
?-3,3

ENTER THE RANGE OF THE X-AXIS.
?-3.1415,3.1415

```
       THE Y-AXIS GOES FROM -3 TO 3 IN INCREMENTS OF .0869565217
------------------------------------------------------------------  X-VALUE
                                  .         *                       -3.1415
                                  .*                                -2.9966
                             *    .                                 -2.8516
                          *       .                                 -2.7067
                    *             .                                 -2.5618
                    *             .                                 -2.4169
                       *          .                                 -2.2719
                          *       .                                 -2.127
                           *.                                       -1.9821
                               *  .                                 -1.8372
                                 *.                                 -1.6922
                                  . *                               -1.5473
                               *  .                                 -1.4024
                            *     .                                 -1.2574
                      *           .                                 -1.1125
                    *             .                                 -.9676
                   *              .                                 -.8227
                  *               .                                 -.6777
                 *                .                                 -.5328
                     *            .                                 -.3879
                       *          .                                 -.2429
                          *       .                                 -.098
.........................................*........................  .0469
                                  .        *                        .1918
                                  .          *                      .3368
                                  .           *                     .4817
                                  .         *                       .6266
                                  .      *                          .7715
                                  .    *                            .9165
                          *       .                                 1.0614
               *                  .                                 1.2063
              *                   .                                 1.3513
           *                      .                                 1.4962
        *                         .                                 1.6411
            *                     .                                 1.786
                *                 .                                 1.931
                          *       .                                 2.0759
                             *    .                                 2.2208
                                *  .                                2.3657
                                  *.                                2.5107
                                  . *                               2.6556
                                  .  *                              2.8005
                                  .*                                2.9455
                              *   .                                 3.0904
```

Example   2:   Graph on the same coordinate system the second degree equation

$$x^2+xy+y^2+x+y=1$$

and the equation obtained by rotating or translating to standard form with center (0,0).

To   be   able   to better distinguish between the graphs of the two curves we will use "*" for the graph of $x^2+xy+y^2+x+y=1$ and "+" for the translated curve.

In section 15.1, example 2, the program SECOND DEGREE was run for $x^2+xy+y^2+x+y=1$ and resulted in the unrotated, untranslated equation

$$\frac{x^2}{0.887} + \frac{y^2}{2.66} = 1$$

which implicitly defines the functions

$$y_1 = \sqrt{2.66-2.66x^2/.887} \text{ and } y_2 = -\sqrt{2.66-2.66x^2/.887}$$

with $-\sqrt{.887} \leq x \leq \sqrt{.887}$.

To   determine   the   implicit   functions   defined   by $x^2+xy+y^2+x+y=1$ we will use the quadratic formula.

$$x^2+xy+y^2+x+y=1$$

$$y^2+y(x+1)+(x^2+x-1)=0$$

$$y = \frac{-(x+1) \pm \sqrt{(x+1)^2-4(x^2+x-1)}}{2}$$

$$= \frac{-(x+1) \pm \sqrt{5-2x-3x^2}}{2}$$

Thus, $y_1 = \dfrac{-(x+1)+\sqrt{5-2x-3x^2}}{2}$ and $y_2 = \dfrac{-(x+1)-\sqrt{5-2x-3x^2}}{2}$

The domain of y_1 and y_2 is for all x such that $0 \le 5-2x-3x^2$ which reduces to $-5/3 \le x \le 1$. As for the range on the y-axis, by inspection of the various implicitly defined functions it appears that a range of -3 to 2 will be sufficient. If not, during execution of the program we will exceed the dimension of the array, and an appropriate error message will be printed. In this case we should rerun the program with a larger range for the y-axis. Also, since the domains of the sets of implicitly defined functions are different we must be sure not to attempt to evaluate a function at a value outside of its domain or execution of the program will terminate.

```
450 DEF FNF(X)=(-(X+1)+SQR(5-2*X-3*X*X))/2
453 DEF FNG(X)=(-(X+1)-SQR(5-2*X-3*X*X))/2
454 LET J=INT((FNG(X)-Y1)/DY+.5)+1
455 LET E$(J)="*"
456 IF X<-SQR(.887) THEN 480
457 IF X>SQR(.887) THEN 480
460 DEF FNH(X)=SQR(2.66-2.66*X*X/.887)
461 LET J=INT((FNH(X)-Y1)/DY+.5)+1
462 LET E$(J)="+"
463 DEF FNL(X)=-SQR(2.66-2.66*X*X/.887)
464 LET J=INT((FNL(X)-Y1)/DY+.5)+1
465 LET E$(J)="+"

RUN
ENTER THE RANGE OF THE Y-AXIS.
?-3,2

ENTER THE RANGE OF THE X-AXIS.
?-1.666,1
```

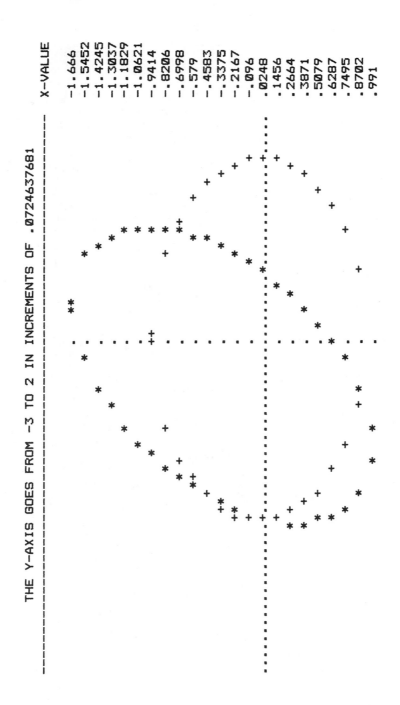

Exercise Set 15.2

In exercises 1-4 use program GRAPH to obtain the graph of each of the following functions. Use the indicated range for the x and y-axes.

		x-axis	y-axis
1.	$f(x)=\cos 2x$	-3.4 to 3.4	-1.1 to 1.1
2.	$f(x)=(\ln(x+1))/(x+1)$	-.7 to 5	-4.1 to .5
3.	$f(x)=e^{-x}$	-1 to 4	-.5 to 3
4.	$f(x)=\tanh x=\dfrac{e^x-e^{-x}}{e^x+e^{-x}}$	-2 to 2	-1 to 1

In exercises 5-8 use program GRAPH to obtain the graph of each of the following relations. Use the indicated range for the x and y-axes.

		x-axis	y-axis
5.	$x^2/9 + y^2/4 = 1$	-3 to 3	-2 to 2
6.	$y^2+2y+x-1=0$	-7 to 2	-4 to 2
7.	$x= y+1$	0 to 5	-6 to 4
8.	$x^2/3 +y^2/3 =1$	-1 to 1	-1 to 1

*
In exercises 9-10 use program GRAPH to graph the given conic section. Also, determine the corresponding conic in standard form and graph that on the same grid as in example 2.

9. $5x^2 -3xy+y^2 +65x-25y+203=0$

10. $1.25x^2 -1.5xy-.75y^2 +42x+6y+153=0$

11. If, during the execution of program GRAPH, a value for y is calculated which falls outside the value of the range entered for the y-values, a fatal error will occur, and the program will stop running.   Two methods for avoiding this type of error are given below.

Method 1: Before placing an "*" in the array check to see if the computed y-value is in the range for the y-values. If not, go to the next x-value without attempting to place the "*" in the array.

Method 2: Instead of entering the range of values on the y-axis, only enter the range of the x-values and then determine the maximum and minimum y-value which will occur for the given x-values.

    a. Modify program GRAPH to include Method 1. As test data use

$$x^2 + y^2 = 9$$

with $-3 \leq y \leq 2$ and $-3 \leq x \leq 3$.

    b. Modify program GRAPH to include Method 2. As test data use

$$2x^2 + y^2 - 2y - 1 = 0$$

with $-1 \leq x \leq 1$.

CHALLENGE ACTIVITY

In the introduction to this chapter we stated that the general second degree equation has a graph which was one of the conics or a degenerate conic section. Yet in writing the program in the first section we did not include any provisions for the degenerate cases. Modify the program in section one to include tests and appropriate messages for all the possible degenerate cases. As test data use the three second degree equations below.

1. $5x^2 + 4xy - y^2 + 24x - 6y - 5 = 0$ two intersecting lines

2. $x^2 - 2xy + 2y^2 - 2x + 2y + 1 = 0$ point (1,0)

3. $4x^2 + 4xy + 5y^2 + 8x + 4y + 5 = 0$ imaginary

Hint:

1. If $B^2 - 4AC = 0$, the equation is a parabola, two parallel lines or imaginary.

2. If $B^2 - 4AC \leq 0$, the equation is an ellipse, a single point, or imaginary.

3. If $B^2 - 4AC > 0$, the equation is a hyperbola or two intersecting lines.

CHAPTER 16
Polar Coordinates

   In the Cartesian coordinate system points are located by their distances from two perpendicular lines, the x and the y-axes. However, there are many instances, such as in the firing of a rocket, where locating points by their distance from a fixed point and their angular displacement as measured from a fixed line is more useful. This scheme for locating points is known as the polar coordinate system. In this chapter we will investigate some applications of polar coordinates as well as develop a program to approximate the graph of a polar equation.

16.1 Applications

   If a region in the plane is bounded by curves given by the polar equations $r=f_1(\theta)$ and $r=f_2(\theta)$, then the area of the region can be found in a manner similar to that of regions bounded by curves given by equations in cartesian coordinates. Namely, we divide the region into subregions whose areas are approximated and summed, forming a Reimann sum which becomes an integral as we pass to the limit.

   However, when using curves given by polar equations it is more convenient to divide the region, not into rectangles, but into sectors of circles as shown in the figure below.

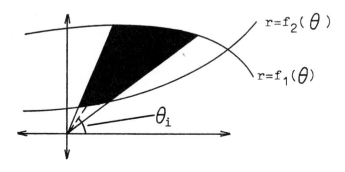

Now to approximate the area of the subregions we make use of the formula for the area of a sector for a circle which subtends an arc of α degrees in a circle of radius r.

$$\text{Area Sector}=\pi\, r^2\,(\alpha/2\pi)=r^2\alpha/2$$

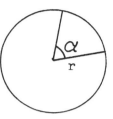

Thus,

Area subregion=Area sector to outer   - Area sector to inner
                 boundary   boundary

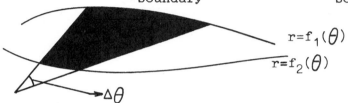

which for the j-th subregion becomes:

$$\approx (f_1(\theta_j))^2\,(\Delta\theta_j/2) - (f_2(\theta_j))^2(\Delta\theta_j/2)$$

Summing over all subregions we have

$$\text{Area}\approx \sum_{j=1}^{n}\left((f_1(\theta_j))^2 - (f_2(\theta_j))^2\right)(\Delta\theta_j/2)$$

which at the limit becomes

$$\text{Area}=(1/2)\int_{\theta_1}^{\theta_2}[(f_1(\theta))^2 - (f_2(\theta))^2]\,d\theta$$

Programming Problem: Approximate the area bounded by two curves described by polar equations.

Output: A table listing the approximation for the area and the number of subdivisions used to obtain the approximation.

Input: Two functions which describe the outer and inner boundary of the area being approximated and the limits of integration, that is, the range on the angle θ.

Strategy: From above we know that if $r = f_1(\theta)$ is the outer boundary and $r = f_2(\theta)$ is the inner boundary, then the area between the curves is given by the integral

$$\text{Area} = (1/2)\int_{\theta_1}^{\theta_2} ([f_1(\theta)]^2 - [f_2(\theta)]^2)\, d\theta$$

for appropriate values of θ_1 and θ_2. Thus, we can obtain an approximation for the area by modifying the program INTEGRAL of section 10.1 . Also, we must convert all angle measures to radians since a computer does not accept angles measured in degrees.

Program:

```
100 REM ** POLAR AREA **
110 REM ** OUTER BOUNDARY **
120 DEF FNF(X)=.....
130 REM ** INNER BOUNDARY **
140 DEF FNG(X)=.....
150 REM ** INPUT **
160 PRINT "ENTER THE RANGE OF"
170 PRINT "THETA IN DEGREES."
180 INPUT A,B
190 PRINT
200 REM ** HEADINGS **
210 PRINT "  N";TAB(10);"APPROXIMATION"
220 REM ** CONVERTING FROM DEGREES TO RADIANS **
230 LET A=A*ATN(1)/45
240 LET B=B*ATN(1)/45
250 REM ** MAIN PROGRAM **
260 LET N=10
270 LET S2=0
280 LET S1=0
290 LET DT=(B-A)/N
300 REM ** LOOP FOR APP.  AREA **
310 FOR I=1 TO N
320 LET L=A+(I-1)*DT
330 LET R=A+I*DT
340 LET M=(L+R)/2
350 LET S1=S1+DT*(FNF(M)*FNF(M)-FNG(M)*FNG(M))
```

```
360 NEXT I
370 LET S1=S1/2
380 PRINT N;TAB(10);S1
390 REM ** TESTING TO END PROGRAM **
400 IF ABS(S1-S2)< .0001 THEN 440
410 LET N=2*N
420 LET S2=S1
430 GO TO 280
440 END
```

Test Data: The polar equations r=5 cos θ and r=2+cos θ intersect at the points (5/2, π/3) and (5/2, -π/3). Thus, the area inside r=5 cos θ and outside r=2+cos θ is given by:

$$(1/2)\int_{-\pi/3}^{\pi/3} ((5\cos\theta)^2 - (2+\cos\theta)^2)\, d\theta \quad =$$

$$(1/2)\int_{-\pi/3}^{\pi/3} (24\cos^2\theta - 4\cos\theta - 4)\, d\theta \quad =$$

$$(1/2)\int_{-\pi/3}^{\pi/3} (12(1+\cos 2\theta) - 4\cos\theta - 4)\, d\theta \quad =$$

$$\int_{-\pi/3}^{\pi/3} (4+6\cos 2\theta - 2\cos\theta)\, d\theta \quad =$$

$$(4\theta + 3\sin 2\theta - 2\sin\theta)\ \Big|_{-\pi/3}^{\pi/3} \quad =$$

$$8/3\, \pi + \sqrt{3} \qquad \approx 10.10963122$$

```
120 DEF FNF(X)=5*COS(X)
140 DEF FNG(X)=2+COS(X)
```

```
RUN
ENTER THE RANGE OF
THETA IN DEGREES.
?-60,60
```

N	APPROXIMATION
10	10.1414752
20	10.1175571
40	10.1116105
80	10.1101259
160	10.1097549
320	10.1096621

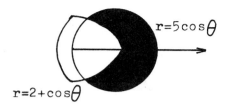

Example 1: Approximate the area bounded by $r=\sqrt{2}(\sin\theta+\cos\theta)$.

Since the area is bounded by a single function, we should use $r=0$ as the function for the inner boundary. As for determining the range on θ, refer to the diagram. As θ goes from 0 to $3/4\pi$, the curve goes form 1 to 0. From $3/4\pi$ to π the value of r is negative and becomes -1 when $\theta=\pi$. Thus, a complete curve is traced out when θ goes from 0 to π.

```
120 DEF FNF(X)=SQR(2)*(SIN(X)+COS(X))
140 DEF FNG(X)=0
```

```
RUN
ENTER THE RANGE OF
THETA IN DEGREES.
?0,180
```

N	APPROXIMATION
10	3.14159265
20	3.14159266

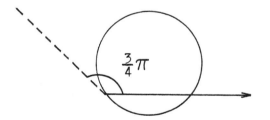

Do you recognize the value for the approximation of the area? To seven decimal places we have π. What figure would you guess the polar equation $r=\sin\theta+\cos\theta$ represents?

Example 2: For any value α the polar equation $r=\alpha\theta$ is known as the spiral of Archimedes. Let A_n be the area bounded by the spiral for $2\pi(n-1)\leq\theta<2\pi n$ and that part of the polar axis joining the endpoints of the spiral for $2\pi(n-1)\leq\theta<2\pi n$. If, for n>2, we let $D_n=A_n-A_{n-1}$, then the following results, which were derived by Archimedes, are valid.

a. $A_1 = 1/3\ \pi(2\pi\ \alpha)^2$

b. $A_2 = 7/12\pi(4\pi\ \alpha)^2$

c. $D_2 = 6A_1$

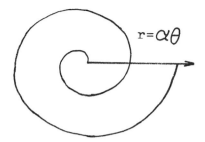

$r = \alpha\theta$

Use program POLAR AREA to illustrate that these results are valid for the spiral $r = \theta\ /3$.

Using a calculator to evaluate the expressions in a, b, and c for $\alpha = 1/3$ gives the following results.

a. $A_1 \approx 4.593522471$
b. $A_2 \approx 32.1546573$
c. $D_2 \approx 27.56113483 = 6(4.593522471)$

Now we verify the above formulas for A_1 and A_2 using the program POLAR AREA.

```
120 DEF FNF(X)=X/3
140 DEF FNG(X)=0
```

RUN
ENTER THE RANGE OF
THETA IN DEGREES.
?0,360

N	APPROXIMATION
10	4.58203867
20	4.59065152
40	4.59280473
80	4.59334304
160	4.59347761
320	4.59351126

Thus, $A_1 \approx 4.59351126$ and parts a and c are verified.

361

```
RUN
ENTER THE RANGE OF
THETA IN DEGREES.
?360,720

    N        APPROXIMATION
   10        32.1431735
   20        32.1517864
   40        32.1539396
   80        32.1544778
  160        32.1546125
  320        32.1546461
```

Thus, $A_2 \approx 32.15464608$

    To   within   the given precision the area determined using the integral agrees with the area resulting from using the formulas.

EXERCISE SET 16.1

In exercises 1-6 use program POLAR AREA to approximate the indicated area.

1. Inside r=3cos θ and outside r=1+cos θ

2. Inside r=9sin θ and outside r=3+3sin θ

3. Bounded by r=3cos 3θ

4. Bounded by r=1/(1+.5cos θ)

5. Inside r=1-cos θ and outside r=cos θ for θ between 0 and 90 degrees.

6. Bounded by the loop of r=3sin θ cos θ /(cos θ +sin θ). (This curve, which in Cartesian coordinates is $x^3+y^3=3xy$, is known as the folium of Descartes. It was submitted to Pierre Fermat by Rene' Descartes as a challenge to determine the tangent line at any point on the curve.)

*
7. From the results of example 1 it appears that the polar equation r=$\sqrt{2}$ (sin θ +cos θ) represents a circle. By converting to Cartesian coordinates determine if this equation actually does represent a circle. If it does, determine its center and radius.

**
8. Write a program which for a function given in polar coordinates will produce a table listing the values of r for θ between 0 degrees and 360 degrees at 15 degree increments. Also, list in the table the corresponding x and y values. Test the program with r=cos θ. A sample output for this program is given below.

RUN

ANGLE	R	X	Y
0	1	1	0
15	.9659	.933	.25
30	.866	.75	.433
45	.7071	.5	.5
60	.5	.25	.433
75	.2588	.067	.25
90	0	0	0
105	-.2588	.067	-.25
120	-.5	.25	-.433
135	-.7071	.5	-.5
150	-.866	.75	-.433
165	-.9659	.933	-.25
180	-1	1	0
195	-.9659	.933	-.25
210	-.866	.75	.433
225	-.7071	.5	.5
240	-.5	.25	.433
255	-.2588	.067	.25
270	0	0	0
285	.2588	.067	.25
300	.5	.25	-.433
315	.7071	.5	-.5
330	.866	.75	-.433
345	.9659	.933	-.25
360	1	1	0

***

9.   a. For the spiral of Archimedes in example 2 sketch the   area represented by D_n.

b. Run program POLAR AREA to approximate the value of D_n for $n=2,3,4$ and 5 for $r=\theta/3$. Make a conjecture about a formula for finding D_n. <u>Hint</u>: Divide each D_n by some integer.

c.   Prove your conjecture from part b.

10. To approximate the arc length of a curve given in polar coordinates consider the following diagram.

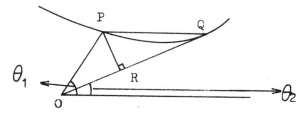

Let the line segment $\overline{PR}$ be perpendicular to line segment $\overline{ORQ}$. The length of the curve from P to Q can be approximated by the hypothenuse $\overline{PQ}$ of the right triangle PQR. Thus, by the Pythagorean Theorem

$$\triangle s(\text{length of curve from P to Q}) \approx \sqrt{RQ^2 + PR^2} \ .$$

Now, the length of RQ is just the change in r, $\triangle r$, and the length of PR can be approximated by the length of the arc PR which is $r\triangle\theta$.

Therefore,

$$\triangle s \approx \sqrt{(\triangle r)^2 + (r\triangle\theta)^2}$$

$$\approx \triangle\theta/\triangle\theta \sqrt{(\triangle r)^2 + (r\triangle\theta)^2}$$

$$\approx \sqrt{(\triangle r/\triangle\theta)^2 + r^2} \ \ \triangle\theta$$

Summing the $\triangle s$ for the entire curve and taking the limit we arrive at the formula for the arc length in polar coordinates.

$$\int_{\theta_1}^{\theta_2} \sqrt{r^2 + (dr/d\theta)^2} \ d\theta$$

Write a program which will approximate the arc length of a curve given in polar coordinates. As test data use the circle r=2.

11. Use your program from exercise 10 to approximate the perimeter of the following curves.

     a. $r=3+3\cos\theta$ (cardioid)

     b. $r=4\cos4\theta$ (eight-leaf rose)

     c. $r=2\sin\theta$ (lemniscate)

     d. $r=4/(1-.5\sin\theta)$ (ellipse)

Ch. 16   POLAR COORDINATES

16.2 Graphing Polar Equations

Programming Problem: Approximate the graph of the polar equation r=f(θ).

Output: A graph of the polar equation r=f(θ).

Input: The function to be graphed, the range on the values of the angle θ, and the maximum value attained by r.

Strategy: We will use a two-dimensional character array to act as a coordinate plane. The dimensions of this array will depend upon what output device is being used and the memory capacity of the computer. For example, monitor screens often display no more than 24 rows and 40 columns at a time. Thus, if the output device is the screen, the dimensions of the array should be no more than 24 by 40. For the programs in this chapter the output device will be a printer, and we will use an array 37 by 37. This is near maximum size which can be dimensioned on a 16K computer.

Since it is impossible to measure angles on a rectangular grid such as our array without the aid of an auxillary device such as a protractor, we will not locate points using their polar coordinates. Instead, we will use their corresponding Cartesian coordinates which can be determined using the formulas:

$$x=r*\cos\theta \quad \text{and} \quad y=r*\sin\theta$$

To locate the points we must remember that an array in BASIC starts with an index (0,0). Therefore, the center of the array, which represents the origin of our Cartesian plane, will have index (18,18) as shown in the figure below. Thus, there will be 18 subdivisions on both the positive and negative x and y-axes. If M is the maximum value of r, then each subdivision will have length D=M/18.

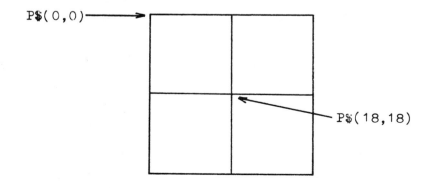

P$(0,0)

P$(18,18)

Although the computer cannot measure distance directly, we can determine the number of subdivisions the x or y-coordinate is from the origin given the length of each subdivision. This value can be computed using the formula:

Number of subdivisions (N) = coordinate/length of subdivision(D)

Note that the value of N, the number of subdivisions, may not be an integer and may be either positive or negative. However, to use N to locate the cell in the array corresponding to a given point we must have an integer. To round N to the nearest integer and also allow for N to be either positive or negative, we will use the BASIC statement:

$$N=SGN(X)*INT(ABS(X)/D+.5)$$

Once the rounding has bee performed, the index for the array can be determined by adding N to 18 for the horizontal component and subtracting N from 18 for the vertical component. We subtract for the vertical component since larger values for y correspond to lower rows in the array.

The final problem that we must resolve is that the line feed and the printer head move different distances as they advance. As was shown in chapter 15 the scale on the x and y-axes can be made approximately the same by multiplying the length of a subdivision by 5/3 on the vertical axis. This compensation can be found in line 240 of the following program.

Program:

```
100 REM ** POLAR PLOT **
110 REM ** FUNCTION **
120 DEF FNF(X)=.....
130 REM ** INPUT **
140 PRINT "ENTER THE RANGE OF"
150 PRINT "THETA IN DEGREES."
160 INPUT A,B
170 PRINT
180 PRINT "ENTER THE MAXIMUM VALUE"
190 PRINT "FOR THE RADIUS."
200 INPUT M
210 PRINT
220 REM ** MAIN PROGRAM **
230 LET D=M/18                          finding scale
240 LET D1=D*5/3
250 DIM P$(36,36)                       setting up plane
260 REM ** INITILIZING PLANE **
270 FOR I=0 TO 36
280 FOR J=0 TO 36
290 LET P$(I,J)=" "
300 NEXT J
310 NEXT I
320 REM ** PLACING POLAR AXIS **
330 FOR I=18 TO 35
340 LET P$(18,I)="-"
350 NEXT I
360 LET P$(18,36)=">"
370 REM ** PLOTTING POLAR POINTS **
380 FOR I=A TO B STEP 5
390 LET T=I*ATN(1)/45                   converting to radians
400 LET R=FNF(T)
410 REM ** CONVERTING TO CARTESIANS **
420 LET X=R*COS(T)
430 LET Y=R*SIN(T)
440 REM ** FINDING LOCATION OF POINT **
450 LET N=SGN(X)*INT(ABS(X)/D+.5)
460 LET X=N+18
470 LET N=SGN(Y)*INT(ABS(Y)/D1+.5)
480 LET Y=18-N
490 REM ** PUTTING POINT IN ARRAY **
500 LET P$(Y,X)="*"
510 NEXT I
520 REM ** PRINTING GRAPH **
530 PRINT
540 PRINT "THE SCALE IS ";D;" UNITS."
550 PRINT
560 FOR I=0 TO 36
570 FOR J=0 TO 36
```

```
580 PRINT P$(I,J);
590 NEXT J
600 PRINT                    advancing to next line
610 NEXT I
620 END
```

Test Data: In polar coordinates the graph of r=3 is a circle of radius 3 centered at the origin.

```
120 DEF FNF(X)=3

RUN
ENTER THE RANGE OF
THETA IN DEGREES.
?0,360

ENTER THE MAXIMUM VALUE
FOR THE RADIUS.
?3


THE SCALE IS .166666667 UNITS.
```

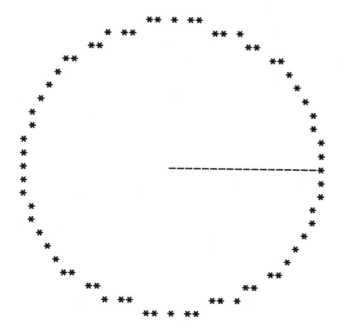

Example 1: Use program POLAR PLOT to graph r=2cos 2θ .

```
120 DEF FNF(X)=2*COS(2*X)
```

RUN
ENTER THE RANGE OF
THETA IN DEGREES.
?0,360

ENTER THE MAXIMUM VALUE
FOR THE RADIUS.
?2

THE SCALE IS .111111111 UNITS.

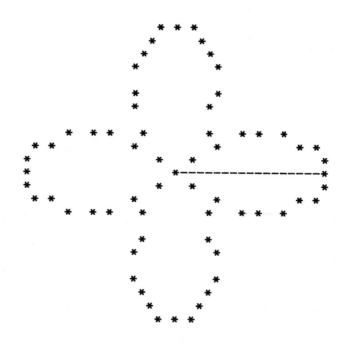

This graph is known as the four-leaf rose.

Example 2: Use program POLAR PLOT to graph the spiral of Archimedes, r= θ/2π .

    120 DEF FNF(X)=X/6.283185307

    RUN
    ENTER THE RANGE OF
    THETA IN DEGREES.
    ?0,720

    ENTER THE MAXIMUM VALUE
    FOR THE RADIUS.
    ?2

    THE SCALE IS .111111111 UNITS.

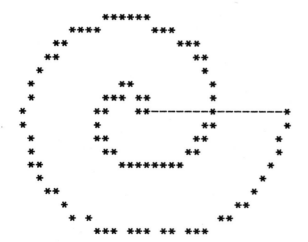

Ch. 16   POLAR COORDINATES

EXERCISE SET 16.2

In exercises 1-8 use program POLAR PLOT to graph each function. Also, whenever possible, identify the curve by name.

1.   r=.3cos3θ

2.   r=2sinθ

3.   r=.25 - .25 cosθ

4.   r=.5 cos3θ + .4 sin4θ

5.   r=θ/3

6.   r=2 cosθ - sinθ

7.   r=1/(1 + .6sinθ)

8.   r=1 - 1.5 sinθ

**
9.   Run program POLAR PLOT a sufficient number of times to be able to make a conjecture about the shape of the graph of the limacon r=a + b cosθ for each of the conditions listed below.

    a.   a/2b ≥ 1

    b.   1/2 < a/2b < 1

    c.   a/2b = 1/2

    d.   a/2b < 1/2

10. Make the necessary modifications to program POLAR PLOT so that the graph of a function in Cartesian coordinates can be estimated. Test your modification with f(x) = 1/3 x - 1. Be sure to include both the x and y-axes on your graph.

11. Use your program from exercise 10 to graph each of the following.

   a. $y = .3 \, x^2$

   b. $y = x$

   c. $y = (x - 1)^2$

12. What are the limitations on using the approach to graphing functions in exercises 10 and 11?

***
13. a. Use program POLAR PLOT to graph $r = 1/(\sin\theta + \cos\theta)$

   b. Explain what went wrong with the graph in part a.

   c. Modify the program so that at least part of the graph of $r=1/(\sin\theta + \cos\theta)$ can be obtained. <u>Hint</u>: Division by zero and plotting points when r is large must be avoided.

CHALLENGE ACTIVITY

Write a program which will:

a.   Graph two polar functions $r=f_1(\theta)$ and $r=f_2(\theta)$ on   the same grid.

b.   Give a list of the appropriate points of intersection of the two functions. Hint: One possible way to   estimate   the possible points of intersection is to   determine   when a location in the array used in part a for the graphing is occupied by both functions.

Test   your   program using $r=-1 + \sin\theta$ and $r=1+(1/2)\sin\theta$ which have points of intersection $(1,0)$ and $(1,\pi)$.

Use your program to estimate the graphs and points of intersection of $r=1/(1-.5\cos\theta)$ and $r=2\cos2\theta$.

CHAPTER 17
Vector Valued Functions

At the end of the nineteenth century the British physicist Lord Kelvin wrote that vectors, "although beautifully ingenious, have been an unmixed evil to those who have touched them in any way" and "have never been of the slightest use to any creature." Yet today many branches of physics are presented in the language of vectors and vector valued functions. In this chapter we investigate vector valued functions and some of their properties.

17.1 THE TRIHEDRAL VECTORS

Although the function $r(t)=x(t)i+y(t)j+z(t)k$ has as its range the set of vectors in three space, we generally think of the function as defining a curve in three space. As with curves in two dimensions, we are interested in both tangent and normal lines to the curve at a point. However, due to the complexity of an equation of a line in three space we do not give the tangent and normal lines but instead the unit tangent and unit normal vectors. These two vectors together with the binormal vector B, defined by B=TXN, form a localized right-handed rectangular coordinate system. This coordinate system is referred to as the triad or trihedral at the given point, or, since it changes for different points on the curve, the moving trihedral.

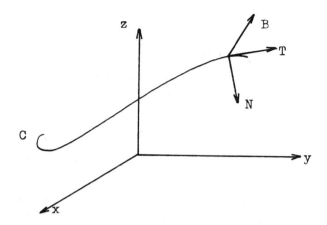

Programming Problem: Approximate for a vector valued function at a given point the position vector, velocity vector, speed, acceleration vector, and trihedral vectors (unit tangent, unit normal, binomial).

Output: The indicated vectors and speed.

Input: The functions which define the i, j, and k components of the vector valued function and the value of t which determines the point on the curve at which the vectors are to be approximated.

Strategy: If r(t)=x(t) i +y(t) j +z(t) k, then we can compute the appropriate vectors and speed at t=t$_0$ using the following formulas.

position vector: $r(t_0)=x(t_0)$ i $+y(t_0)$ j $+z(t_0)$ k

velocity vector: $r'(t_0)=$ V

$$=\frac{dx(t_0)}{dt} \text{ i } +\frac{dy(t_0)}{dt} \text{ j } +\frac{dz(t_0)}{dt} \text{ k}$$

speed: $\frac{ds}{dt}$ = V = r'(t$_0$) =

$$\sqrt{\frac{dx(t_0)}{dt}^2 + \frac{dy(t_0)}{dt}^2 + \frac{dz(t_0)}{dt}^2}$$

acceleration vector: $r''(t_0) = A =$

$$\frac{d^2 x(t_0)}{dt^2} \, i + \frac{d^2 y(t_0)}{dt^2} \, j + \frac{d^2 z(t_0)}{dt^2} \, k$$

unit tangent vector: $T(t_0) = \dfrac{r'(t_0)}{|r'(t_0)|} = \dfrac{V}{|V|}$

unit normal vector: $N(t_0) = \dfrac{T'(t_0)}{|T'(t_0)|}$

binomial vector: $B(t_0) = T \, X \, N$

Except for the normal and binomial vectors the remaining vectors and speed involve only the first and second derivatives of the single variable functions $x(t)$, $y(t)$, and $z(t)$ which can be approximated using the concepts presented in chapter 7. To approximate the normal vector it will be necessary to rewrite its formula.

$$\frac{ds}{dt} = |r'(t)| = |V|$$

$$= \sqrt{(x'(t))^2 + (y'(t))^2 + (z'(t))^2}$$

Then,

$$T(t) = \frac{x'(t)}{ds/dt} \, i + \frac{y'(t)}{ds/dt} \, j + \frac{z'(t)}{ds/dt} \, k$$

Therefore,

$$T'(t) = \frac{(x''(t)s'(t) - x'(t)s''(t))i + (y''(t)s'(t) - y'(t)s''(t))j}{(s'(t))^2}$$

$$+ \frac{(z''(t)s'(t) - z'(t)s''(t))k}{(s'(t))^2}$$

and

$$\frac{d^2 s}{dt^2} = \frac{x'(t)x''(t)+y'(t)y''(t)+z'(t)z''(t)}{\sqrt{(x'(t))^2+(y'(t))^2+(z'(t))^2}} = \frac{r'(t) \cdot r''(t)}{|r'(t)|}$$

Although it is a much more complex formula, $T'(t_0)$, and, thus, $N(t_0)$, is now expressed in terms of first and second derivatives of $x(t)$, $y(t)$, and $z(t)$. The binormal vector can then be approximated using the cross product of T and N, namely,

$$B(t_0)=(T_2 N_3 - N_2 T_3)i+(N_1 T_3 - T_1 N_3)j+(T_1 N_2 - T_2 N_1)k$$

where $T = T_1 i + T_2 j + T_3 k$ and $N = N_1 i + N_2 j + N_3 k$.

Finally, due to round-off error in computing the first and second derivative and also to make the output easier to read, all values will be rounded to three decimal places.

Program:

```
100 REM ** VECTOR FUNCTION **
110 REM ** I-COMPONENT **
120 DEF FNX(T)=.....
130 REM ** J-COMPONENT **
140 DEF FNY(T)=.....
150 REM ** K-COMPONENT **
160 DEF FNZ(T)=.....
170 REM ** INPUT **
180 PRINT "ENTER THE VALUE OF T."
190 INPUT T
200 REM ** SETTING UP ARRAYS **
210 DIM V(3),A(3),TA(3),N(3),B(3)
220 REM ** MAIN PROGRAM **
230 REM ** FINDING FIRST DERIVATIVES **
240 FOR I=1 TO 20
250 LET H=(.3)^I
260 LET R=(FNX(T+H)-FNX(T))/H
270 LET L=(FNX(T-H)-FNX(T))/(-H)
280 IF ABS(R-L)<.0001 THEN 300
290 NEXT I
300 LET V(1)=(R+L)/2
310 FOR I=1 TO 20
320 LET H=(.3)^I
330 LET R=(FNY(T+H)-FNY(T))/H
340 LET L=(FNY(T-H)-FNY(T))/(-H)
350 IF ABS(R-L)<.0001 THEN 370
```

```
360 NEXT I
370 LET V(2)=(R+L)/2
380 FOR I=1 TO 20
390 LET H=(.3)^I
400 LET R=(FNZ(T+H)-FNZ(T))/H
410 LET L=(FNZ(T-H)-FNZ(T))/(-H)
420 IF ABS(R-L)<.0001 THEN 440
430 NEXT I
440 LET V(3)=(R+L)/2
450 REM ** FINDING SECOND DERIVATIVE **
460 FOR I=1 TO 20
470 LET H=(.3)^I
480 LET R=(FNX(T+2*H)-2*FNX(T+H)+FNX(T))/(H*H)
490 LET L=(FNX(T-2*H)-2*FNX(T-H)+FNX(T))/(H*H)
500 IF ABS(R-L)<.001 THEN 520
510 NEXT I
520 LET A(1)=(R+L)/2
530 FOR I=1 TO 20
540 LET H=(.3)^I
550 LET R=(FNY(T+2*H)-2*FNY(T+H)+FNY(T))/(H*H)
560 LET L=(FNY(T-2*H)-2*FNY(T-H)+FNY(T))/(H*H)
570 IF ABS(R-L)<.001 THEN 590
580 NEXT I
590 LET A(2)=(R+L)/2
600 FOR I=1 TO 20
610 LET H=(.3)^I
620 LET R=(FNZ(T+2*H)-2*FNZ(T+H)+FNZ(T))/(H*H)
630 LET L=(FNZ(T-2*H)-2*FNZ(T-H)+FNZ(T))/(H*H)
640 IF ABS(R-L)<.001 THEN 660
650 NEXT I
660 LET A(3)=(R+L)/2
670 REM ** FINDING SPEED **
680 LET S=SQR(V(1)*V(1)+V(2)*V(2)+V(3)*V(3))
690 REM ** FINDING UNIT TANGENT **
700 FOR I=1 TO 3
710 LET TA(I)=V(I)/S
720 NEXT I
730 REM ** FINDING UNIT NORMAL **
740 LET M=0
750 FOR I=1 TO 3          finding dot product
760 LET M=M+V(I)*A(I)     of velocity and
770 NEXT I                acceleration
780 LET M=M/S
790 FOR I=1 TO 3
800 LET N(I)=(A(I)*S-V(I)*M)/(S*S)
810 NEXT I
820 LET L=SQR(N(1)*N(1)+N(2)*N(2)+N(3)*N(3))
830 FOR I=1 TO 3
840 LET N(I)=N(I)/L
850 NEXT I
```

```
860 REM ** FINDING BINORMAL VECTOR **
870 LET B(1)=TA(2)*N(3)-N(2)*TA(3)
880 LET B(2)=N(1)*TA(3)-TA(1)*N(3)
890 LET B(3)=TA(1)*N(2)-TA(2)*N(1)
900 REM ** ROUNDING **
910 DEF FNR(T)=SGN(T)*INT(ABS(T)*1000+.5)/1000
920 LET X1=FNR(FNX(T))
930 LET X2=FNR(FNY(T))
940 LET X3=FNR(FNZ(T))
950 LET S=FNR(S)
960 FOR I=1 TO 3
970 LET V(I)=FNR(V(I))
980 LET A(I)=FNR(A(I))
990 LET TA(I)=FNR(TA(I))
1000 LET N(I)=FNR(N(I))
1010 LET B(I)=FNR(B(I))
1020 NEXT I
1030 REM ** OUTPUT **
1040 PRINT
1050 PRINT "POSITION VECTOR:";TAB(17);
1060 PRINT X1;"I+";Y1;"J+";Z1;"K"
1070 PRINT "VELOCITY: ";TAB(17);
1080 PRINT V(1);"I+";V(2);"J+";V(3);"K"
1090 PRINT "SPEED: ";TAB(17);S
1100 PRINT "ACCELERATION: ";TAB(17);
1110 PRINT A(1);"I+";A(2);"J+";A(3);"K"
1120 PRINT "UNIT TANGENT: ";TAB(17);
1130 PRINT TA(1);"I+";TA(2);"J+";TA(3);"K"
1140 PRINT "UNIT NORMAL: ";TAB(17);
1150 PRINT N(1);"I+";N(2);"J+";N(3);"K"
1160 PRINT "BINORMAL: ";TAB(17);
1170 PRINT B(1);"I+";B(2);"J+";B(3);"K"
1180 END
```

Test Data: For the function r(t)=3cos t i + 3sin t j + 4t k at
t=π/2

$r(\pi/2)=3j+2\pi$ k=3j+6.283k

$r'(\pi/2)=-3i+4k$

ds/dt = 5

$r''(\pi/2)=-3j$

$T(\pi/2)=-.6i+.8k$

$N(\pi/2) = -j$

$B(\pi/2) = .8i + .6k$

```
120 DEF FNX(T)=3*COS(T)
140 DEF FNY(T)=3*SIN(T)
160 DEF FNZ(T)=4*t
```

```
RUN
ENTER THE VALUE OF T.
?1.5707963
```

POSITION:	0I+3J+6.283K
VELOCITY:	-2.955I+0J+4K
SPEED:	4.973
ACCELERATION:	2E-03I+-2.845J+0K
UNIT TANGENT:	-.594I+0J+.804K
UNIT NORMAL:	0I+-1J+0K
BINORMAL:	.804I+1E-03J+.594K

By comparing the exact values with those computed note the effect of the round-off error in the computation of the derivatives.

Example 1: Show that the acceleration vector at the point t=1 for $r(t)=e^t i + e^t \cos t \, j + e^t \sin t \, k$ can be written as a sum of multiples of the unit tangent and unit normal vector.

```
120 DEF FNX(T)=EXP(-T)
140 DEF FNY(T)=EXP(-T)*COS(T)
160 DEF FNZ(T)=EXP(-T)*SIN(T)
```

```
RUN
ENTER THE VALUE OF T.
?1
```

POSITION:	.368I+.199J+.31K
VELOCITY:	-.368I+-.508J+-.111K
SPEED:	.637
ACCELERATION:	.368I+.619J+-.4K
UNIT TANGENT:	-.577I+-.798J+-.174K
UNIT NORMAL:	1E-03I+.212J+-.977K
BINORMAL:	.816I+-.564J+-.122K

Now we want to find numbers E and F such that a(t)=E*T+f*N.

0.368i+.619j-.398k=E(-.577i-.798j-.174k)+F(.213j-.977k)

Therefore,

0.368=-.577E, .619=-.798E+.213F, -.398=-.174E-.977F

or

E≈-.638 and F≈.517

Example 2: For the trihedral coordinate system the coordinate planes are referred to as the osculating plane, normal plane, and rectifying plane. The osculating plane contains the unit tangent and unit normal vectors; the normal plane contains the unit normal and binomial vectors; the rectifying plane contains the unit tangent and binormal vectors.

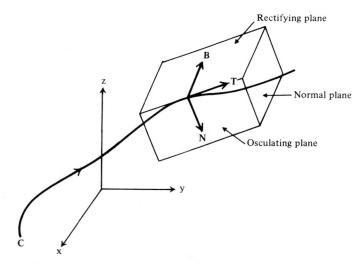

Approximate the osculating, normal and rectifying planes for the twisted cubic $r(t)=ti+t^2j+t^3k$ at t=2.

```
120 DEF FNX(T)=T
140 DEF FNY(T)=T*T
160 DEF FNZ(T)=T*T*T

RUN
ENTER THE VALUE OF T.
?2
```

POSITION:	2I+4J+8K
VELOCITY:	1I+4J+11.999K
SPEED:	12.688
ACCELERATION:	0I+2J+12.004K
UNIT TANGENT:	.079I+.315J+.946K
UNIT NORMAL:	-.445I+-.838J+.316K
BINORMAL:	.892I+-.446J+.074K

From the position vector the point on the curve at which we want to find the three planes is (2,4,8). By definition the osculating plane contains the unit normal and unit tangent vectors, and, thus, is perpendicular to the binormal vector. Therefore, using the point-normal form for a plane we obtain the following.

> Point-Normal Form: If the vector n=ai+bj+ck is normal (perpendicular) to a plane containing the point (x_0, y_0, z_0) the the equation of the plane is:
>
> $$a(x-x_0)+b(y-y_0)+c(z-z_0)=0$$

Thus, the equation for the osculating plane is:

$$.892(x-2)-.446(y-4)+.074(z-8)=0$$

The normal plane is perpendicular to the unit tangent vector and has equation

$$.079(x-2)+.315(y-4)+.946(z-8)=0$$

Finally, the rectifying plane is perpendicular to the normal vector and has equation

$$-.445(x-2)-.838(y-4)+.316(z-8)=0$$

EXERCISE SET 17.1

In exercises 1-5 use program VECTOR FUNCTION to approximate the position vector, velocity vector, speed, acceleration, and the trihedral vectors for each function at the given value of t.

1. $r(t)=t$ i+ln(t) j+e^t k; t=2

2. $r(t)=\cos 2t$ i+sin 2t j+3t k; t= $\pi/4$

3. $r(t)=2/(2+t)$ i+$t^2/(1+t)$ j+3t k; t=1

4. $r(t)=\tan t$ i+cos t j+sin t k; t=$\pi/4$

5. $r(t)=(t \cos t -\sin t)$i+$(\cos t +t\sin t)$j+t^2k; t=1

*

In exercises 6-10 use program VECTOR FUNCTION to find the osculating, normal and rectifying planes for each function at the given value of t.

6. $r(t)=e^{-2t}$ i+e^{-2t} j+t k; t=1

7. $r(t)=\sin t$ i+t j+cos t k; t=$-\pi/4$

8. $r(t)=3\cos^3 t$ i+$3\sin^3 t$ j+k; t=0

9. $r(t)=e^t \sin t$ i+$e^t \cos t$ j+e^t k; t=$\pi/2$

10. $r(t)=\ln(t+2)$ i+$t/(t^3+1)$ j-$3t^2$k; t=3

11. In the function used for the test data
$r'(t)=-3\sin t\ i+3\cos t\ j+4k$, and $|r'(t)| =\sqrt{9\sin^2 t +9\cos^2 t +16}$
$=\sqrt{9+16} = 5$ for all values of t. Thus, the speed is a constant.
An interesting fact about a particle which moves along a curve at
a constant speed is that the acceleration and velocity vectors are
always perpendicular. Note that this is true for the test data.
Verify that the velocity and acceleration vectors are
perpendicular for the function $r(t)=3\cos t\ i+3\sin t\ j+4tk$ used in
the test data for t=1, $\pi/4,\pi$, and 4 by running the program VECTOR
FUNCTION for these values of t.

**

12. a. The angular momentum, L(t), of a particle of mass m with
position vector r(t) is $L(t)=m(r(t)$ X $V(t))$, and the torque, N(t),
is $N(t)=m(r(t)$ X $A(t))$. Use program VECTOR FUNCTION to
approximate the angular momentum and torque of a particle with
unit mass at $t=\pi$ if its position at any time is given by
$r(t)=e^t i+e^t\cos t\ j-e^t\sin t\ k$. Note that N(t) in this
problem stands for the torque, not the unit normal vector.

    b. Show that $dL(t)/dt = N(t)$. Thus, if the torque, which is
a force causing rotation, is zero, then the angular momentum is a
constant. This is the law of conservation of angular momentum.

13. In the program VECTOR FUNCTION it is necessary to divide by
the speed to find several of the vectors. However, no test was
included in the program to check for a speed of 0. Modify the
program to include such a test with appropriate messages for the
affected vectors.

14. In example 1 we showed that at t=1 the acceleration vector
for $r(t)=e^{-t} i+e^{-t}\cos t\ j+e^{-t}\sin t\ k$ could be written as a
sum of multiples of the unit tangent and unit normal vectors.
This illustrates the fact that the acceleration vector at any
point on a curve can always be written as a linear combination of
the unit tangent and the unit normal vectors. Modify program
VECTOR FUNCTION to include the computation of the tangential and
normal components of acceleration.

$$\text{tangential component} = \frac{|r'(t)\ r''(t)|}{|r'(t)|} =r''(t)\ T(t)$$

$$\text{normal component} = \frac{|r'(t)\ X\ r''(t)|}{|r'(t)|} =r''(t)\ N(t)$$

As test data use the function of example 1.

15.   In chapter 7 alternate difference quotients for approximating the first and second derivatives were given which generally reduce round-off error.   Modify program VECTOR FUNCTION to compute the first and second derivatives using these alternate difference quotients.   As test data use r(t)=3cos t i+3sin t j+4t k at t=π/2. Compare your results to those obtained in this section for the same function.

16.   Write a program which will find:

a.   the length of a vector.

b.   the dot (scalar) product of two vectors.

c.   the cross product of two vectors.

d.   the angle between two vectors.

e.   the scalar and vector projection of one vector on another.

Your program should be set up so that the user can choose what is to be done--this is referred to as a MENU.   Use appropriate test data in checking your program.   A sample output is given below.

RUN

1. LENGTH OF V.
2. DOT PRODUCT OF V AND W.
3. CROSS PRODUCT OF V AND W.
4. ANGLE BETWEEN V AND W.
5. PROJECTION OF V ON W, SCALAR.
6. PROJECTION OF V ON W, VECTOR.
7. QUIT.

WHAT DO YOU WANT TO DO?(1-7)
?6

ENTER THE COMPONENTS OF V, ALL ON THE SAME LINE.
?1,2,3

ENTER THE COMPONENTS OF W, ALL ON THE SAME LINE.
?4,5,6

THE VECTOR PROJECTION OF 1I+2J+3K ON 4I+5J+6K IS
1.66I+2.078J+2.494K

17.2 CURVATURE

Curvature is a measure of how much a curve is bending. Small values indicate that the curve is nearly flat, whereas large values indicate a sharp bend. For curves in the plane given by $y=f(x)$ the curvature at the point $(x_0,f(x_0))$ is given by:

$$C= \frac{|f''(x_0)|}{(1+(f'(x_0))^2)^{3/2}}$$

Also, if the curvature is nonzero at a point there is associated a circle of curvature or osculating circle. This circle is tangent to the curve at the point and has the same concavity which makes it the circle which best approximates the curve at the point.

Programming Problem: Approximate the curvature of a function at a point when the function is given in rectangular coordinates, $y=f(x)$. Also, estimate the center of the osculating circle at the given point and the corresponding radius of this circle.

Output: The functional value, the curvature, radius of curvature, and the center of the osculating circle.

Input: The function and the value of x at which the values are to be approximated.

Strategy: As stated above the curvature of the function $y=f(x)$ at $x=x_0$ is given by:

$$C= \frac{|f''(x_0)|}{(1+(f'(x_0))^2)^{3/2}}$$

Also the radius of curvature, which we will denote by **p**, is $p=1/C$. Finally, the center of the osculating circle is given by:

$$\left(x_0 - \frac{f'(x_0)(1+(f'(x_0))^2)}{f''(x_0)} \ , \ f(x_0) + \frac{1+(f'(x_0))^2}{f''(x_0)} \right)$$

Since all the formulas involve first and second derivatives of the single variable function $f(x)$, we can again use the concepts of chapter 7 to approximate the derivatives, and, thus, evaluate the formulas given above.

Program:

```
100 REM ** CURVATURE RECTANGULAR **
110 REM ** FUNCTION **
120 DEF FNF(X)=.....
130 REM ** INPUT **
140 PRINT "ENTER THE VALUE OF X."
150 INPUT X
160 PRINT
170 REM ** MAIN PROGRAM **
180 REM ** FINDING FIRST DERIVATIVE **
190 FOR I=1 TO 20
200 LET H=(.3)^I
210 LET R=(FNF(X+H)-FNF(X))/H
220 LET L=(FNF(X-H)-FNF(X))/(-H)
230 IF ABS(R-L)<.0001 THEN 250
240 NEXT I
250 LET D1=(R+L)/2
260 REM ** FINDING SECOND DERIVATIVE **
270 FOR I=1 TO 20
280 LET H=(.3)^I
290 LET R=(FNF(X+2*H)-2*FNF(X+H)+FNF(X))/(H*H)
300 LET L=(FNF(X-2*H)-2*FNF(X-H)+FNF(X))/(-H*-H)
310 IF ABS(R-L)<.001 THEN 330
320 NEXT I
330 LET D2=(R+L)/2
340 REM ** FINDING CURVATURE AND RADIUS **
350 LET C=ABS(D2)/(1+D1*D1) (1.5)
360 PRINT "AT X=";X;" ,F(X)=";FNF(X)
370 PRINT "AND CURVATURE IS ";C
380 IF C<>0 THEN 420
390 PRINT "THE RADIUS OF CURVATURE IS"
400 PRINT "INFINITE."
410 GO TO 490
420 LET P=1/C
430 PRINT "THE RADIUS OF CURVATURE IS ";P
440 REM ** FINDING CENTER OF OSCULATING CIRCLE **
450 LET A=X-D1*(1+D1*D1)/D2
460 LET B=FNF(X)+(1+D1*D1)/D2
470 PRINT "THE CENTER OF THE OSCULATING CIRCLE"
480 PRINT "IS (";A;",";B;")."
490 END
```

Test Data:

At x=0 the function y=x² has curvature 2,   radius
of curvature .5 and an osculating circle with center (0,.5).

```
120 DEF FNF(X)=X*X

RUN
```

ENTER THE VALUE OF X.
?0

AT X=0 ,F(X)=0
AND THE CURVATURE IS 2
THE RADIUS OF CURVATURE IS .5
THE CENTER OF THE OSCULATING CIRCLE
IS (0,.5).

Example 1: Use program CURVATURE RECTANGULAR to aid in finding the osculating circle at the point $(3\sqrt{3}/8,\ 1/8)$ on the hypocycloid $x^{2/3}+y^{2/3}=1$. Sketch a graph.

In order to use our program we must have y as a function of x.

$$x^{2/3}+y^{2/3}=1$$

$$y^{2/3}=1-x^{2/3}$$

$$y=+(1-x^{2/3})^{3/2}$$

We choose the positive root since we are interested in a point in the first quadrant. Also, it will be necessary to change the range of the FOR-NEXT loop used in calculating the second derivative on line 270 to I=2 to 20. This is necessary since, for I=1 and H=.3, X+2*H=1.1, which when substituted into the function requires taking the square root of a negative number.

```
120 DEF FNF(X)=(1-(X*X)^(1/3))^(1.5)
270 FOR I=2 TO 20

RUN
```
ENTER THE VALUE OF X.
?.64951905

AT X=.64951905 ,F(X)=.125000001
AND THE CURVATURE IS .769837197
THE RADIUS OF CURVATURE IS 1.29897594
THE CENTER OF THE OSCULATING CIRCLE
IS (1.29900807,1.24994557).

Considering round-off error we will use p=1.3 and center (1.3, 1.25).   Therefore, the osculating circle is $(x-1.3)^2+(y-1.25)^2=(1.3)^2$

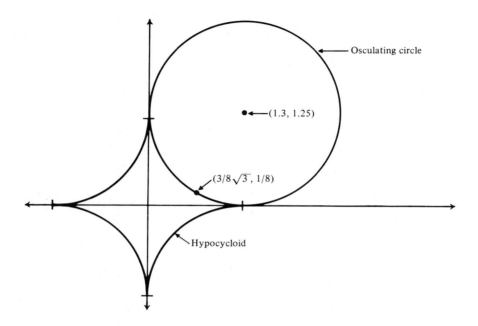

Osculating circle

(1.3, 1.25)

$(3/8\sqrt{3}, 1/8)$

Hypocycloid

Programming Problem: Approximate the curvature of a space curve which is defined parametrically.

Output: The point on the curve and the curvature.

Input: The three component functions used to define the curve and the value of t used to determine the point on the curve.

Strategy: If the curve is given by r(t)=x(t)i+y(t)j+z(t)k, then the curvature at t=to is given by:

$$C=\frac{|r'(to) \ X \ r''(to)|}{|r'(to)^3|}$$

Again this formula involves only first and second derivatives of the single variable functions x(t), y(t), and z(t) and can be approximated as in the program VECTOR FUNCTION of section 1.

Program:

```
100 REM ** CURVATURE SPACE **
110 REM ** I-COMPONENT **
120 DEF FNX(T)=.....
130 REM ** J-COMPONENT **
140 DEF FNY(T)=.....
150 REM ** K-COMPONENT **
160 DEF FNZ(T)=.....
170 REM ** INPUT **
180 PRINT "ENTER THE VALUE OF T."
190 INPUT T
200 REM ** SETTING UP ARRAYS **
210 DIM V(3),A(3)
220 REM ** MAIN PROGRAM **
230 REM ** FINDING FIRST DERIVATIVES **
240 FOR I=1 TO 20
250 LET H=(.3)^I
260 LET R=(FNX(T+H)-FNX(T))/H
270 LET L=(FNX(T-H)-FNX(T))/(-H)
280 IF ABS(R-L)<.0001 THEN 300
290 NEXT I
300 LET V(1)=(R+L)/2
310 FOR I=1 TO 20
320 LET H=(.3)^I
330 LET R=(FNY(T+H)-FNY(T))/H
340 LET L=(FNY(T-H)-FNY(T))/(-H)
350 IF ABS(R-L)<.0001 THEN 370
360 NEXT I
370 LET V(2)=(R+L)/2
380 FOR I=1 TO 20
390 LET H=(.3)^I
400 LET R=(FNZ(T+H)-FNZ(T))/H
410 LET L=(FNZ(T-H)-FNZ(T))/(-H)
420 IF ABS(R-L)<.0001 THEN 440
430 NEXT I
440 LET V(3)=(R+L)/2
450 REM ** FINDING SECOND DERIVATIVES **
460 FOR I=1 TO 20
470 LET H=(.3)^I
480 LET R=(FNX(T+2*H)-2*FNX(T+H)+FNX(T))/(H*H)
490 LET L=(FNX(T-2*H)-2*FNX(T-H)+FNX(T))/(H*H)
500 IF ABS(R-L)<.01 THEN 520
510 NEXT I
520 LET A(1)=(R+L)/2
530 FOR I=1 TO 20
540 LET H=(.3)^I
550 LET R=(FNY(T+2*H)-2*FNY(T+H)+FNY(T))/(H*H)
560 LET L=(FNY(T-2*H)-2*FNY(T-H)+FNY(T))/(H*H)
570 IF ABS(R-L)<.01 THEN 590
```

```
580 NEXT I
590 LET A(2)=(R+L)/2
600 FOR I=1 TO 20
610 LET H=(.3)^I
620 LET R=(FNZ(T+2*H)-2*FNZ(T+H)+FNZ(T))/(H*H)
630 LET L=(FNZ(T-2*H)-2*FNZ(T-H)+FNZ(T))/(H*H)
640 IF ABS(R-L)<.01 THEN 660
650 NEXT I
660 LET A(3)=(R+L)/2
670 REM ** FINDING MAGNITUDE OF VELOCITY **
680 LET S=SQR(V(1)*V(1)+V(2)*V(2)+V(3)*V(3))
690 IF S=0 THEN 790          testing for infinite curvature
700 REM ** FINDING V CROSS A **
710 LET K1=V(2)*A(3)-V(3)*A(2)
720 LET K2=V(3)*A(1)-V(1)
730 LET K3=V(1)*A(2)-V(2)*A(1)
740 REM ** FINDING MAGNITUDE **
750 LET C=SQR(K1*K1+K2*K2+K3*K3)
760 REM ** FINDING CURVATURE **
770 LET C=C/(S*S*S)
780 REM ** ROUNDING **
790 DEF FNR(T)=SGN(T)*INT(ABS(T)*1000+.05)/1000
800 LET X1=FNR(FNX(T))
810 LET Y1=FNR(FNY(T))
820 LET Z1=FNR(FNZ(T))
830 REM ** OUTPUT **
840 PRINT "AT T=";T;" THE POINT ON THE CURVE IS"
850 PRINT "(";X;",";Y;",";Z;")."
860 IF S=0 THEN 890
870 PRINT "THE CURVATURE IS ";C
880 GO TO 900
890 PRINT "THE CURVATURE IS INFINITE."
900 END
```

Test Data:
At t=1 for the twisted cubic $r(t)=ti+t^2 j+t^3$, $r'(1)=i+2j+3k$ and $r''(1)=2j+6k$.

$$|r'(1) \text{ X } r''(1)| = \begin{vmatrix} i & j & k \\ 1 & 2 & 3 \\ 0 & 2 & 6 \end{vmatrix} = 6i-6j+2k$$

$$r'(1) \text{ X } r''(1) = \sqrt{76} \text{ and } r'(1) = \sqrt{14}$$

Thus, $C = \sqrt{76}(\sqrt{14})^3 \approx .16642353$

```
120 DEF FNX(T)=T
140 DEF FNY(T)=T*T
160 DEF FNZ(T)=T*T*T

RUN
ENTER THE VALUE OF T.
?1
AT T=1 THE POINT ON THE CURVE IS
(1,1,1).
THE CURVATURE IS .166399458
```

Example 2: In example 1 of section 17.1 we showed that the acceleration could be written as a sum of multiples of the unit tangent and unit normal for $r(t)=e^{-t}i+e^{-t}\cos t \; j+e^{-t}\sin t \; k$ at t=1. Also, in exercise 14 of that section we gave formulas for decomposing the acceleration into tangential and normal components. An alternate set of formulas for these components are

$$\text{tangential component} = \frac{d^2 s}{dt^2} \quad \text{and} \quad \text{normal component} = C\left(\frac{ds}{dt}\right)^2$$

where the derivatives are evaluated at t_0.

Run program CURVATURE SPACE to approximate C and use the results of example 1, section 17.1, to verify that the normal component of acceleration of $r(t)=e^{-t}i+e^{-t}\cos t \; j+e^{-t}\sin t \; k$ at t=1 is $C(ds/dt)^2$.

```
120 DEF FNX(T)=EXP(-T)
140 DEF FNY(T)=EXP(-T)*COS(T)
160 DEF FNZ(T)=EXP(-T)*SIN(T)

RUN
ENTER THE VALUE OF T.
?1
AT T=1 THE POINT ON THE CURVE IS
(.368,.199,.31).
THE CURVATURE IS 1.28149599
```

From example 1, section 17.1, ds/dt $\approx$ .637 and the normal component $\approx$ .517 Thus, $C(ds/dt)^2$ = (1.28149599)(.637)^2 $\approx$ .5199513

Example 3: The curvature of a plane curve given parametrically by x=x(t) and y=y(t) at t=to is given by:

$$C = \frac{x'(t_o)y''(t_o)-x''(t_o)y'(t_o)}{((x'(t_o))^2+(y'(t_o))^2)^{3/2}}$$

Illustrate that the formulas for calculating curvature in three dimensions holds for two dimensions by comparing the results of calculating the curvature of $x=t+t^3$, $y=t^2+1/t$ at t=1 by the formulas mentioned above and by running the program CURVATURE SPACE.

x'(t)=1+3t	y'(t)=2t-1/t²
x''(t)=6t	y''(t)=2+2/t³
x'(1)=4	y'(1)=1
x''(1)=6	y''(1)=4

$$C= \frac{4(4)-6(1)}{(16+1)^{3/2}} = \frac{10}{17^{3/2}} \approx .14266801$$

```
120 DEF FNX(T)=T+T*T*T
140 DEF FNY(T)=T*T+1/T
160 DEF FNZ(T)=0
```

```
RUN
ENTER THE VALUE FOR T.
?1
AT T=1 THE POINT ON THE CURVE IS
(2,2,0)
THE CURVATURE IS .14267707
```

The results agree to four decimal places.

EXERCISE SET 17.2

In exercises 1-4 use program CURVATURE RECTANGULAR to aid in finding the equation of the osculating circle to the curve at the indicated point. Sketch a graph of both the function and the osculating circle.

1. $y=x^2-.5$; $x=0$

2. $y=e^x$; $x=0$

3. $y=1/x$; $x=1$

4. $y=\cos x$; $x=\pi$

In exercises 5-9 use program CURVATURE SPACE to approximate the curvature at the given value of t.

5. $r(t)=(3t+1)i+(2t^2-1)j$; $t=2$

6. $r(t)=ti+4tj+3t^2k$; $t=1$

7. $r(t)=3t/(1+t^3)\ i+3t^2/(1+t^3)\ j$; $t=1$ (folium of Descartes)

8. $r(t)=t^2i+t^4k$; $t=2$

9. $r(t)=4\cos t\ i+4\sin t\ j+tk$; $t=3\pi/4$ (helix)

10. The curvature of a straight line should be 0--no bending, while that of a circle should be constant since it bends at a constant rate. Illustrate these facts by running program CURVATURE RECTANGULAR for the functions below at different values of x.

   a. $y=5x+7$; $x=-2,0,2$

   b. $x^2+y^2=25$; $x=-1,0,1$ <u>Hint</u>:Solve for y in terms of x.

What is the constant curvature for a circle?

Ch. 17   VECTOR VALUED FUNCTIONS

11. In example 1 show that the osculating circle passes through the point $(3\sqrt{3/8}, 1/8)$.

**

12. Write a program to approximate the curvature of a curve in the plane given parametrically by the formula

$$C = \frac{|x'(t)y''(t) - x''(t)y'(t)|}{((x'(t))^2 + (y'(t))^2)^{3/2}}$$

As test data use $x = t + t^3$, $y = t^2 + 1/t$ at $t = 1$. Compare your results to example 3.

13. Write a program to approximate the tangential and normal components of acceleration using

tangential component$= d^2 s/dt^2$, normal component$= C(ds/dt)^2$

As test data use $r(t) = e^{-t} i + e^{-t} \cos t \, j + e^{-t} \sin t \, k$ at $t = 1$. Compare your results to example 1, section 17.1.

14. For a curve given in polar coordinates $r = f(\theta)$ the curvature is

$$C = \frac{|r^2 + 2(dr/d\theta)^2 - r(d^2 r/d\theta^2)|}{(r^2 + (dr/d\theta)^2)^{3/2}}$$

Write a program to approximate the curvature of a curve given in polar coordinates. As test data use $r = 1 + \sin\theta$ which has curvature 0.8117942 at $\theta = \pi/4$.

***
15. A remarkable fact about curvature is that the shape of a curve in a plane is completely determined knowing the curvature. Thus, two plane curves which have the same curvature for every value of x or t are identical except for their locations. A similar statement holds in three dimensions but requires the introduction of a new quantity, Tr, called the torsion. Torsion basically measures movement of the curve in and out of the osculating plane while curvature measures movement in the osculating plane. The relations between torsion, curvature, and the trihedral vectors are known as the Frenet-Serret formulas given by

$$\frac{dT}{ds} = CN, \quad \frac{dN}{ds} = TrB-CT, \quad \frac{dB}{ds} = -TrN.$$

If r(t)=x(t) i+y(t) j+z(t) k, then the torsion at t=to can be computed using

$$Tr= \frac{r'(to) \ (r''(to) \ X \ r'''(to))}{|r'(to) \ X \ r''(to)|^2} \quad .$$

a. Write a program to approximate the torsion of a curve in R which is defined parametrically. As test data use the twisted cubic r(t)=ti+t²j+t³k at t=1 which has Tr=.15789473 .

b. Run your program for r(t)=t²i+t³j at t=2.

c. Run your program for r(t)=cos t i+sin t j at t=π/4.

d. In general, what is the torsion of a plane curve?

CHALLENGE ACTIVITY

Write a program which will approximate the graph of a function $y=f(x)$ and, at a specified point of the curve, graph the osculating circle. Be sure that the point at which the osculating circle is to be drawn is included in your graph. As test data use $y=x$ $-.5$ at $x=0$. An example of what the output for the program could look like is given below.

"*"--graph of function

"+"--graph of osculating circle

"#"--where function and osculating circle overlap

y-axis--horizontal

x-axis--vertical

INPUT X
? 0
AT X= O F(X)=-.5
THE CURVATURE IS 2 .
THE RADIUS OF CURVATURE IS .5 .
CENTER OF OSCULATING CIRCLE:(0 , 0)

SCALE ON Y-AXIS: .0144927536

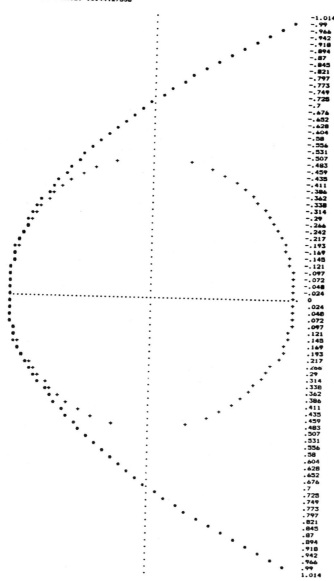

-1.014
-.99
-.966
-.942
-.918
-.894
-.87
-.845
-.821
-.797
-.773
-.749
-.728
-.7
-.676
-.652
-.628
-.604
-.58
-.556
-.531
-.507
-.483
-.459
-.435
-.411
-.386
-.362
-.338
-.314
-.29
-.266
-.242
-.217
-.193
-.169
-.145
-.121
-.097
-.072
-.048
-.024
0
.024
.048
.072
.097
.121
.145
.169
.193
.217
.266
.29
.314
.338
.362
.386
.411
.435
.459
.483
.507
.531
.556
.58
.604
.628
.652
.676
.7
.725
.749
.773
.797
.821
.845
.87
.894
.918
.942
.966
.99
1.014

CHAPTER 18
Functions Of Several Variables

For many physical applications the quantity under study is not dependent on a single variable but rather on several independent variables, such as position in space and time. Thus, to investigate such quantities it is necessary to study what are referred to as functions of several variables. In this chapter we will explore the concepts of limits and derivatives as they pertain to these functions.

18.1 LIMITS AND CONTINUITY

Programming Problem: Approximate
$$\lim_{(x,y) \to (a,b)} f(x,y)$$

Output: A table listing the points used to approximate the limit as well as the approximation.

Input: The point at which the limit is to be taken and the function $f(x,y)$.

Strategy: To approximate
$$\lim_{(x,y) \to (a,b)} f(x,y)$$
we will select points (x,y) with decreasing distance from the point (a,b) and evaluate $f(x,y)$ at these points. If the evaluations appear to "cluster" about a particular value, we will use it to approximate the limit; otherwise, we will assume that the limit does not exist.

In approximating the $\lim_{x \to a} f(x)$, a function of a single variable, the values of x could approach a from only one of two directions, either the left or the right.

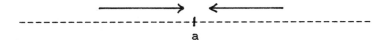

However, for a function of two variables there are an infinite number of paths along which the points (x,y) could approach (a,b).

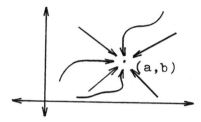

Since it would be impossible to select points along all these paths, we will try to go across as many of these paths as we can by approaching (a,b) along a spiral path.

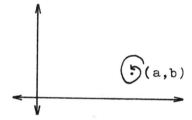

To do this we will use polar coordinates and compute the point (x,y) by

$$x = a + r \cos \theta \text{ and } y = b + r \sin \theta$$

where, for successive points, θ is incremented by 20 degrees and r is decreased so that the points approach (a,b). Also, to improve readability we will round the values of x and y to five decimal places.

Finally, in order to write this program it is necessary to understand how the DEF statement works in BASIC. For most computers the DEF statement allows only one argument on the left-hand side of the assignment statement; for example, DEF FNF(X) is a statement where X is the argument. However, there is no limitation on the number of variables which may occur in the right-hand side of the statement. When called, the function is evaluated using the value assigned to the argument in the calling statement and the current value of all other variables. For example, the following statements could be used to evaluate f(x,y)=2x+3y at (1,2).

401

```
10 DEF FNF(X)=2*X+3*Y
20 LET Y=2
30 LET A=FNF(1)
```

Thus, we can use the DEF statement in dealing with functions of
several variables as long as all the variables are assigned the
correct value before referring to the function.

Program:

```
100 REM ** LIMIT TWO VARIABLES **
110 REM ** FUNCTION **
120 DEF FNF(X)=.....
130 REM ** ROUNDING FUNCTION **
140 DEF FNR(X)=SGN(X)*INT(ABS(X)*100000+.5)/100000
150 REM ** INPUT **
160 PRINT "ENTER THE POINT AT WHICH"
170 PRINT "THE LIMIT IS TO BE TAKEN."
180 PRINT "FIRST THE X THEN THE Y VALUE."
190 INPUT X1,Y1
200 PRINT
210 REM ** HEADINGS **
220 PRINT "X-VALUE";TAB(12);"Y-VALUE";
230 PRINT TAB(23);"APP.  LIMIT"
240 REM ** MAIN PROGRAM **
250 FOR I=0 TO 360 STEP 20
260 LET R=.7 (I/10)
270 LET T=I*ATN(1)/45
280 LET X=X1+R*COS(T)
290 LET Y=Y1+R*SIN(T)
300 LET A=FNF(X)
310 REM ** ROUNDING X AND Y **
320 LET X=FNR(X)
330 LET Y=FNR(Y)
340 PRINT X;TAB(12);Y;TAB(23);A
350 NEXT I
360 END
```

Test Data: As test data we will approximate two limits, one for a function which has a limit and a second for a function which does not have a limit.

a. $\lim(x,y)\to(1,2)$ $(3-x+y)/2xy=1$

```
120 DEF FNF(X)=(3-X+Y)/(2*X*Y)
```

```
RUN
ENTER THE POINT AT WHICH
THE LIMIT IS TO BE TAKEN.
FIRST THE X THEN THE Y VALUE.
?1,2
```

X-VALUE	Y-VALUE	APP. LIMIT
2	2	.375
1.46045	2.16759	.585524977
1.18393	2.15433	.7783358
1.05882	2.10189	.90833722
1.01001	2.05677	.974014826
0.99509	2.02782	.999251732
0.99308	2.01199	1.00570108
0.9948	2.00436	1.00543225
0.99688	2.00114	1.00363049
0.99837	2	1.00203884
0.99925	1.99973	1.00100627
0.9997	1.99975	1.00043737
0.9999	1.99983	1.00016124
0.99998	1.99991	1.00004349
1.00001	1.99995	1.00000134
1.00001	1.99998	.999990793
1.00001	1.99999	.999991199
1.00001	2	.999994106
1	2	.999996685

b. $\lim_{(x,y)\to(0,0)} (x^2+yx^3)/(x^2+y)$ does not exist

120 DEF FNF(X)=(X*X+Y*X*X*X)/(X*X+Y)

```
RUN
ENTER THE POINT AT WHICH
THE LIMIT IS TO BE TAKEN.
FIRST THE X THEN THE Y VALUE.
?0,0
```

X-VALUE	Y-VALUE	APP. LIMIT
1	0	1
0.46045	.16759	.601611984
0.18393	.15433	.184890788
0.05882	.10189	.0330436528
0.01001	.05677	1.76300752E-03
-4.91E-03	.02782	8.64041646E-04
-6.92E-03	.01199	3.97940375E-03
-5.2E-03	4.36E-03	6.15350295E-03
-3.12E-03	1.14E-03	8.50700497E-03
-1.63E-03	0	1
-7.5E-04	-2.7E-04	-2.06431724E-03
-3E-04	-2.5E-04	-3.57069208E-04
-1E-04	-1.7E-04	-5.53077979E-05
-2E-05	-9E-05	-2.87434765E-06
1E-05	-5E-05	-1.40842831E-06
1E-05	-2E-05	-6.50658952E-06
1E-05	-1E-05	-1.0082822E-05
1E-05	0	-1.39720259E-05
0	0	1

Since the approximations are not "clustering" about a single value, we should conclude that the limit does not exist.

Example 1: Investigate
$$\lim_{(x,y) \to (1,-1)} e^{xy} (\sqrt{x} + y) / (y^4 - 2xy^2 + x^2)$$

```
120 DEF FNF(X)=(EXP(X*Y)*(SQR(X)+Y))/(Y*Y*Y*Y/2*X*Y*Y+X*X)
```

```
RUN
```
ENTER THE POINT AT WHICH
THE LIMIT IS TO BE TAKEN.
FIRST THE X THEN THE Y VALUE.
?1,-1

X-VALUE	Y-VALUE	APP. LIMIT
2	-1	.0560577098
1.46045	-.83241	.189280925
1.18393	-.84567	.405336648
1.05882	-.89811	.794931186
1.01001	-.94323	1.64528888
0.99509	-.97218	3.86232654
0.99308	-.98801	11.1709364
0.9948	-.99564	53.1747723
0.99688	-.99886	-217.295295
0.99837	-1	-113.186567
0.99925	-1.00027	-142.033587
0.9997	-1.00025	-229.34904
0.9999	-1.00017	-431.19861
0.99998	-1.00009	-907.485441
1.00001	-1.00005	-2308.85672
1.00001	-1.00002	-4841.36765

?DIVISION BY ZERO ERROR IN 300

Since we are dividing by zero and from the trend of the output it appears that as (x,y) approaches (-1,1) the function is approaching minus infinity, we should conclude that the limit does not exist.

Example 2: By definition a function $f(x,y)$ is continuous at (a,b) if and only if

$$\lim_{(x,y)\to(a,b)} f(x,y)=f(a,b).$$

Is it possible to assign a value to C so that the function

$$f(x,y)=\begin{cases} \dfrac{x^2+y^2}{|x|+|y|} & , (x,y)\neq(0,0) \\[2ex] C & , (x,y)=(0,0) \end{cases}$$

is continuous for all (x,y)?

```
120 DEF FNF(X)=(X*X+Y*Y)/(ABS(X)+ABS(Y))
```

```
RUN
ENTER THE POINT AT WHICH
THE LIMIT IS TO BE TAKEN.
FIRST THE X THEN THE Y VALUE.
?0,0
```

X-VALUE	Y-VALUE	APP. LIMIT
1	0	1
0.46045	.16759	.382300945
0.18393	.15433	.170424856
0.05882	.10189	.0861250457
0.01001	.05677	.0497627995
-4.91E-03	.02782	.0243837718
-6.92E-03	.01199	.0101325255
-5.2E-03	4.36E-03	4.8140804E-03
-3.12E-03	1.14E-03	2.59285323E-03
-1.63E-03	0	1.62841361E-03
-7.5E-04	-2.7E-04	6.22544061E-04
-3E-04	-2.5E-04	2.77522156E-04
-1E-04	-1.7E-04	1.40247197E-04
-2E-05	-9E-05	8.10344199E-05
1E-05	-5E-05	3.97068658E-05
1E-05	-2E-05	1.64999425E-05
1E-05	-1E-05	7.83931404E-06
1E-05	0	4.22223749E-06
0	0	2.65173089E-06

It appears that $\lim_{(x,y)\to(0,0)} (x^2+y^2)/(|x|+|y|)=0$. Thus, by letting C=0 the function will be continuous for all (x,y).

EXERCISE SET 18.1

In exercises 1-10 use program LIMIT TWO VARIABLES to approximate the given limit. If possible , determine the exact value of the limit and compare this value to the approximation.

1. $\lim_{(x,y)\to(0,0)} \dfrac{\cos 2xy}{x^2+y^2}$

2. $\lim_{(x,y)\to(0,0)} \dfrac{xy}{\sqrt{x^2+y^2}}$

3. $\lim_{(x,y)\to(0,0)} \dfrac{y^2-x^2}{\sqrt{x^2+y^2}}$

4. $\lim_{(x,y)\to(-3,9)} \dfrac{y-9}{y-x^2}$

5. $\lim_{(x,y)\to(1,-1)} \dfrac{\sin(x+y)}{x+y}$

6. $\lim_{(x,y)\to(0,0)} \dfrac{e^{xy+1}\sin xy}{xy}$

7. $\lim_{(x,y)\to(2,-2)} \dfrac{\ln|x^2-4|}{y+2}$

8. $\lim_{(x,y)\to(0,0)} \dfrac{y^2-x^2}{y^2+x^2}$

9. $\lim_{(x,y)\to(0,0)} \dfrac{\tan(xy+\pi/2)}{\ln(x^2+y^2)}$

10. $\lim_{(x,y)\to(1,1)} \dfrac{\sqrt{x}-\sqrt{y}}{x^2-2xy+y^2}\ \sin(x+y)$

*
In exercises 11-13 use program LIMIT TWO VARIABLES to determine, if possible, a value of C so that the function will be continuous

for all (x,y).

11.
$$f(x,y) = \begin{cases} \dfrac{x^3-y}{x^2+y^2} & \text{if } (x,y) \neq (0,0) \\[2ex] C & \text{if } (x,y)=(0,0) \end{cases}$$

12.
$$f(x,y) = \begin{cases} \dfrac{10xy}{\sqrt{x^2+y^2}} & \text{if } (x,y) \neq (0,0) \\[2ex] C & \text{if } (x,y)=(0,0) \end{cases}$$

13.
$$f(x,y) = \begin{cases} \dfrac{\cos(x^2+y^2) - 1}{x^2+y^2} & \text{if } (x,y) \neq (0,0) \\[2ex] C & \text{if } (x,y)=(0,0) \end{cases}$$

**

14. a.   Approximate $\lim_{(x,y) \to (0,0)} (x^2-3xy+y^2)/(3x-2y)$ using program LIMIT TWO VARIABLES.

b.   Evaluate the function in part a for the point (.027735009,.041602514).

c.   Do the results of parts a and b agree? Explain.

d.   What is the value of the limit in part a along the path $y=(3/2)x$?

e.   Explain why program LIMIT TWO VARIABLES did not give the correct answer.

f.   Can our program be modified to overcome this difficulty for any function? Explain.

***

15. Write a program which will approximate the limit for a function of three variables. As test data use both

$$\lim_{(x,y,z) \to (1,1,1)} x/(x-y+z) = 1 \text{ and}$$

$$\lim_{(x,y,z) \to (0,0,0)} (x^2y)/(x^3+y^3+z^3)$$

which does not exist. <u>Hint</u>: Use spherical coordinates.

18.2 PARTIAL DERIVATIVES

Programming Problem: Approximate the partial derivatives of a function f(x,y) at a given point.

Output: An approximation for both partial derivatives at the given point.

Input: The function, the point at which the partial derivatives are to be evaluated, and the precision to which the approximation will be calculated.

Strategy: The partial derivatives of a function f(x,y) are given by:

$$\frac{\partial f}{\partial x} = \lim_{\triangle x \to 0} \frac{f(x+\triangle x, y) - f(x,y)}{\triangle x}$$

$$\frac{\partial f}{\partial y} = \lim_{\triangle y \to 0} \frac{f(x, y+\triangle y) - f(x,y)}{\triangle y}$$

Since both formulas involve only taking a limit of a difference quotient of one variable , with the other variable being held constant, we can use the same techniques as used in chapter 7 to approximate derivatives for functions of a single variable. That is, we will compute a left-hand, (f(x-h,y)-f(x,y))/(-h), and a right-hand, (f(x+h,y)-f(x,y))/h, difference quotient for x with decreasing values of h until the difference between the two values for the same h is within the desired precision. When this happens, we approximate the partial derivative by the average of the left and right-hand difference quotient. On the other hand, if the difference of the two values never gets within the desired precision, we will conclude that the partial derivative does not exist. The partial derivatives with respect to y can be approximated in the same manner.

Program:

```
100 REM ** PARTIAL DERIVATIVE **
110 REM ** FUNCTION **
120 DEF FNF(X)=.....
130 REM ** INPUT **
140 PRINT "ENTER THE POINT AT WHICH THE"
150 PRINT "DERIVATIVE IS TO BE EVALUATED"
160 PRINT "FIRST THE X THEN THE Y VALUE."
170 INPUT X,Y1
180 PRINT
190 PRINT "ENTER THE DESIRED PRECISION"
200 INPUT P
210 PRINT
220 REM ** HEADINGS **
230 PRINT "AT (";X;",";Y1;") THE PARTIAL"
240 PRINT "DERIVATIVES ARE:"
250 REM ** MAIN PROGRAM **
260 REM ** FINDING X-PARTIAL **
270 LET Y=Y1
280 FOR I=1 TO 20
290 LET H=.3^I
300 LET R=(FNF(X+H)-FNF(X))/H
310 LET L=(FNF(X-H)-FNF(X))/(-H)
320 IF ABS(R-L)<P THEN 360
330 NEXT I
340 PRINT "WITH RESPECT TO X: DOES NOT EXIST."
350 GO TO 370
360 PRINT "WITH RESPECT TO X: ";(R+L)/2
370 REM ** FINDING Y-PARTIAL **
380 FOR I=1 TO 20
390 LET H=.3^I
400 LET Y=Y1+H
410 LET R=FNF(X)
420 LET Y=Y1-H
430 LET L=FNF(X)  440 LET Y=Y1
450 LET R=(R-FNF(X))/H
460 LET L=(L-FNF(X))/(-H)
470 IF ABS(R-L)<P THEN 510
480 NEXT I
490 PRINT "WITH RESPECT TO Y: DOES NOT EXIST."
500 GO TO 520
510 PRINT "WITH RESPECT TO Y: ";(R+L)/2
520 END
```

Test Data: For $f(x,y)=x^2-xy-2y^2+3,$ $f_x(3,2)=4$ and $f_y(3,2)=-11.$

```
120 DEF FNF(X)=X*X-X*Y-2*Y*Y+3
```

```
RUN
ENTER THE POINT AT WHICH THE
DERIVATIVE IS TO BE EVALUATED.
FIRST THE X THEN THE Y VALUE.
?3,2
```

```
ENTER THE DESIRED PRECISION.
?.0001
```

```
AT (3,2) THE PARTIAL
DERIVATIVES ARE:
WITH RESPECT TO X: 3.99991287
WITH RESPECT TO Y: -10.9994134
```

Example 1: The range of a projectile fired at an angle of θ with the horizontal and initial speed v_0 is a function of the variables θ, v_0 and is given by

$$R(v_0, \theta) = \frac{v_0^2 \sin 2\theta}{9.807}$$

if v_0 is in m/sec and θ in radians. To hit a target 15000 meters away with a projectile which is fired with an initial speed of 425 m/sec requires an angle of elevation about 27° 16'.

$$15000 = (425) \sin 2\theta / 9.807$$
$$\sin 2\theta = .814422145$$
$$2\theta = .951732665$$
$$\theta = .475866332 \quad \text{or} \quad 27° 16'$$

Now, if the measurement of the initial speed can be in error by ±.7 m/sec and the angle by ±1/2° , will the projectile land within 200 meters of the target?

We will approximate the error in the range by the total differential

$$dR = \frac{\partial R}{\partial v_0} \triangle v_0 + \frac{\partial R}{\partial \theta} \triangle \theta$$

411

and use program PARTIAL DERIVATIVE to approximate the partial
derivatives.

```
120 DEF FNF(X)=X*X*SIN(2*Y)/9.807

RUN
ENTER THE POINT AT WHICH THE
DERIVATIVE IS TO BE EVALUATED.
FIRST THE X THEN THE Y VALUE.
?425,.475866332

ENTER THE DESIRED PRECISION.
?.0001

AT (425,.475866332) THE PARTIAL
DERIVATIVES ARE:
WITH RESPECT TO X: 70.5883826
WITH RESPECT TO Y: 21374.6179
```

Then,

$$dR=70.5883826(\pm.7)+21374.6179(\pm.5\pi /180)$$

$$=\pm49.411867 \pm 186.52872$$

$$=\pm235.94058$$

Thus, the projectile may not land within 200 meters of the target.

Example 2: Approximate the <u>equation</u> of the tangent plane and
normal line to the surface $z=\sqrt{x^3+y^2+y^3}$ (cos x)/ln(x+y)   at
the point (1.3,.6,.694).

The equation of the tangent plane is given by

$$f_x(x_0,y_0)(x-x_0)+f_y(x_0,y_0)(y-y_0)-(z-z_0) = 0$$

and the normal line by

$$\frac{x-x_0}{f_x(x_0,y_0)} = \frac{y-y_0}{f_y(x_0,y_0)} = \frac{z-z_0}{-1}$$

We will use program PARTIAL DERIVATIVE to approximate the partial derivatives.

```
120 DEF FNF(X)=(SQR(X*X*X+Y*Y+Y*Y*Y)*COS(X))/LOG(X+Y)

RUN
ENTER THE POINT AT WHICH THE
DERIVATIVE IS TO BE EVALUATED.
FIRST THE X THEN THE Y VALUE.
?1.3,.6

ENTER THE DESIRED PRECISION.
?.0001

AT (1.3,.6) THE PARTIAL
DERIVATIVES ARE:
WITH RESPECT TO X: -2.43449083
WITH RESPECT TO Y: -.283767004
```

Thus, the tangent plane is

$$-2.43(x-1.3)-.284(y-.6)-(z-.694)=0$$

and the normal line is

$$\frac{x-1.3}{-2.43} = \frac{y-.6}{-.284} = \frac{z-.694}{-1}$$

EXERCISE SET 18.2

In exercises 1-5 use program PARTIAL DERIVATIVE to approximate both partial derivatives at the given point. Also, compute the exact value of the partial derivatives and determine the percent of error in the approximation using the program.

$$\text{PERCENT ERROR} = \frac{\text{approximation-exact value}}{\text{exact value}} \times 100$$

1. $f(x,y) = 3x^2 - 4xy + 7y^3$ at $(-3,2)$

2. $f(x,y) = \dfrac{4x - 3x - y^3 - 4xy}{2x^2 - 3xy + 8}$ at $(5,-3)$

3. $f(x,y) = \dfrac{2x(x^2 - y^2)^3 + 3y}{\sqrt{x^2 + y^2 - 1}}$ at $(5,5)$

4. $f(x,y) = \dfrac{3x \sin xy + 2y}{4 \cos x^2 - 3xy}$ at $(4,7)$

5. $f(x,y) = \dfrac{1}{2\ln x + 3\ln y}$ at $(6,8)$

*
6.   The volume of a cone is given by $(1/3)\pi r^2 h$. A cone of height .15 cm and radius .05 cm is to be constructed. However, the accuracy of both dimensions can only be guaranteed to ±.00025 cm. Using program PARTIAL DERIVATIVE estimate the maximum error in the volume.

7.   For a simple pendulum the period of oscillation is given by $T = 2\pi \sqrt{L/g}$, where L is the length of the pedulum. If L is measured at 0.137 meters and g at 10.27 meters/second, estimate the error in the period of oscillation if L can be in error by ±.0015 meters and g=±.01 m/sec using program PARTIAL DERIVATIVE.

In exercises 8-10 use program PARTIAL DERIVATIVE to aid in approximating the equation of the tangent plane and normal line to the given surface at the given point.

8. $z = \dfrac{2\ln(x+y^2)}{3x^2+4y^2}$ at $(3,5)$

9. $z = \dfrac{\sec xy - \tan y}{2 \tan x}$ at $(-1,2)$

10. $z = \dfrac{e^{xy}\cos xy}{2x^2+2y^2}$ at $(0,3)$

**
11. In exercise 12 of section 7.1 the formula

$$\frac{f(a+h)-f(a-h)}{2h}$$

was derived for approximating the derivative of a function at the value $x=a$. At that time it was stated that this formula was more accurate than the normal difference quotient. Modify program PARTIAL DERIVATIVE to use the appropriate form of this formula to compute both partial derivatives.   As test data use $f(x,y) = \sqrt{x^3+y^2+y^3}\ \cos x\ /\ln(x+y)$ for $(x,y)=(1.3,.6)$. Compare your results to those obtained in Example 2 and to the actual values of the partial derivatives. Decide if the formula is more accurate for approximating partial derivatives.

12. Write a program to approximate the three partial derivatives of a function $w=f(x,y,z)$. As test data use $w=4x^2y-3xyz+2yz$ at $(1,2,1)$. The values of the partial derivatives are $w_x=10$, $w_y=3$, and $w_z=-2$.

***
13. Write a program to approximate the partial derivatives f_{xx} and f_{yy} for a function $z=f(x,y)$. As test data use $f(x,y)=3x^2y-4x^3y^2+3y^2$ at $(1,2)$. $f_{xx}=-84$ and $f_{yy}=-2$.
<u>Hint</u>: See Section 7.2

14.   a.   For a function $z=f(x,y)$ can a program be written to approximate f_{xy}? Explain.

b.   If the answer of part a is yes, then write a program to approximate f_{xy}.   As test data use $f(x,y)=3x^2y-4x^3y^2+3y^2$
$(1,2)$. $f_{xy}=-42$.

c.   Is a second program needed to approximate f_{yx}? Explain.

CHALLENGE ACTIVITY

Write a program to approximate the graph of the level curves of a function $f(x,y)$. Since there are an infinite number of level curves for a function, it will not be possible to use a different symbol for each level curve. Instead, use the same symbol for all level curves for which the value of $f(x,y)$ is within a predetermined interval. As test data use $f(x,y)=2x^2-y^2$, which has two families of hyperbolas as level curves, those opening horizontally for positive values of $f(x,y)$ and vertically for negative values of $f(x,y)$. A possible example of the output for this function is given below.

```
ENTER RANGE OF X-AXIS
?-5.5

ENTER RANGE OF Y-AXIS
?-5.5

+*))('’&&%%$$$###""""""""!!""""""##$$$%%&&’’())*+
+**)(('’&&%%$$$####"""""""""""""##$$$%%&&’’(()**+
.+**))(('’&&%%$$$###"""""""""""##$$$%%%&&’’(())*+.
.+**)(('’’&&%%$$$#######################$$$%%%&&’’(()**+.
.++**))(('’’&&%%$$$$###############$$$$$%%%&&’’(())***+.
-.+**))('’’&&&%%%$$$$$###########$$$$$%%%&&’’())**+.-
-.++**))(('’’&&&%%%$$$$$$$$$$$$$$$$$%%%&&&’’(())***+.-
-..+**))(('’’&&%%%%$$$$$$$$$$$$$$$$$%%%%&&’’(())**+..-
--.++**))(('’’&&&%%%%%$$$$$$$$$$$$$%%%%%&&&’’(())**+.--
-.++**))(('’’&&&%%%%%%$$$$$$$$$%%%%%%&&&’’(())***+.-.
-..++**))(('’’’&&&%%%%%%%%%%%%%%%%%&&&’’’(())***+..-.
.--.++**))((('’’&&&&%%%%%%%%%%%%%%%%&&&&’’’(())**++.--.
..-.++**))((('’’’&&&&%%%%%%%%%%%%%&&&’’’’(())***+..-.
..-..++**))((('’’’’&&&&%%%%%%%%%%%%&&&&’’’’(()))***+..-.
/.-.++**))((('’’’&&&&&%%%%%%%%%%%&&&&&’’’’(())***++..-./
/.--.++**))((('’’’’&&&&&&%%%%%%%%&&&&&&’’’’(())***+.--./
/.--.++**))((('’’’’&&&&&&&&&&&&&&&&&&&’’’’(())***+..--./
/.--..++**))((('’’’’&&&&&&&&&&&&&&&&’’’’(()))***+..-../
//.-.++**))((('’’’’&&&&&&&&&&&&&&&&’’’’(()))***++..-.//
//.--.++**))((('’’’’&&&&&&&&&&&&&’’’’(()))****+..--.//
//.--.++**))))((('’’’’&&&&&&&&&&&’’’’’(()))****+..--.//
//.--.++**))))(((('’’’’&&&&&&&&&&&’’’’(()))****+..--.//
//.--.++**))((('’’’’&&&&&&&&&&&’’’’(()))****+..--.//
//.--.++**))((('’’’’&&&&&&&&&&&’’’’(()))****+..--.//
//.--.++**))((('’’’’&&&&&&&&&&&’’’’(()))****+..--.//
//.--.++**))((('’’’’&&&&&&&&&&&’’’’(()))****+..--.//
//.--..++**))((('’’’’&&&&&&&&&&&’’’’(()))****+..--.//
/.--..++**))((('’’’’&&&&&&&&&&&&’’’’(()))***+..-../
/.--..++**))((('’’’’&&&&&&&&&&&&&’’’’(()))***+..--./
/.--.++**))((('’’’’&&&&&&&&&&&&&&’’’’(()))***+..--./
/.-..++**))((('’’’’&&&&&&%%%%%%&&&&&&’’’’(()))***+..-./
..-..++**))((('’’’’&&&&&%%%%%%%%%%%&&&&’’’’(()))***+..-.
..-.++**))(('’’’&&&&%%%%%%%%%%%%%%&&&&’’’(())***+..-.
.--.++**))(('’’’&&&&%%%%%%%%%%%%%%%%&&&’’’(())***+.--.
.-..++**))(('’’’&&&%%%%%%%%%%%%%%%%%&&&’’’(())***+..-.
.-.++**))(('’’’&&&%%%%%$$$$$$$$%%%%%&&&’’’(())***+..-.
--.++**))(('’’&&&%%%%%$$$$$$$$$$%%%%%&&&’’’(())***+.--
-..++**))(('’’&&%%%%$$$$$$$$$$$$$$%%%%&&’’(())***+..-
-.++**))(('’’&&&%%$$$$$$$$$$$$$$$$$$%%&&&’’(())***+.-
-.+**))(('’’&&&%%$$$###########$$$$$%%&&&’’(())**+.-
.+**))(('’’&&%%$$$####################$$$%%&&&’’(()**+.
.+**)(('’’&&%%$$$###################$$$%%&&&’’(())**+.
.+*))(('’’&&%%%$$$####"""""""""##$$$%%%&&’’(())*+.
+**)(('’’&&%%$$$###""""""""""""""##$$$%%&&’’(()**+
+*))('’’&&%%$$$###""""""""!!""""""##$$$%%&&’’())*+
```

418

CHAPTER 19
Applications Of Partial Derivatives

Partial derivatives are used to describe the change in a function of several variables when one of its independent variables is changed. For example, if $z=f(x,y)$ and if one of its variables, say y, is treated as a constant, then we can calculate the derivative of f with respect to x alone. Symbolically we use $\partial f/\partial x$ of f_x to denote the partial derivative of f with respect to x.

For example, if,

$$z=f(x,y)=x^3y^2-2x\sin x^2y$$

then

$$\frac{\partial f}{\partial x} = \frac{\partial(x^3y^2-\cos x^2y)}{\partial x} = 3x^2y^2+2x\sin x^2y$$

Although partial derivatives have geometric interpretations such as slope, they can also be applied to finding roots of two equations in two unknowns and in the determination of the gradient of a function. We discuss both of these applications in this chapter.

19.1 NEWTON'S METHOD FOR FINDING ROOTS IN TWO VARIABLES

In Chapter 9 we approximated roots of an equation $f(x)=0$ by Newton's method which utilizes the iterative formula

$$x_{n+1}=x_n - \frac{f(x_n)}{f'(x_n)} .$$

For functions of two variables there are similar formulas. However, they cannot be used to solve for roots of a single equation, $f(x,y)=0$, but must be used for solving a system of two equations, $f(x,y)=0$ and $g(x,y)=0$, simultaneously. This is analogous to needing two equations in two unknowns when solving a system of two linear equations in algebra. Using Newton's method to approximate the solutions of the two equations f(x,y)=0 and g(x,y)=0 simultaneously yields the iterative formulas below.

$$x_{n+1} = x_n - \frac{\begin{vmatrix} f(x_n,y_n) & f_y(x_n,y_n) \\ g(x_n,y_n) & g_y(x_n,y_n) \end{vmatrix}}{\begin{vmatrix} f_x(x_n,y_n) & f_y(x_n,y_n) \\ g_x(x_n,y_n) & g_y(x_n,y_n) \end{vmatrix}}$$

$$y_{n+1} = y_n - \frac{\begin{vmatrix} f_x(x_n,y_n) & f(x_n,y_n) \\ g_x(x_n,y_n) & g(x_n,y_n) \end{vmatrix}}{\begin{vmatrix} f_x(x_n,y_n) & f_y(x_n,y_n) \\ g_x(x_n,y_n) & g_y(x_n,y_n) \end{vmatrix}}$$

The notation of the bars in the numerators and denominators refers to taking determinants. The determinate in the denominators occurs quite frequently in applications involving functions of several variables and is known as the determinate of the Jacobian matrix.

Programming Problem: Approximate solutions to the simultaneous system of equations f(x,y)=0 and g(x,y)=0 using Newton's method.

Output: A table listing the approximations for the solution and the sum of the absolute value of the corresponding functional values for f(x,y) and g(x,y).

Input: The functions f(x,y), g(x,y), the four partial derivatives $f_x(x,y)$, $f_y(x,y)$, $g_x(x,y)$, $g_y(x,y)$, an initial guess to the solution and the desired precision to which the solution is to be determined.

Strategy: The actual computation of successive approximations is straightforward, involving only the evaluation of the three determinants and simple division and subtraction. Deciding when to terminate the program, however, is more complicated. Three possibilities exist:

1. A solution is determined.

2. Successive approximations do not converge to a solution.

3. The Jacobian is zero, and the next approximation cannot be determined.

19.1 NEWTON'S METHOD FOR FINDING ROOTS IN TWO VARIABLES

For this program we will consider a solution to have been determined if the distance, using the distance formula, between two successive approximations is less than the precision specified by the user. If, on the other hand, after twenty iterations no solution has been found, we will assume that the approximations are not going to converge to a solution. A simple IF-THEN statement will be used to test for the Jacobian being zero.

A final difficulty is that with a 40 column display there will not be sufficient space to print both the values of x and y as well as the functional values for f(x,y) and g(x,y). Since the only purpose of the functional values is to see how close they are to zero, instead of printing each value separately we will print the sum of the absolute values of the functional values.

Program:

```
100 REM ** ROOT **
110 REM ** FIRST FUNCTION **
120 DEF FNF(X)=.....
130 REM ** SECOND FUNCTION **
140 DEF FNG(X)=.....
150 REM ** F PARTIAL-X **
160 DEF FNH(X)=.....
170 REM ** F PARTIAL-Y **
180 DEF FNJ(X)=.....
190 REM ** G PARTIAL-X **
200 DEF FNK(X)=.....
210 REM ** G PARTIAL-Y **
220 DEF FNL(X)=.....
230 REM ** INPUT **
240 PRINT "ENTER THE INITIAL GUESS FOR"
250 PRINT "COMMON ROOT."
260 INPUT X,Y
270 PRINT
280 PRINT "ENTER THE DESIRED PRECISION."
290 INPUT P
300 PRINT
310 REM ** HEADINGS **
320 PRINT "APP.   X";TAB(13);"APP.   Y";TAB(26);"ABS(F)+ABS(G)"
330 PRINT X;TAB(13);Y;TAB(26);ABS(FNF(X))+ABS(FNG(X))
340 REM ** MAIN PROGRAM **
350 FOR I=1 TO 20
360 REM ** FINDING NEXT X **
370 LET A=FNF(X)*FNL(X)-FNJ(X)*FNG(Y)
380 LET B=FNH(X)*FNL(X)-FNJ(X)*FNK(X)
390 IF B=0 THEN 540
400 LET X1=X-(A/B)
410 REM ** FINDING NEXT Y **
420 LET C=FNG(X)*FNH(X)-FNK(X)*FNF(X)
430 LET Y1=Y-(C/B)
440 LET D1=X1-X
450 LET D2=Y1-Y
460 LET X=X1
470 LET Y=Y1
480 PRINT X;TAB(13);Y;TAB(26);ABS(FNF(X))+ABS(FNG(X))
490 REM ** TESTING TO END PROGRAM **
500 IF SQR(D1*D1+D2*D2)<P THEN 550
510 NEXT I
520 PRINT "NO ROOT FOUND AFTER 20 ITERATIONS."
530 GO TO 550
540 PRINT "NO ROOT FOUND-DENOMINATOR ZERO."
550 END
```

19.1 NEWTON'S METHOD FOR FINDING ROOTS IN TWO VARIABLES

Test Data: The equations

$$x+y^2-16=0 \text{ and } x+y-1=0$$

have common solutions $(-2.283882182, 3.283882182)$ and $(3.283882182, -2.283882182)$.

```
120 DEF FNF(X)=X*X+Y*Y-16
140 DEF FNG(X)=X+Y-1
160 DEF FNH(X)=2*X
180 DEF FNJ(X)=2*Y
200 DEF FNK(X)=1
220 DEF FNL(X)=1

RUN
ENTER THE INITIAL GUESS FOR
THE COMMON ROOT.
?1,2

ENTER THE DESIRED PRECISION.
?.0001
```

APP. X	APP. Y	ABS(F)+ABS(G)
1	2	13
-8.5	9.5	146.5
-4.43055556	5.43055556	33.1207562
-2.75119327	3.75119327	5.64051537
-2.31746672	3.31746672	.376237437
-2.28408235	3.28408235	2.2290349E-03
-2.28388219	3.28388219	8.19563866E-08
-2.28388218	3.28388218	0

```
RUN
ENTER THE INITIAL GUESS FOR
THE COMMON ROOT.
?-1,-2

ENTER THE DESIRED PRECISION.
?.0001
```

APP. X	APP. Y	ABS(F)+ABS(G)
-1	-2	15
12.5	-11.5	272.5
6.82291667	-5.82291667	64.4585504
4.27430842	-3.27430842	12.990808
3.141383236	-2.41383236	1.48083809
3.28677992	-2.28677992	.0322846398
3.28388369	-2.28388369	1.67787075E-05
3.28388218	-2.28388218	0

Example 1: Determine the intersection of the two curves $4x^2+7y^2=8$ and $x^3+y^2=2$.

Rewriting the equations we have

$$\frac{x^2}{2} + \frac{y^2}{8/7} = 1 \text{ and } y^2 = 2-x^3 .$$

The first is an ellipse, and the second will have a graph essentially parabolic.

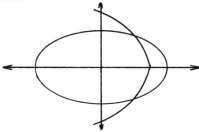

From the graph we see that there should be two points of intersection, and that they should be symmetric with respect to the x-axis since y occurs only to the second power in both equations.   Also, we will use (1,1) as an initial guess since it satisfies the second equation $x^3+y^2=2$.

```
120 DEF FNF(X)=4*X*X+7*Y*Y-8
140 DEF FNG(X)=X*X*X+Y*Y-2
160 DEF FNH(X)=8*X
180 DEF FNJ(X)=14*Y
200 DEF FNK(X)=3*X*X
220 DEF FNL(X)=2*Y

RUN
ENTER THE INITIAL GUESS FOR
THE COMMON ROOT.
?1,1

ENTER THE DESIRED PRECISION.
?.0001
```

APP. X	APP. Y	ABS(F)+ABS(G)
1	1	3
1.23076923	.653846154	1.34365044
1.18563495	.587494149	.0507987915
1.18344584	.585281033	7.53868371E-05
1.18344082	.585280307	0

Thus, the points of intersection are approximated by (1.18344082,.585280307) and (1.18344082,-.585280307).

Example 2: Solve the system of equations $y=\sin(x+y)$ and $x=\cos(y-x)$ simultaneously.

```
120 DEF FNF(X)=Y-SIN(X+Y)
140 DEF FNG(X)=X-COS(Y-X)
160 DEF FNH(X)=-COS(X+Y)
180 DEF FNJ(X)=1-COS(X+Y)
200 DEF FNK(X)=1-SIN(Y-X)
220 DEF FNL(X)=SIN(Y-X)
```

RUN
ENTER THE INITIAL GUESS FOR
THE COMMON ROOT.
?1,0

ENTER THE DESIRED PRECISION.
?.0001

APP. X	APP. Y	ABS(F)+ABS(G)
1	0	1.30116868
2.26763943	3.3203984	5.73336058
6.61071388	.624350805	5.84459958
1.4051396	-6.1159076	8.19422817
4.30039228	.983797071	7.10984352
-.153267886	-8.24532151	7.47255248
1.86462939	-4.06600132	4.18444876
-1.23795752	-.865449393	2.17333996
2.9850502	-2.28502334	5.38508876
-1.4892334	-4.38040606	5.30268982
-11.5848846	-57.3154855	68.9368168
-73.6778141	-194.083023	268.943683
2.61565323	-115.336472	118.177889
-121.896327	-116.562326	239.338801
-266.833852	-590.449062	855.920031
1.89396268	-414.574849	417.580473
126.674075	-160.125217	286.515507
5885.82481	-1831.7234	7718.7645
3791.29306	-35.9279161	3825.57935
1790.55572	558.28316	2348.99811
-603.387116	-2705.03732	3307.74883

NO ROOT FOUND AFTER 20 ITERATIONS.

However, selecting a different initial guess yields a much different result.

```
RUN
ENTER THE INITIAL GUESS FOR
THE COMMON ROOT.
?0,1

ENTER THE DESIRED PRECISION.
?.0001

APP. X          APP. Y          ABS(F)+ABS(G)
0               1               .698831321
0.723707722     1.50574962      .72916884
11.4922288      -3.02628325     15.7092027
7.25817084      .960547497      6.28460484
1.11168707      2.55716878      4.04703908
-.308528277     1.57389492      .622097807
1.54095104      1.48232708      1.90695731
1.00520419      1.064723388     .193700339
1.0057868       .938255435      .015172947
0.998032297     .935119685      6.5937089E-05
0.998020058     .935082065      1.94449967E-09
```

As can be seen from the output, using (1,0) as the initial guess does not result in the process converging to a solution. Instead, we obtain an erratic sequence of values. However, if we select (0,1), just as likely a choice for the initial guess, we obtain the correct solution (.998,.935). This illustrates just one of the difficulties of Newton's method, a difficulty similar to that illustrated in chapter 9 for functions of one variable. Since an analysis of these difficulties and their causes are beyond the scope of this text, it is a good procedure to run the program for different initial guesses when erratic results are obtained.

19.1 NEWTON'S METHOD FOR FINDING ROOTS IN TWO VARIABLES

EXERCISE SET 19.1

In exercises 1-6 use program ROOT to approximate points of intersection of the two curves. If possible, sketch both curves and use your sketch to aid in approximating all points of the intersection and in making the initial guess.

1. $2x-7y=6$ and $3x+17y=9$

2. $4x^2+3y^2=16$ and $5x^2y+2xy=14$

3. $x=\cos y$ and $y=\sin x$

4. $\cos(x+y) + \sin(x+y) = 0$ and $\tan x - \tan y = 0$

5. $\ln(x+y) - y = 0$ and $e^{x-y}+x^2=5$

6. $\tan^{-1}(x+2) -\cos y =0$ and $\sin(x-y)+xy^2=0$

*
7. a.   Run program ROOT to solve the system of equations $2x^2+y^2=9$ and $3y^2+6x^2=27$ simultaneously.

b.   In part a you should not have obtained a solution.   Is this due to the failure of the method or is it a correct solution? Explain.

8. Same as exercise 7 except use the equations $yx^4+x^2y+y-1=0$ and $x^2+x^2y-6x+11=0$.

**
9. In determining the maximum and minimum values of a function $f(x,y)$ it is necessary to solve the two equations $f_x(x,y)=0$ and $f_y(x,y)=0$ simultaneously in order to find critical points. Using program ROOT to aid in determining critical points find the extrema of $f(x,y)=\cos x + \cos y + \cos(x+y)$.

10.   Same as exercise 9 for $f(x,y)=e^{xy}\sin(x+y)$.

***
11. Modify program ROOT so that the four partial derivatives
f_x, f_y, g_x, and g_y are approximated during the execution
of the program instead of being entered by a DEF statement. As
test data use $x^2+y^2-16=0$ and $x+y-1=0$ which were used as test
data in the section. <u>Hint</u>: See chapter 18 for methods to
approximate the partial derivatives. All the partial derivatives
will have to be approximated during each iteration.

12.   One drawback to Newton's method presented in this section is
the large number of computations needed in computing all the
partial derivatives and determinants for each iteration. An
easier method to apply, though slower and not as dependable, is
known as the modified Newton's method. As before, we are trying
to solve the two equations $f(x,y)=0$ and $g(x,y)=0$ simultaneously,
but in this method we use each function individually to
approximate either x or y. Specifically, two iteration formulas
can be obtained.

Formula I   Formula II

$$x_{n+1} = x_n - \frac{f(x_n,y_n)}{f_x(x_n,y_n)} \qquad x_{n+1} = x_n - \frac{g(x_n,y_n)}{g_x(x_n,y_n)}$$

$$y_{n+1} = y_n - \frac{g(x_{n+1},y_n)}{g_y(x_{n+1},y_n)} \qquad y_{n+1} = y_n - \frac{f(x_{n+1},y_n)}{f_y(x_{n+1},y_n)}$$

Note that for both formulas x_{n+1} must be computed before
computing y_{n+1}.

   a. Write two programs, one for each formula, to approximate
roots by the modified Newton's method. As test data use the
equations used in the test data for the section.

   b. Explain why this method is easier to apply than the
method presented in the section.

   c.   For any particular set of equations one of the formulas
for the modified Newton's method usually diverges while the other
converges. To illustrate, run both programs from part a for
y=sin x and x=cos y. Compare the results with exercise 3.

19.2 THE GRADIENT

An important and interesting vector which can be defined from a function of several variables is the gradient, denoted by ∇f. One property that it has is that it points in the direction in which a path through $(x,y,f(x,y))$ on the graph of $z=f(x,y)$ will rise most rapidly. A second application of ∇f concerns particles which move so as to maximize an attribute of the medium through which they are moving, for example, a heat seeking missle. In this section we use partial derivatives to compute the gradient.

Programming Problem: Approximate the gradient of a function of three variables, $f(x,y,z)$, at a specified point.

Output: The approximation of the gradient of $f(x,y,z)$ at the specified point.

Input: The function $f(x,y,z)$, the point at which the gradient is to be approximated, and the precision to which the components of the gradient are to be approximated.

Strategy: The gradient of the function $f(x,y,z)$ at the point (x_0,y_0,z_0) is given by

$$\nabla f(x_0,y_0,z_0)=f_x(x_0,y_0,z_0)i+f_y(x_0,y_0,z_0)j+$$

$$f_z(x_0,y_0,z_0)k.$$

Since this definition involves approximating the three partial derivatives f_x, f_y, f_z at (x_0,y_0,z_0), we can use the same procedure for approximating partial derivatives presented in chapter 18.

Program:

```
100 REM ** GRADIENT **
110 REM ** FUNCTION **
120 DEF FNF(X)=.....
130 REM ** INPUT **
140 PRINT "ENTER THE POINT AT WHICH"
150 PRINT "THE GRADIENT IS TO BE EVALUATED."
160 INPUT X1,Y1,Z1
170 PRINT
180 PRINT "ENTER THE DESIRED PRECISION."
190 INPUT P
```

```
200 PRINT
210 REM ** MAIN PROGRAM **
220 REM ** I-COMPONENT **
230 LET Y=Y1
240 LET Z=Z1
250 FOR I=1 TO 25
260 LET H=.3^I
270 LET R=(FNF(X1+H)-FNF(X1))/H
280 LET L=(FNF(X1-H)-FNF(X1))/(-H)
290 IF ABS(R-L)<P THEN 330
300 NEXT I
310 PRINT "NO GRADIENT-X PARTIAL DOES NOT EXIST."
320 GO TO 640
330 LET IC=(R+L)/2
340 REM ** J-COMPONENT **
350 LET X=X1
360 LET N=FNF(X)
370 FOR I=1 TO 25
380 LET H=.3^I
390 LET Y=Y1+H
400 LET R=(FNF(X)-N)/H
410 LET Y=Y1-H
420 LET L=(FNF(X)-N)/(-H)
430 IF ABS(R-L)<P THEN 470
440 NEXT I
450 PRINT "NO GRADIENT-Y PARTIAL DOES NOT EXIST."
460 GO TO 640
470 LET JC=(R+L)/22
480 REM ** K-COMPONENT **
490 LET Y=Y1
500 FOR I=1 TO 25
510 LET H=.3^I
520 LET Z=Z1+H
530 LET R=(FNF(X)-N)/H
540 LET Z=Z1-H
550 LET L=(FNF(X)-N)/(-H)
560 IF ABS(R-L)<P THEN 600
570 NEXT I
580 PRINT "NO GRADIENT-Z PARTIAL DOES NOT EXIST."
590 GO TO 640
600 LET KC=(R+L)/2
610 REM ** OUTPUT **
620 PRINT "THE GRADIENT AT(";X1;",";Y1;",";Z1;") IS "
630 PRINT IC;"I+";JC;"J+";KC;"K"
640 END
```

Test Data: The gradient of

$$f(x,y,z)=x^2+y^2+z^2-50$$

at $(2,1,3)$ is $4i+2j+6k$.

```
120 DEF FNF(X)=X*X+Y*Y+Z*Z-50

RUN
ENTER THE POINT AT WHICH
THE GRADIENT IS TO BE EVALUATED.
?2,1,3

ENTER THE DESIRED PRECISION.
?.0001

THE GRADIENT AT (2,1,3) IS
3.99998858I+1.99999429J+5.99998286K
```

Example 1: Approximate the directional derivative of $f(x,y)=xy^3-e^{xy}$ at $(1.25,2.37)$ in the direction $u=3i-6j$.

By treating the function xy^3-e^{xy} as a function of three variables with $z=0$, we can use program GRADIENT to approximate the gradient.

```
120 DEF FNF(X)=X*X*Y*Y*Y-EXP(X*Y)

RUN
ENTER THE POINT AT WHICH
THE GRADIENT IS TO BE EVALUATED.
?1.25,2.37,0

ENTER THE DESIRED PRECISION.
?.001

THE GRADIENT AT (1.25,2.37,0) IS
-12.5703864I+2.14642102J+0K
```

Now by definition the directional derivative in the direction of the vector V is given by:

$$Dv=\nabla f \frac{V}{V}$$

Thus, $Dvf(x,y)=(-12.570i+2.146j)(1/\sqrt{45})(3i-6j)$
$$=-7.541$$

431

Example 2: Approximate the tangent plane and normal line to the surface 0.5xz sin x + ln xyz = 0 at the point where x=.6 and y=1.59.

The gradient of F(x,y,z)=.5xz sin x + ln xyz is normal to the surface f(x,y,z)=0.

Thus, we can use program GRADIENT to determine the normal vector. However, first we must determine the z-coordinate on the surface for x=.6 and y=1.59. By substitution, z must satisfy

0.3 sin(.6) z + ln(.954z) = 0

Since this equation is not solvable by elementary means, we will use Newton's method in one variable to approximate the value for z. Using program NEWTON'S METHOD from chapter 9, we obtain the following.

```
110 DEF FNF(X)=.3*SIN(.6)*X+LOG(.954*X)
120 DEF FNG(X)=.3*SIN(.6)+(1/X)
```

RUN
**ENTER THE INITIAL GUESS FOR THE
ROOT OF THE FUNCTION.**
?1

ENTER THE DESIRED PRECISION.
?.0001

ITERATION	APPROXIMATION
1	.895414833
2	.899988796
3	.89999895

**AT X=.89999895 A ROOT OF THE
FUNCTION TO WITHIN A PRECISION OF 1E-04
HAS BEEN FOUND. THE FUNCTIONAL VALUE
IS -2.19188223E-10**

Thus, we want to approximate the gradient at $(.6, 1.59, .89999895)$.

```
120 DEF FNF(X)=.5*X*Z*SIN(X)+LOG(X*Y*Z)

RUN
```
ENTER THE POINT AT WHICH
THE GRADIENT IS TO BE EVALUATED.
?.6,1.59,.89999895

ENTER THE DESIRED PRECISION.
?.0001

THE GRADIENT AT (.6,1.59,.89999895) IS
2.1436034I+.628930258J+1.2805028K

Using $2.144i+.629j+1.281k$ as the normal vector we have the following results.

Tangent plane: $2.144(x-.6)+.629(y-1.59)+1.281(z-.9)=0$

Normal line: $\dfrac{x-.6}{2.144} = \dfrac{y-1.59}{.629} = \dfrac{z-.9}{1.281}$

Example 3: Approximate the direction and magnitude of the greatest rate of increase of $f(x,y,z)=(3x-4e^{xy}-3ye^z)/\sqrt{x^2+y^2+z^3}$ at $(1,1,1)$.

We know that the greatest rate of change occurs in the direction of the gradient and has magnitude equal to the length of the gradient.

```
120 DEF FNF(X)=(3*X*X-4*EXP(X*Y)-3*Y*EXP(Z))
    /(SQR(X*X+Y*Y+Z*Z*Z))

RUN
```
ENTER THE POINT AT WHICH
THE GRADIENT IS TO BE EVALUATED.
?1,1,1

ENTER THE DESIRED PRECISION.
?.001

THE GRADIENT AT (1,1,1) IS
0.271085133I+-7.901211808J+-.0813278396K

Thus, the direction of the greatest rate of change is 0.271i-7.901j-.083k and the magnitude is $\sqrt{(.271)^2 +(-7.901)^2 +(-.083)^2} \approx 7.906$

Example 4: Illustrate that the curves

$$x^2 +y^2 +z^2 =25 \text{ and } z^2 =4x^2 +9y^2$$

are perpendicular at any of their points of intersection.

The two curves intersect, by substitution, when

$$x^2 +y^2 +4x^2 +9y^2 =25$$

$$5x^2 +10y^2 =25$$

$$x^2 +2y^2 =5.$$

Letting x=1 we obtain $y=\sqrt{2}$ and $z=\sqrt{22}$.

Now we will use program GRADIENT to approximate the gradient to both surfaces at $(1,\sqrt{2},\sqrt{22})$. Since the gradient is normal to the surface at a given point, we can show that the surfaces are perpendicular by showing that the normal vectors, or gradients, are perpendicular at a point.

For the sphere $x^2 +y^2 +z^2 =25$:

```
120 DEF FNF(X)=X*X+Y*Y+Z*Z-25

RUN
ENTER THE POINT AT WHICH
THE GRADIENT IS TO BE EVALUATED.
?1,1.4142135,4.690415

ENTER THE DESIRED PRECISION.
?.001

THE GRADIENT AT (1,1.4142135,4.690415) IS
1.99998861I+2.82842708J+9.38086693K
```

For the cone $z^2 =4x^2 +9y^2$:

```
120 DEF FNF(X)=4*X*X+9*Y*Y-Z*Z
```

RUN
**ENTER THE POINT AT WHICH
THE GRADIENT IS TO BE EVALUATED.**
?1,1.4142135,4.690415

ENTER THE DESIRED PRECISION.
?.001

**THE GRADIENT AT (1,1.4142135,4.690415) IS
7.99997716I+25.4558664J+-9.38085641K**

Now, the two normal vectors and, thus, the surfaces will be normal
if their dot product is 0.

$$(2i+2.828j+9.381k) \cdot (8i+25.456j-9.381k) \approx .0136$$

Considering round-off error we can assume the dot product is
zero and thus the surfaces are perpendicular or normal at the
point of intersection. A similar result will hold for any other
point of intersection.

Example 5: Illustrate the validity of the theorem $\nabla(fg)=f\nabla g+g\nabla f$
using $f(x,y,z)=ye^{xz}$ and $g(x,y,z)=z\sin(x+y)$ at $(1,1,2)$.

Finding ∇f at $(1,1,2)$:

```
120 DEF FNF(X)=Y*EXP(X*Z)
```

RUN
**ENTER THE POINT AT WHICH
THE GRADIENT IS TO BE EVALUATED.**
?1,1,2

ENTER THE DESIRED PRECISION.
?.001

**THE GRADIENT AT (1,1,2) IS
14.7781393I+7.38905611J+7.38903182K**

435

Finding ∇g at (1,1,2):

```
120 DEF FNF(X)=Z*SIN(X+Y)

RUN
ENTER THE POINT AT WHICH
THE GRADIENT IS TO BE EVALUATED.
?1,1,2

ENTER THE DESIRED PRECISION.
?.001

THE GRADIENT AT (1,1,2) IS
-.832294495I+-.832294495J+.909297426K
```

Finding ∇(fg) at (1,1,2):

```
120 DEF FNF(X)=Y*EXP(X*Z)*Z*SIN(X+Y)

RUN
ENTER THE POINT AT WHICH
THE GRADIENT IS TO BE EVALUATED.
?1,1,2

ENTER THE DESIRED PRECISION.
?.0001

THE GRADIENT AT (1,1,2) IS
20.7254973I+7.28781324J+20.1563699K
```

At (1,1,2), f=7.389056099 and g=1.818594854.

Now,

$$f\nabla g \approx 7.3891(-.8323i-.8323j+.9093k)$$
$$\approx -6.1499479i-6.1499479j+6.7189086k$$

$$g\nabla f \approx 1.8186(14.7781i+7.389j+7.389k)$$
$$\approx 26.875452i+13.437817j+13.437635k$$

Adding, we obtain

$$f\nabla g+g\nabla f=20.725504i+7.2878701j+20.156626k$$

which agrees with ∇fg considering round-off error.

EXERCISE SET 19.2

In exercises 1-5 use program GRADIENT to approximate the gradient of the given function at the indicated point.

1. $f(x,y)=3xe^y + \cos xy$ at $(1,2)$

2. $f(x,y)=3e^{xy}/(x^2+y^2)$ at $(1,1)$

3. $f(x,y,z)=\ln xyz + x^2 - \cos yz$ at $(1,1,1)$

4. $f(x,y,z)=ze^{xy}+xe^{yz}+ye^{xz}$ at $(-1.5,2.76,3.81)$

5. $f(x,y,z)=3 \ln xy - e^{5z}+z^x$ at $(1,1,2.5)$

In exercises 6-8 use program GRADIENT to aid in approximating the tangent plane and normal line to the surface at the given point.

6. $4xyz - 2x^2 + 3y^2 = 5$ at $(1.5,2.6)$

7. $4 \sin xy - e^{2y} + 5 \cos y = 0$ at $(1.6,-1,0)$

8. $5xz \cos z + 2 \ln xy - 5z = 0$ at $(2.1,4)$

In exercises 9-10 use program GRADIENT to approximate the direction and magnitude of the greatest rate of change of the function at the given point.

9. $f(x,y,z) = \dfrac{e^x xz+yz}{\sqrt{2x^2+3y^2+4z}}$ at $(1,1,3)$

10. $f(x,y,z) = \dfrac{\sin xy+\sin yz+\cos xz}{e^{xyz}}$ at $(2,1,3)$

*
In exercises 11-12 illustrate the validity of the given theorem using the functions and point of example 5.

11. $\triangledown(f+g)= \triangledown f+\triangledown g$

12. $\triangledown(f/g)=(g\triangledown f-f\triangledown g)/g^2$

13. Determine another point of intersection of the sphere and cone of example 4 and verify with the help of program GRADIENT that the curves are perpendicular at the point of intersection.

**

14. a. Let (x_0,y_0,z_0) be in the intersection of $f(x,y,z)=0$ and $g(x,y,z)=0$. If all partial derivatives of f and g exist, then show that the vector $\triangledown f \times \triangledown g$ is tangent to the curve of the intersection of the two surfaces at (x_0,y_0,z_0).

b. Use the method of part a to find the tangent to the curve which is the intersection of $z=x+y^2$ and $x^2+y^2+z^2=12$ at $(1, 2,3)$.

***

15. Since the gradient of $w=f(x,y,z)$ points in the direction of the maximum rate of increase of the function, its negation will be in the direction of maximum decrease or "steepest descent". We can make use of this fact to approximate the minimum value of the function $f(x,y,z)$. If (x_n,y_n,z_n) is the nth approximation for the location of the minimum, then we can use as the next approximation

$$x_{n+1}=x_n - kf_x(x_n,y_n,z_n)$$

$$y_{n+1}=y_n - kf_y(x_n,y_n,z_n)$$

$$z_{n+1}=z_n - kf_z(x_n,y_n,z_n)$$

where the value of k is arbitrarily chosen, usually the value of the precision to which the point is to be determined.

Write a program to approximate the minimum of a function $f(x,y,z)$ using the method of steepest descent. Terminate your program when two successive approximations are within the desired precision, using the final approximation as the location of the minimum value. If no location for the minimum has been found after thirty iterations, assume that the process is not going to converge to an answer and terminate the program with an appropriate message. As test data use $f(x,y,z)=x^2+y^2+z^2$ which has a minimum at $(0,0,0)$.

16.   Use the program from exercise 15 to approximate the simultaneous solution for the system:

$$3x+3y^2+3z^2=4$$

$$-3x^2+2y+z^3=-4$$

$$4x+2xy-3z^2=3.7$$

<u>Hint</u>: If (x_0,y_0,z_0) is a root of the functions $f(x,y,z)$, $g(x,y,z)$ and $h(x,y,z)$, then it is a minimum value of $F(x,y,z)=f^2(x,y,z)+g^2(x,y,z)+h^2(x,y,z)$

CHALLENGE ACTIVITY

Newton's method in two variables of section 19.1 can be extended to solve systems of equations simultaneously for any number of variables as long as the number of variables and the number of equations in the system are the same. If the system of equations is expressed as

$$f_1(x_1, x_2, \ldots x_n) = 0$$
$$\cdot$$
$$\cdot$$
$$\cdot$$
$$f_n(x_1, x_2, \ldots x_n) = 0$$

then the iteration formula for approximating x_i is given by

$$_{m+1}x_i = {}_mx_i - \frac{|J_i({}_mx_1, {}_mx_2, \ldots, {}_mx_n)|}{|J({}_mx_1, {}_mx_2, \ldots, {}_mx_n)|} \quad .$$

where $J({}_mx_1, {}_mx_2, \ldots, {}_mx_n)$ stands for the nth order Jacobian matrix

$$\begin{bmatrix} \dfrac{\partial f_1}{\partial x_1} & \dfrac{\partial f_1}{\partial x_2} & . & . & \dfrac{\partial f_1}{\partial x_n} \\ . & . & & & . \\ . & . & & & . \\ . & . & & & . \\ \dfrac{\partial f_n}{\partial x_1} & \dfrac{\partial f_n}{\partial x_2} & . & . & \dfrac{\partial f_n}{\partial x_n} \end{bmatrix}$$

evaluated at the m-th approximation, and $J_i({}_mx_1, {}_mx_2, \ldots, {}_mx_n)$ is the same Jacobian matrix except the i-th column is replaced by

$$\begin{bmatrix} f_1(mX_1, mX_2, \ldots, mX_n) \\ \cdot \\ \cdot \\ \cdot \\ f_n(mX_1, mX_2, \ldots, mX_n) \end{bmatrix}$$

As in section 19.1 the bars around J and J_i refer to taking determinants.

Write a program which will solve a system of four equations in four unknowns. The input should consist of the four equations, the initial guess for the root, and the precision to which we wish to approximate the answer. Be sure to include routines for terminating the program if a root is not found, successive approximations do not converge, or if the Jacobian is zero. As test data approximate the simultaneous solutions to the system

$$x + y + z + w - 4 = 0$$

$$x^2 - y^2 + z^2 - w^2 = 0$$

$$x^2 - y^2 = 0$$

$$z^2 + w^2 - 2 = 0$$

which are $(1,1,1,1)$, $(3,3,-1,-1)$, $(2,2,-1,1)$ and $(2,2,1,-1)$.

A useful application of this method for solving systems of equations is in determining maximum and minimum by Lagrange multipliers. Use your program from above to determine the extrema of $f(x,y,z) = x - y + z$ subject to the constraint $x^2 - y^2 - z^2 = -1$.

CHAPTER 20
Multiple Integration

Multiple integrals have a number of practical applications. For example, double integrals are used to calculate area and volume. Some physical applications include the moment of inertia, important in connection with the knetic energy of rotating bodies. The triple integral can be used to determine the volume enclosed between two surfaces. However, mass, center of gravity, and moments of inertia of a mass distributed over a region of three space also can be expressed as triple integrals. In this chapter we will write computer programs to approximate double and triple integrals.

20.1 DOUBLE INTEGRALS

Programming Problem: Approximate the value of the double integral

$$\int_L^U \int_{g(x)}^{h(x)} f(x,y) \, dy \, dx$$

Output: The approximation of the double integral.

Input: The function being integrated, the upper and lower bounds of integration for the inner and outer integrals and the number of subdivisions to be used in approximating the integral.

Strategy: The double integral $\int_L^U \int_{g(x)}^{h(x)} f(x,y) \, dy \, dx$ can be approximated by the double summation

$$\sum_{i=1}^{M} \sum_{j=1}^{N} f(x_i, y_j) \, \triangle y_j \, \triangle x_i$$

with the approximation becoming more accurate as M and N increase. For the purposes of this program we will assume that M=N. Now the evaluation of a double summation can be handled in much the same manner as a double integral, namely, an iterated summation

$$\sum_{i=1}^{N} \left(\sum_{j=1}^{N} f(x_i, y_j) \, \triangle y_j \right) \triangle x_i$$

Thus, for each value of i we compute the inner summation, multiply by $\triangle x_i$ and add the result to a "running" sum. In BASIC this can be accomplished by two nested FOR-NEXT loops.

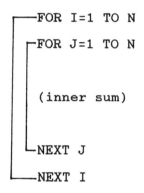

```
FOR I=1 TO N

  FOR J=1 TO N



    (inner sum)



  NEXT J

NEXT I
```

In computing the sums we need to determine the points (x_i, y_j) and the lengths of the subdivisions $\triangle y_j$ and $\triangle x_i$. Since the program is being written to approximate the double integral with the order of integration dy dx, the limits of integration on x will always be constants, U and L, and, thus, the length of each subdivision $\triangle x_i$ will be

$$H=(U-L)/N$$

For   each x_i we   will   use   the   midpoint   of   the   i-th
subdivision just as we did in chapter 10 in approximating a single
integral.   Determining   the   values   for $\triangle y_j$ and y_j is   more
difficult since the limits of integration for y may both be
functions   of   x.   However, for each subdivision $\triangle x_i$ the interval
over which y varies   can   be   approximated   by $[G(x_i),H(x_i)]$.
Thus, the length of $\triangle y_j$ will be

$$K=(H(x_i)-G(x_i))/N$$

and y_j the midpoint of each subdivision of $[G(x_i),H(x_i)]$.

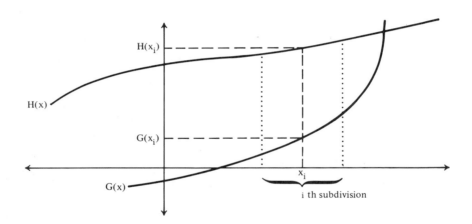

Program:

```
100 REM ** DOUBLE INTEGRATION **
110 REM ** FUNCTION TO BE INTEGRATED **
120 DEF FNF(X)=.....
130 REM ** LOWER LIMIT-INNER INTEGRAL **
140 DEF FNG(X)=.....
150 REM ** UPPER LIMIT-INNER INTEGRAL **
160 DEF FNH(X)=.....
170 REM ** LIMITS-OUTER INTEGRAL **
180 LET L=.....
190 LET U=.....
200 PRINT "ENTER THE NUMBER OF SUBDIVISIONS."
210 INPUT N
220 PRINT
230 REM ** MAIN PROGRAM **
240 LET S=0
250 LET H=(U-L)/N
260 REM ** LOOP-OUTER SUM **
270 FOR I=1 TO N
280 LET XL=L+(I-1)*H
290 LET XR=L+I*H
300 LET X=(XL+XR)/2              finding midpoint
310 LET S1=0
320 REM ** FINDING DELTA-YI **
330 LET YL=FNG(X)               finding lower limit
340 LET YU=FNH(X)               finding upper limit
350 LET K=(YU-YL)/N
360 REM ** LOOP-INNER SUM **
370 FOR J=1 TO N
380 LET Y1=YL+(J-1)*K
390 LET Y2=YL+J*K
400 LET Y=(Y1+Y2)/2             finding midpoint
410 LET S1=S1+FNF(X)*K
420 NEXT J
430 LET S=S+S1*H
440 NEXT I
450 PRINT "THE APPROXIMATION OF THE";
460 PRINT "INTEGRAL IS ";S
470 END
```

In running the program the value for N should not be chosen too large since the inner loop is executed N^2 times.

Test Data:

$$\int_0^1 \int_0^{x^2} x+y \ dy \ dx = \int_0^1 \left. xy + 1/2 \ y^2 \right|_0^{x^2} dx$$

$$= \int_0^1 x^3 + 1/2 \ x^4 \ dx$$

$$= \left. 1/4 \ x^4 + 1/10 \ x^5 \ \right|_0^1$$

$$= 1/4 + 1/10 = 7/20 = .35$$

```
120 DEF FNF(X)=X+Y
140 DEF FNG(X)=0
160 DEF FNH(X)=X*X
180 LET L=0
190 LET U=1
```

```
RUN
ENTER THE NUMBER OF SUBDIVISIONS.
?25

THE APPROXIMATION OF THE
INTEGRAL IS .3494666704
```

Example 1: Approximate the volume bounded by

$$z=9-(x^2+y^2) \text{ and the xy-plane where } x \geq 0.$$

Sketches of the surface and the region over which the integration is to be performed are shown below. Thus, the volume can be approximated using program DOUBLE INTEGRATION for the integral

$$\int_0^3 \int_{-\sqrt{9-x^2}}^{\sqrt{9-x^2}} 9-(x^2+y^2) \ dy \ dx.$$

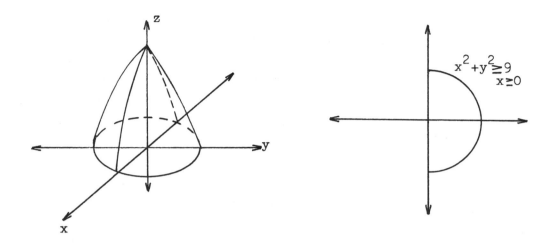

```
120 DEF FNF(X)=9-(X*X+Y*Y)
140 DEF FNG(X)=-SQR(9-X*X)
160 DEF FNH(X)=SQR(9-X*X)
180 LET L=0
190 LET U=3


RUN
ENTER THE NUMBER OF SUBDIVISIONS.
?25

THE APPROXIMATION OF THE
INTEGRAL IS 63.6697773
```

Thus, the volume is approximately 63.7

Example 2: Approximate the double integral

$$\int_0^4 \int_{\sqrt{x}}^2 (\sin y)/y \ dy \ dx$$

by changing the order of integration and using program DOUBLE INTEGRATION.

$$\int_0^4 \int_{\sqrt{x}}^2 (\sin y)/y \ dy \ dx = \int_0^2 \int_0^{y^2} (\sin y)/y \ dx \ dy$$

447

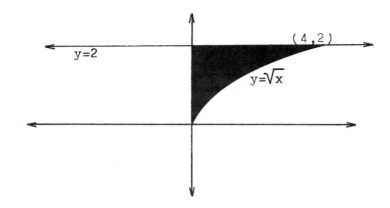

The program DOUBLE INTEGRATION was written with the order of integration dy dx. To approximate a double integral dx dy using program DOUBLE INTEGRATION we must interchange x and y and approximate

$$\int_0^2 \int_0^{x^2} (\sin\ x)/x\ dy\ dx$$

```
120 DEF FNF(X)+(SIN(X))/X
140 DEF FNG(X)=0
160 DEF FNH(X)=X*X
180 LET L=0
190 LET U=2
```

RUN
ENTER THE NUMBER OF SUBDIVISIONS.
?25

THE APPROXIMATION OF THE
INTEGRAL IS 1.74157047

Thus, $\int_0^4 \int_{\sqrt{x}}^2 (\sin\ y)/y\ dy\ dx \approx 1.74$

Example 3: Approximate the integral

$$\int\int x^2 + y^2\ dx\ dy$$

over the region bounded by $(x^2+y^2)^2=x^2-y^2$ using program DOUBLE INTEGRATION.

The occurrence of x^2 and y^2 in both the double integral and the boundary suggests the use of polar coordinates. Thus letting $x=r\cos\theta$ and $y=r\sin\theta$ we have

$$(x^2+y^2)^2=x^2-y^2$$
$$(r^2\cos\theta + r^2\sin\theta)^2=r^2\cos\theta - r^2\sin\theta$$
$$r^4(\cos^2\theta+\sin^2\theta)^2=r^2(\cos^2\theta-\sin^2\theta)$$
$$r^2=\cos 2\theta$$

In the simplification we have used the identities $\cos^2\theta+\sin^2\theta=1$ and $\cos 2\theta=\cos^2\theta-\sin^2\theta$. Thus, the boundary is the lemniscate sketched below and the double integral can be written as

$$4\int_0^{\pi/4}\int_0^{\sqrt{\cos 2\theta}} r^2 (r\ dr\ d\theta)$$

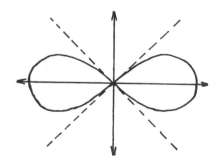

In writing the double integral in polar coordinates we are using the fact that $dx\ dy = r\ dr\ d\theta$, and the symmetry of both the boundary and the function being integrated allows us to integrate over that portion of the region in the first quadrant and multiply by 4.

```
120 DEF FNF(X)=Y*Y*Y     letting r=y and θ=x
140 DEF FNG(X)=0
160 DEF FNH(X)=SQR(COS(2*X))
180 LET L=0
190 LET U=.785398163
```

RUN
ENTER THE NUMBER OF SUBDIVISIONS.
?25

THE APPROXIMATION OF THE
INTEGRAL IS .0980962307

$$\iint_R x+y^2\, dx\ dy \approx 4(.0980962307) \approx .3923849228$$

where R is bounded by $(x^2+y^2)^2=x^2-y^2$.

Example 4: Approximate the total area of the surface z=xy above/below the unit circle using program DOUBLE INTEGRATION.

If z=f(x,y) is the equation of the surface, then its area is given by

$$S = \iint_R \sqrt{1+(f_x)^2+(f_y)^2}\ dx\ dy$$

where R is the region in the xy-plane over which the integration is to be performed. For z=xy, f_x =y and f_y =x, and the surface area above/below the unit circle is

$$\int_{-1}^{1}\int_{-\sqrt{1-X^2}}^{\sqrt{1-X^2}} \sqrt{1+x^2+y^2}\ dy\ dx$$

```
120 DEF FNF(X)=SQR(1+X*X+Y*Y)
140 DEF FNG(X)=-SQR(1-X*X)
160 DEF FNH(X)=+SQR(1-X*X)
180 LET L=-1
190 LET U=1
```

RUN
ENTER THE NUMBER OF SUBDIVISIONS.
?25

THE APPROXIMATION OF THE
INTEGRAL IS 3.83988598

Total surface area is approximately 3.83988598

EXERCISE SET 20.1

In exercises 1-5 use program DOUBLE INTEGRATION to approximate the double integral.

1. $\displaystyle\int_{-1}^{2}\int_{-\sqrt{1+x}}^{\sqrt{1+x}} 4xy^2 \ dy \ dx$

2. $\displaystyle\int_{0}^{3}\int_{0}^{2x} e^{y*y} \ dy \ dx$

3. $\displaystyle\int_{0}^{4}\int_{0}^{\sqrt{x}} y \sin xy \ dy \ dx$

4. $\displaystyle\int_{0}^{2}\int_{0}^{\sqrt{1-x^2/4}} (x-y)/(1+x+y) \ dy \ dx$

5. $\displaystyle\int_{0}^{2}\int_{0}^{3\sqrt{8-x^3}} 8-x^3-y^3 \ dy \ dx$

In exercises 6-8 use program DOUBLE INTEGRATION to approximate the double integral obtained after converting the given integral to polar coordinates.

6. $\displaystyle\iint_{R} x+y^2 \ dy \ dx$; R bounded by r=3+sin θ

7. $\displaystyle\iint_{R} (dx \ dy)/(1+x^2+y^2)^2$; R the region in the first quadrant bounded by $x^2+y^2=4$, $y=\sqrt{3} \ x$ and the x-axis.

8. $\displaystyle\iint_{R} \sqrt{x^2+y^2} \ dy \ dx$; R bounded by $(x-1)^2+y^2=1$

In exercises 9-11 use program DOUBLE INTEGRATION to approximate the indicated surface area.

9.   The surface area of $z=\sqrt{1+x+y}$ above the unit circle in the first quadrant.

10.   The surface area of the sphere $x^2+y^2+z^2=16$ inside the cylinder $x^2+y^2=4y$.

11.   The surface area of $z=y \sin x + x \cos y$ inside the cylinder bounded by $y=x$, $x=2$ and the x-axis.

*
12.   The mass of a thin sheet of material or lamina with density $p(x,y)$ is given by

$$\iint_R p(x,y) \ dx \ dy$$

where R is the region in the xy-plane which corresponds to the lamina. Use program DOUBLE INTEGRATION to approximate the mass of the given lamina.

    a.   lamina: ellipse $4x^2+9y^2=36$; density: $p(x,y)=x^2+y^2$

    b.   lamina: bounded by $y=3x-x^2$ and $y=x^2-3x$; density: $p(x,y)=x^2 y^2$

13.   The center of mass of a lamina R of mass M is the point (x,y) such that:

$$\overline{x} = \frac{\iint_R x \ p(x,y) \ dx \ dy}{M} \quad \text{and} \quad \overline{y} = \frac{\iint_R y \ p(x,y) \ dx \ dy}{M}$$

for density function $p(x,y)$.

    a.   Approximate the center of mass of the lamina of exercise 12a using program DOUBLE INTEGRATION.

    b.   Same as part a except use the lamina of exercise 12b.

**
14. By converting to polar coordinates and using program DOUBLE INTEGRATION approximate the center of mass of one leaf of the four leaf rose r=4 sin 2 θ .

***
15. Determine a formula for the surface area of the ellipsoid:

$$\frac{x^2}{a^2} + \frac{y^2}{b^2} + \frac{z^2}{c^2} = 1$$

<u>Hint</u>: Run program DOUBLE INTEGRATION for different values of a,b,c and generalize.

16. The Theorem of Pappus states:

> If a plane region is revolved about an axis in its plane which does not intersect the interior of the region, then the volume, V, of the solid of revolution is the product of the area A of R and the distance travelled by the centroid (center of mass).

For example, the volume of the torus is $V = 2\pi \bar{y} A$ where A is the area of the generating circle, since the centroid travels around a circle of radius $\bar{y}$.

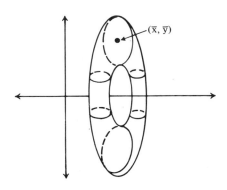

a. Show that if R is a region in the xy-plane with the x-axis not intersecting its interior (the x-axis may be part of the boundary of R), then the volume obtained by revolving the region about the x-axis is

$$\iint_R 2\pi y \; dx \; dy$$

b.   Approximate the volume obtained by revolving the cardiod $r=2+2\cos\theta$ about the x-axis using the Theorem of Pappus and program DOUBLE INTEGRATION.

17. Modify program DOUBLE INTEGRATION so that the double integral is approximated for larger and larger values of N until two successive approximations are within a specified precision which is part of the input. As test data use the same integral as in the test data for the section.

18. Write a program to approximate the improper integral

$$\int_a^\infty \int_{g(x)}^\infty f(x,y) \; dy \; dx.$$

As test data use

$$\int_0^\infty \int_x^\infty 1/ye^y \; dy \; dx = 1.$$

20.2 TRIPLE INTEGRALS AND APPLICATIONS

Programming Problem: Approximate the value of the triple integral

$$\int_{L}^{U} \int_{G(x)}^{H(x)} \int_{U(x,y)}^{V(x,y)} f(x,y,z) \; dz \; dy \; dx$$

Output: The approximation of the triple integral.

Input: The function being integrated, the various limits of integration, and the number of subdivisions to be used in approximating the triple integral.

Strategy: The triple integral can be approximated using the triple summation

$$\sum_{i=1}^{L} \sum_{j=1}^{M} \sum_{k=1}^{N} f(x_i, y_j, z_k) \; \triangle z_k \; \triangle y_j \; \triangle x_i$$

As in section 20.1 we will assume L=M=N and evaluate the summation by using an iterated summation

$$\sum_{i=1}^{N} \left(\sum_{j=1}^{N} \left(\sum_{k=1}^{N} f(x_i, y_j, z_k) \; \triangle z_k \right) \triangle y_j \right) \triangle x_i$$

In BASIC this will require the use of three nested FOR-NEXT loops, one for each of the summations. The values of x_i, y_j, z_k, $\triangle x_i$, $\triangle y_j$, and $\triangle z_k$ used in the summation will be determined in a manner similar to that used for the double integral of section 20.1 Specifically , we will write the program so that the order of integration is dz dy dx. Thus, the limits for x will always be constants, L and U, and $\triangle x_i = (U-L)/N$ with x_i the midpoint of the i-th subinterval. For each i, the limits for y will be approximated by $G(x_i)$ and $H(x_i)$, and we will let $\triangle y_j = (H(x_i)-G(x_i))/N$ with y_j the midpoint of successive subintervals of $[g(x_i),H(x_i)]$. Finally, for each i and j we can approximate the limits for z by $U(x_i,y_j)$ and $V(x_i,y_j)$ and let $\triangle z_k = (V(x_i,y_j)-U(x_i,y_j))/N$ with z_k the midpoint of successive subintervals of $[U(x_i,y_j),V(x_i,y_j)]$.

Program:

```
100 REM ** TRIPLE INTEGRATION **
110 REM ** FUNCTION TO BE INTEGRATED **
120 DEF FNF(X)=.....
130 REM ** LIMITS-INNER INTEGRAL **
140 DEF FNU(X)=.....
150 DEF FNV(X)=.....
160 REM ** LIMITS-MIDDLE INTEGRAL **
170 DEF FNG(X)=.....
180 DEF FNH(X)=.....
190 REM ** LIMITS-OUTER INTEGRAL **
200 LET L=.....
210 LET U=.....
220 REM ** NUMBER OF SUBDIVISIONS **
230 PRINT "ENTER THE NUMBER OF SUBDIVISIONS."
240 INPUT N
250 PRINT"
260 REM ** MAIN PROGRAM **
270 LET S=0
280 LET DX=(U-L)/N
290 REM ** LOOP-OUTER INTEGRAL **
300 FOR I=1 TO N
310 LET XL=L+(I-1)*DX
320 LET XR=L+I*DX
330 LET X=(XR+XL)/2
340 LET Y1=FNG(X)
350 LET Y2=FNH(X)
360 LET DY=(Y2-Y1)/N
370 LET S1=0
380 REM ** LOOP-MIDDLE INTEGRAL **
390 FOR J=1 TO N
400 LET YL=Y1+(J-1)*DY
410 LET YR=Y1+J*DY
420 LET Y=(YR+YL)/2
430 LET Z1=FNU(X)
440 LET Z2=FNV(X)
450 LET DZ=(Z2-Z1)/N
460 LET S2=0
470 REM ** LOOP-INNER INTEGRAL **
480 FOR K=1 TO N
490 LET ZL=Z1+(K-1)*DZ
500 LET ZR=Z1+K*DZ
510 LET Z=(ZR+ZL)/2
520 LET S2=S2+FNF(X)*DZ
530 NEXT K
540 LET S1=S1+S2*DY
550 NEXT J
560 LET S=S+S1*DX
570 NEXT I
```

```
580 REM ** OUTPUT **
590 PRINT "THE APPROXIMATION OF THE"
600 PRINT "INTEGRAL IS ";S
610 END
```

Again do not choose too large a value for N since, for three nested loops, the inner loop will be performed N^3 times.

Test Data:
$$\int_0^1 \int_0^x \int_0^{x+y} x+y+z \ dz \ dy \ dx =$$

$$\int_0^1 \int_0^x (3/2)(x^2+2xy+y^2) \ dy \ dx=$$

$$\int_0^1 (7/2) \ x^3 \ dx = .875$$

```
120 DEF FNF(X)=X+Y+Z
140 DEF FNU(X)=0
150 DEF FNV(X)=X+Y
170 DEF FNG(X)=0
180 DEF FNH(X)=X
200 LET L=0
210 LET U=1
```

```
RUN
ENTER THE NUMBER OF SUBDIVISIONS.
?15

THE APPROXIMATION OF THE
INTEGRAL IS .872916977
```

Example 1: Approximate the volume inside both the sphere $x^2+y^2+z^2=4$ and the cylinder $(x-1)^2+y^2=1$ using program TRIPLE INTEGRAL.

The volume is given by the triple integral

$$\int_0^2 \int_{-\sqrt{1-(x-1)^2}}^{\sqrt{1-(x-1)^2}} \int_{-\sqrt{4-x^2-y^2}}^{\sqrt{4-x^2-y^2}} dz \ dy \ dx$$

```
120 DEF FNF(X)=1
140 DEF FNU(X)=-SQR(4-X*X-Y*Y)
150 DEF FNV(X)=SQR(4-X*X-Y*Y)
170 DEF FNG(X)=-SQR(1-(X-1)*(X-1))
180 DEF FNH(X)=SQR(1-(X-1)*(X-1))
200 LET L=0
210 LET U=2
```

RUN
ENTER THE NUMBER OF SUBDIVISIONS.
?15

THE APPROXIMATION OF THE
INTEGRAL IS 9.68751955

The volume is approximately 9.69

Example 2: Use program TRIPLE INTEGRATION to approximate the mass of the region in the first octant bounded by $x^2+y^2=16$ and $z=x+y$ if the density function is $p(x,y,z)=x+y+z$.

The mass is given by the triple integral

$$\iiint p(x,y,z)\ dz\ dy\ dx = \int_0^4 \int_0^{\sqrt{16-x^2}} \int_0^{x+y} x+y+z\ dz\ dy\ dx$$

```
120 DEF FNF(X)=X+Y+Z
140 DEF FNU(X)=0
150 DEF FNV(X)=X+Y
170 DEF FNG(X)=0
180 DEF FNH(X)=SQR(16-X*X)
200 LET L=0
210 LET U=4
```

RUN
ENTER THE NUMBER OF SUBDIVISIONS.
?15

THE APPROXIMATION OF THE
INTEGRAL IS 247.478634

Thus, the mass is approximately 247.5

Example 3: The average value of the function $f(x,y,z)$ on a region R in R^3 is given by

$$(\iiint_R f(x,y,z) \; dz \; dy \; dx)/(\text{Volume of R})$$

Approximate the average value of $f(x,y,z)=z^2$ over the unit sphere using program TRIPLE INTEGRATION.

Since the volume of the unit sphere iS $4\pi/3$, we only need to approximate the integral

$$\int_{-1}^{1} \int_{-\sqrt{1-x^2}}^{\sqrt{1-x^2}} \int_{-\sqrt{1-x^2-y^2}}^{\sqrt{1-x^2-y^2}} z^2 \; dz \; dy \; dx$$

```
120 DEF FNF(X)=Z*Z
140 DEF FNU(X)=-SQR(1-X*X-Y*Y)
150 DEF FNV(X)=SQR(1-X*X-Y*Y)
170 DEF FNG(X)=-SQR(1-X*X)
180 DEF FNH(X)=SQR(1-X*X)
200 LET L=-1
210 LET U=1
```

```
RUN
ENTER THE NUMBER OF SUBDIVISIONS.
?15

THE APPROXIMATION OF THE
INTEGRAL IS .834495447
```

Average value $\approx (.834495447)/(4\pi/3) \approx .1992211$

Example 4: By converting the integral to spherical coordinates and using program TRIPLE INTEGRATION approximate the value of the integral

$$\iiint_R dV/(x+y+z)$$

where R is the region between the spheres $x^2+y^2+z^2=16$ and $x^2+y^2+z^2=9$.

For spherical coordinates

$$x=p \sin \phi \cos \theta$$

459

$$y = p \sin \phi \sin \theta$$

$$z = p \cos \phi$$

$$dV = p^2 \sin \phi \, dp \, d\phi \, d\theta$$

Thus, $\iiint_R dV/(x+y+z) =$

$$\int_0^{2\pi} \int_0^{\pi} \int_3^4 \frac{p^2 \sin \phi \, dp \, d\phi \, d\theta}{p \sin \phi \cos \theta + p \sin \phi \sin \theta + p \cos \phi} =$$

$$\int_0^{2\pi} \int_0^{\pi} \int_3^4 \frac{p \sin \phi \, dp \, d\phi \, d\theta}{\sin \phi \cos \theta + \sin \phi \sin \theta + \cos \phi}$$

Now to use program TRIPLE INTEGRATION we must use the variables x, y, z with the order of integration dz dy dx. Thus, we will let p=z, ϕ=y, and θ=x. Then the integral becomes

$$\int_0^{2\pi} \int_0^{\pi} \int_3^4 \frac{z \sin y}{\sin y \cos x + \sin y \sin x + \cos y} \, dz \, dy \, dx$$

```
120 DEF FNF(X)=(Z*SIN(Y))/(SIN(Y)
    *COS(X)+SIN(Y)*SIN(X)+COS(Y))
140 DEF FNU(X)=3
150 DEF FNV(X)=4
170 DEF FNG(X)=0
180 DEF FNH(X)=3.1415926
200 LET L=0
210 LET U=6.2831853


RUN
ENTER THE NUMBER OF SUBDIVISIONS.
?15

THE APPROXIMATION OF THE
INTEGRAL IS 17.1016182
```

EXERCISE SET 20.2

In exercises 1-5 use program TRIPLE INTEGRATION to approximate each integral.   Also determine the exact value of each integral and compare to the estimated value.

1. $\displaystyle\int_{0}^{4}\int_{0}^{2}\int_{0}^{x^2} xyz\ dz\ dy\ dx$

2. $\displaystyle\int_{0}^{\pi}\int_{0}^{x/2}\int_{0}^{1/x} \sin x\ dz\ dy\ dx$

3. $\displaystyle\int_{0}^{2}\int_{0}^{\sqrt{4-x^2}}\int_{0}^{\sqrt{4-x^2-y^2}} xyz\ dz\ dy\ dx$

4. $\displaystyle\int_{0}^{1}\int_{0}^{\sqrt{1-x^2}}\int_{0}^{x} zy\ dz\ dy\ dx$

5. $\displaystyle\int_{-1}^{1}\int_{-\sqrt{1-x^2}}^{\sqrt{1-x^2}}\int_{0}^{x+y} zy\ dz\ dy\ dx$

In exercises 6-8 use program TRIPLE INTEGRATION to approximate the volume of the region described in each exercise.

6.   Bounded by $x^2+y^2=9$, $z=-x^2+9$, and $z=0$.

7.   Bounded by $z=4-y^2-x^2$ and the xy-plane.

8.   Bounded by $x^2+y^2=z^2$, $z=0$, $z=9$.

In exercises 9-11 approximate the average value of the indicated function over the given region using program TRIPLE INTEGRATION.

9. The function $f(x,y,z)=2xyz$ over the region bounded by $z=9-3x^2$ and $z=6x^2+y^2$.

10.   The   function $f(x,y,z)=xy$ cos zy over the region bounded by
the tetrahedron formed by   the   plane $3x+2y+4z=12$ in   the   first
octant.

11.   The   function $f(x,y,z)=5x^2+5y^2$ over the region bounded by
the part of the sphere $x^2+y^2+z^2=16$,   z=1,   and z=3.

*
In exercises 12-14 approximate each triple integral by   converting
to cylindrical coordinates and using program TRIPLE INTEGRATION.

12. $\int_0^1 \int_0^{\sqrt{1-x^2}} \int_0^{\sqrt{x^2+y^2}} xy\ dz\ dy\ dx$

13. $\int_0^4 \int_{-\sqrt{16-x^2}}^{\sqrt{16-x^2}} \int_{-\sqrt{16-x^2-y^2}}^{\sqrt{16-x^2-y^2}} x^2+y^2+z\ dz\ dy\ dx$

14. $\int_{-1}^1 \int_{-\sqrt{1-x^2}}^{\sqrt{1-x^2}} \int_{-\sqrt{4-x^2-y^2}}^{\sqrt{4-x^2-y^2}} z^4\ dz\ dy\ dx$

In   exercises 15-16 approximate each triple integral by converting
to spherical coordinates and using program TRIPLE INTEGRATION.

15. $\int_{-5}^5 \int_{-\sqrt{25-x^2}}^{\sqrt{25-x^2}} \int_{-\sqrt{25-x^2-y^2}}^{\sqrt{25-x^2-y^2}} (x^2+y^2+z^2)^{5/2}\ dz\ dy\ dx$

16. $\int_0^3 \int_0^{\sqrt{9-x^2}} \int_0^{\sqrt{x^2+y^2}} xz\ dz\ dy\ dx$

In exercises 17-19 use program TRIPLE INTEGRATION   to   approximate
the mass, M, of the region if the density function is $p(x,y,z)$.

17. $p(x,y,z)=5\sqrt{x^2+y^2+z^2}$ and   the   region is bounded by the
hemisphere $z=\sqrt{49-x^2-y^2}$.

18. $p(x,y,z)=\ln z$ and R is the region bounded by $z^2=x^2+y^2$, $z=4$, and $z=5$.

19. $p(x,y,z)=x^2+y^2+z^2$ and R is the region inside both the sphere $x^2+y^2+z^2=16$ and the cone $z^2=3x^2+3y^2$.

**

For exercises 20-22 use program TRIPLE INTEGRATION to approximate the center of mass of the regions in exercises 17-19. The center of mass of a region is the point (x,y,z) where

$$x= \frac{\iiint_R x\, p(x,y,z)\ dV}{M}$$

$$y= \frac{\iiint_R y\, p(x,y,z)\ dV}{M}$$

$$z= \frac{\iiint_R z\, p(x,y,z)\ dV}{M}$$

***
23. Write a program to approximate the value of the integral

$$\int_L^U \int_{G(x)}^{H(x)} \int_{U(x,y)}^{V(x,y)} \int_{S(x,y,z)}^{T(x,y,z)} f(x,y,z,w)\ dw\ dz\ dy\ dx.$$

As test data use

$$\int_0^1 \int_0^1 \int_0^x \int_0^{x+y} z\ dw\ dz\ dy\ dx = 5/24 = .2083333$$

24. Approximate the volume of the sphere $x^2+y^2+z^2+w^2=4$ in R^4 using the program form exercise 23.

CHALLENGE ACTIVITY

Note: This is a very difficult challenge activity.

The trapezoidal rule and Simpson's rule for approximating integrals can both be extended to double and triple integrals. To illustrate we will derive the trapezoidal rule for the double integral

$$\int_L^U \int_{g(x)}^{h(x)} f(x,y) \, dy \, dx$$

If N is the number of subdivisions, then we can subdivide [L,U] into N subintervals with the endpoints $L = x_0, x_1, \ldots, x_N = U$, and

$$\int_L^U \int_{g(x)}^{h(x)} f(x+y) \, dy \, dx =$$

$$\int_{x_0}^{x_1} \int_{g(x)}^{h(x)} f(x,y) \, dy \, dx + \ldots + \int_{x_{N-1}}^{x_N} \int_{g(x)}^{h(x)} f(x,y) \, dy \, dx$$

Now, each integral on the right side of the above equation can be approximated by approximating the limits of integration for y with $g(\bar{x}_i)$ to $h(\bar{x}_i)$ where $\bar{x}_i$ is the midpoint of the i-th interval.

$$\int_{x_{i-1}}^{x_i} \int_{g(x)}^{h(x)} f(x,y) \, dy \, dx \approx \int_{x_{i-1}}^{x_i} \int_{g(\bar{x}_i)}^{h(\bar{x}_i)} f(x,y) \, dy \, dx$$

and

$$\int_L^U \int_{g(x)}^{h(x)} f(x,y) \, dy \, dx \approx \sum_{i=1}^N \int_{x_{i-1}}^{x_i} \int_{g(x_i)}^{h(x_i)} f(x,y) \, dy \, dx.$$

To approximate each double integral in the summation we apply the trapezoidal rule twice. Let $g(\bar{x}_i) = y_{i-1}$ and $h(\bar{x}_i) = y_i$. In applying the trapezoidal rule to the inner integral we must subdivide the interval $[y_{i-1}, y_i]$ into N subdivisions.

Let the endpoints of the subinterval be $y_{i-1} = y_{i,0}, y_{i,1}, \ldots, y_{i,N} = y_i$ and the length of each subdivision k_i, that is $k_i = (y_i - y_{i-1})/N$. Thus, using the trapezoidal rule on the inner integral,

$$\int_{x_{i-1}}^{x_i} \int_{y_{i-1}}^{y_i} f(x,y) \, dy \, dx = \int_{x_{i-1}}^{x_i} \left(\int_{y_{i-1}}^{y_i} f(x,y) \, dy \right) dx$$

$$\approx \int_{x_{i-1}}^{x_i} \left[(k_i/2)(f(x,y_{i,0}) + f(x,y_{i,N}) + 2 \sum_{j=1}^{N-1} f(x,y_{i,j}) \right] dx$$

Since the $[L,U]$ has already been subdivided into N subdivisions of length $h = (U-L)/N$, we approximate the outer integral using the trapezoidal rule wit one subdivision.

$$\int_{x_{i-1}}^{x_i} \int_{y_{i-1}}^{y_i} f(x,y) \, dy \, dx \approx$$

$$(h/2)[(k_i/2)(f(x_{i-1},y_i,0) + f(x_{i-1},y_i,N) +$$

$$2 \sum_{j=1}^{N-1} f(x_{i-1},y_i,j) + f(x_i,y_i,0) + f(x_i,y_i,N) +$$

$$2 \sum_{j=1}^{N-1} f(x_i,y_i,j))]$$

The approximation of the original integral is now obtained by summing for i=1 to N.

a. Write a program to approximate double integrals using the trapezoidal rule. As test data use

$$\int_0^1 \int_0^{x^2} x+y \, dy \, dx = .35$$

b. In a manner similar to that presented above, derive a formula for approximating a double integral using Simpson's rule.

   c.   Write a program to approximate the value of a double
integral using the formula you developed in part b. Use the same
test data as in part a.

   d.   Run the programs of parts a and c and DOUBLE INTEGRATION
to approximate the same integral for several different integrals.
Compare the results and decide which method is most accurate.

CHAPTER 21
Vector Calculus

George Green (1793-1841) was a self-taught mathematical physicist and one of the first to treat static electricity and magnetism in mathematical terms. The theorem that bears his name transforms a line integral around the boundary of a plane region into a double integral over that region and conversely. One of the most important physical applications of line integrals is in calculating work done in a force field. The other major topic we consider in this chapter is the Divergence Theorem, an extension of Green's Theorem related to the flow of a fluid through a region in space.

21.1 LINE INTEGRALS AND GREEN'S THEOREM

Programming Problem: Approximate the value of a line integral in the plane given in differential form

$$\int_C M(x,y) \ dx + N(x,y) \ dy$$

Output: The approximation of the line integral.

Input: The functions $M(x,y)$, $N(x,y)$, the parametric representation of x, y, dx, and dy along the curve C, the limits of integration on the parameter, and the precision to be used in approximating the line integral.

Strategy: The standard method for evaluating a line integral in differential form,

$$\int_C M(x,y)\ dx\ +\ N(x,y)\ dy$$

involves the following steps.

1.   Determine parametric equations g(t) and h(t)   so that C={(x,y):x=g(t),y=h(t),t$_1$≤t≤t$_2$}.

2.   Express dx and dy in terms of the parameter t, where dx=g'(t)dt and dy=h'(t)dt.

3.   Substitute into the line integral to obtain the equivalent integral.

$$\int_{t_1}^{t_2} [M(g(t),h(t))\ g'(t)\ +\ N(g(t),h(t))\ h'(t)]dt$$

This final integral is now in terms of the single variable t and can be approximated using one of the methods presented earlier. For this program we will use the midpoint rule developed in chapter 10.

Program:

```
100 REM ** LINE INTEGRAL **
110 REM ** M-FUNCTION **
120 DEF FNM(X)=.....
130 REM ** N-FUNCTION **
140 DEF FNN(X)=.....
150 REM ** PARAMETERIZATION OF X **
160 DEF FNC(T)=.....
170 REM ** PARAMETERIZATION OF Y **
180 DEF FND(T)=.....
190 REM ** PARAMETERIZATION OF DELTA X **
200 DEF FNA(T)=.....
210 REM ** PARAMETERIZATION OF DELTA Y **
220 DEF FNB(T)=.....
230 REM ** INPUT **
240 PRINT "ENTER THE RANGE ON THE PARAMETER T."
250 PRINT "ENTER BOTH ON ONE LINE."
260 INPUT T1,T2
270 PRINT
280 PRINT "ENTER THE DESIRED PRECISION."
290 INPUT P
300 PRINT"
310 REM ** MAIN PROGRAM **
320 LET S2=0
```

```
330 LET N=10
340 LET S1=0
350 LET DT=(T2-T1)/N
360 FOR I=1 TO N
370 LET TL=T1+(I-1)*DT
380 LET TR=T1+I*DT
390 LET TM=(TL+TR)/2
400 LET X=FNC(TM)
410 LET Y=FND(TM)
420 LET S1=S1+(FNM(X)*FNA(TM)+FNN(X)*FNB(TM))*DT
430 NEXT I
440 REM ** TESTING TO END **
450 IF ABS(S1-S2) < P THEN 530
460 LET N=2*N
470 LET S2=S1
480 REM ** TESTING FOR NO SOLUTION **
490 IF N < 20480 THEN 340
500 PRINT "NO VALUE OF THE LINE INTEGRAL"
510 PRINT "AFTER 20480 SUBDIVISIONS."
520 GO TO 550
530 PRINT "THE VALUE OF THE LINE"
540 PRINT "INTEGRAL IS ";S1
550 END
```

Test Data:

$$\int_C xy \, dx + x^2 y \, dy$$

where C is the unit circle in the first quadrant.

$$x=\cos t \qquad y=\sin t \qquad 0 \leq t \leq \pi/2$$
$$dx=-\sin t \, dt \qquad dy=\cos t \, dt$$

$$\int_C xy \, dx + x^2 y \, dy = \int_0^{\pi/2} (-\cos t \sin^2 t + \cos^3 t \sin t) \, dt$$

$$= [-(\sin^3 t)/3 - (\cos^4 t)/4] \Big|_0^{\pi/2}$$

$$= (-1/3)-(-1/4)$$

$$=-1/12$$

$$\approx -.08333333$$

```
120 DEF FNM(X)=X*Y
140 DEF FNN(X)=X*X*Y
160 DEF FNC(T)=COS(T)
180 DEF FND(T)=SIN(T)
200 DEF FNA(T)=-SIN(T)
220 DEF FNB(T)=COS(T)
```

RUN
ENTER THE RANGE ON THE PARAMETER T.
ENTER BOTH ON ONE LINE.
?0,1.5707963

ENTER THE DESIRED PRECISION.
?.0001

THE VALUE OF THE LINE
INTEGRAL IS -.0833334723

Example 1: Illustrate that the value of the line integral is independent of the parametric representation of the curve C by running program LINE INTEGRAL to approximate

$$\int_C x^2 \cos xy \, dx + y^3 \sin yx^2 \, dy$$

where C is the line segment from (0,0) to (1,1) using:

   a.   the parametric representation x=t, y=t, $0 \leq t \leq 1$

   b.   the parametric representation x=ln t, y=ln t, $1 \leq t \leq 2.7182818$

For the parameterization of part a we have x=t, y=t, dx=dt, and dy=dt.

```
120 DEF FNM(X)=X*X*COS(X*Y)
140 DEF FNN(X)=Y*Y*Y*SIN(Y*X*X)
160 DEF FNC(T)=T
180 DEF FND(T)=T
200 DEF FNA(T)=1
220 DEF FNB(T)=1
```

```
RUN
ENTER THE RANGE ON THE PARAMETER T.
ENTER BOTH ON ONE LINE.
?0,1

ENTER THE DESIRED PRECISION.
?.0001

THE VALUE OF THE LINE
INTEGRAL IS .396045651
```

For part b x=ln t, y=ln t, dx=1/t dt, dy=1/t dt.

```
160 DEF FNC(T)=LOG(T)
180 DEF FND(T)=LOG(T)
200 DEF FNA(T)=1/T
220 DEF FNB(T)=1/T

RUN
ENTER THE RANGE ON THE PARAMETER T.
ENTER BOTH ON ONE LINE.
?0,2.7182818

ENTER THE DESIRED PRECISION.
?.0001

THE VALUE OF THE LINE
INTEGRAL IS .396046261
```

Considering round-off and discretization error, the values of the line integrals are equal.

Example 2: Approximate the work done by the force

$$F(x,y)=3x^2yi+4xy^3j$$

on an object moving along the hypocycloid $x^{2/3}+y^{2/3}=1$ in a counterclockwise direction using program LINE INTEGRAL.

By definition the work done by a force $F(x,y)$ acting on a particle along a curve C is

$$\int_C F\cdot dr \text{ where } dr=dxi+dyj$$

471

Thus, the work in this instance is given by:

$$\int_C (3x^2yi+4xy^3j)\ (dxi+dyj)=\int_C 3x^2y\ dx+4xy^3\,dy$$

Also, a parametric representation of the hypocycloid is $x=\cos^3 t$, $y=\sin^3 t$, $0 \leq t \leq 2\pi$.

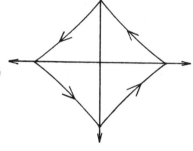

```
120 DEF FNM(X)=3*X*X*Y
140 DEF FNN(X)=4*X*Y*Y*Y
160 DEF FNC(T)=COS(T)*COS(T)*COS(T)
180 DEF FND(T)=SIN(T)*SIN(T)*SIN(T)
200 DEF FNA(T)=-3*COS(T)*COS(T)*SIN(T)
220 DEF FNB(T)=3*SIN(T)*SIN(T)*COS(T)

RUN
ENTER THE RANGE ON THE PARAMETER T.
ENTER BOTH ON ONE LINE.
?0,6.2831853

ENTER THE DESIRED PRECISION.
?.0001

THE VALUE OF THE LINE
INTEGRAL IS -.386563159
```

The work done by the force is approximately -.38653159 . What does the negative value of the line integral indicate?

Example 3: Illustrate the validity of Green's Theorem:

Let R be a region bounded by a simple closed path C. If M,N, $\partial M/\partial y$, and $\partial N/\partial x$ are all continuous on R then,

$$\int_C M(x,y)\ dx\ +\ N(x,y)\ dy\ =\ \iint_R [(\partial N/\partial x)-(\partial M/\partial y)\ dy\ dx$$

By using programs LINE INTEGRAL and DOUBLE INTEGRATION with $M(x,y)=e^x \sin y$, $N(x,y)=e^y \cos x$, and R the region bounded by $y=x^2$ and $y=1$.

To approximate the line integral C must be broken into two pieces, the parabola $y=x^2$ from $(-1,1)$ to $(1,1)$ and the line segment $y=1$ from $(1,1)$ to $(-1,1)$.

For the parabola $x=t$, $y=t^2$, $-1 \le t \le 1$, $dx=dt$, $dy=2t\ dt$.

```
120 DEF FNM(X)=EXP(X)*SIN(Y)
140 DEF FNN(X)=EXP(Y)*COS(X)
160 DEF FNC(T)=T
180 DEF FND(T)=T*T
200 DEF FNA(T)=1
220 DEF FNB(T)=2*T
```

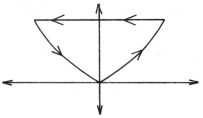

RUN
ENTER THE RANGE ON THE PARAMETER T.
ENTER BOTH ON ONE LINE.
?-1,1

ENTER THE DESIRED PRECISION.
?.0001

THE VALUE OF THE LINE
INTEGRAL IS .813604877

On the line segment $x=t$, $y=1$, $dx=dt$, $dy=0$. The limits of integration on the parameter t must be from 1 to -1 to maintain the orientation of the curve C.

```
160 DEF FNC(T)=T
180 DEF FND(T)=1
200 DEF FNA(T)=1
220 DEF FNB(T)=0
```

RUN
ENTER THE RANGE ON THE PARAMETER T.
ENTER BOTH ON ONE LINE.
?1,-1

ENTER THE DESIRED PRECISION.
?.0001

THE VALUE OF THE LINE
INTEGRAL IS -1.97778254

Thus, $\int_C$ e^x sin y dx + e^y cos x dy $\approx$

$$.813259191-1.977663565 \approx -1.164404374$$

Now to approximate

$$\iint_R [(\partial N/\partial x)-(\partial M)/\partial y)] \; dy \; dx$$

we will use program DOUBLE INTEGRATION from chapter 20.

$$\partial N/\partial x = -e^y \sin x \qquad \partial M/\partial y = e^x \cos y$$

Thus, $\qquad \partial N/\partial x - \partial M/\partial y = -e^y \sin x - e^x \cos y$.

Hence,

$$\iint [(\partial N/\partial x) - (\partial M/\partial y) dy \; dx = \int_{-1}^{1}\int_{x^2}^{1} -e^y \sin x - e^x \cos y \; dy \; dx.$$

```
120 DEF FNF(X)=-EXP(Y)*SIN(X)-EXP(X)*COS(Y)
140 DEF FNG(X)=X*X
160 DEF FNH(X)=1
180 LET L=-1
190 LET U=1
```

```
RUN
ENTER THE NUMBER OF SUBDIVISIONS.
?25

THE APPROXIMATION OF THE
INTEGRAL IS -1.16512636
```

Considering round-off error and discretization error the two integrals are equal, and Green's Theorem is verified.

EXERCISE SET 21.1

In exercises 1-4 approximate each line integral using program LINE INTEGRAL.

1. $\displaystyle\int_C 2xy^2\,dx + 3x^2y^3\,dy$; C:x=t-1, y=t²; 0≤t≤1

2. $\displaystyle\int_C 3x\,dx + 4y\,dy$; C:x=t, y=$\sqrt{4-t^2}$; -2≤t≤2

3. $\displaystyle\int_C x\cos y\,dx - x^2\sin y\,dy$; C:x=t, y=(t-1)²; 0≤t≤1

4. $\displaystyle\int_C x\sin x\,dx + \cos y\,dy$; C:x=t, y=t³; 0≤t≤1

In exercises 5-8 use program LINE INTEGRAL to approximate the work done by the force F(x,y) acting on an object moving along the curve C.

5. F(x,y)=5x²y⁴i + 7y³x³j;   C:unit   circle   moving counterclockwise.

6. F(x,y)=3 cos xy i + 2 sin x² j; C:semicircle with radius 4 and center the origin.

7. F(x,y)=eˣsin y i + eˣ⁺ʸln x j; C:the ellipse x²/4 + y²/9 = 1 in a counterclockwise direction from ($\sqrt{3}$,3/2) to ($\sqrt{2}$,3$\sqrt{2}$/2).

8. F(x,y)=x³sin xy i + eʸcos x²j; C:the triangle with vertices (1,1), (5,7),(-1,10) starting at (1,1) going to (5,7).

*

9. Illustrate that $\displaystyle\int_{-C} M(x,y)\,dx + N(x,y)\,dy =$

$$-\int_C M(x,y) \; dx \; + \; N(x,y) \; dy$$

by running program LINE INTEGRAL twice to approximate

$$\int_C x^2 y \; dx \; + \; 3xye^y \, dy$$

where C is:

     a.   the unit circle in a counterclockwise direction.

     b.   the unit circle in a clockwise direction.

In exercises 10-11 use programs LINE INTEGRAL and DOUBLE INTEGRATION to verify Green's Theorem for the indicated functions $M(x,y)$, $N(x,y)$, and region R.

10. $M(x,y)=2 \cos x^2 y$; $N(x,y)=4 \sin y^2 x^3$; R bounded by $y=x^3$, $x=0$, and $y=8$

11. $M(x,y)=e^{x \ast x + y}$; $N(x,y)=\ln(x+y^2)$; R bounded by $y=x^2$, $y=4$, and $x=4$

**
In exercises 12-14 use program LINE INTEGRAL to evaluate the line integral

$$\int_C M(x,y) \; dx \; + \; N(x,y) \; dy \quad \text{on the curve}$$

     a.   from (0,0) to (.5,.125) parameterized by $x=t$, $y=t^3$, $0 \le t \le .5$

     b.   from (0,0) to (.5,.125) parameterized by $x=t$, $y=.5t^2$, $0 \le t \le .5$

If the line integral appears to be path independent, determine a function $f(x,y)$ such that the gradient of $f(x,y)$ is $M(x,y) \; i + N(x,y) \; j$ and decide if the line integral equals $f(.5,.125)-f(0,0)$. If it does not appear to be path independent, show that $\partial N/\partial x \neq \partial M/\partial y$.

12. $\int_C (e^y + x^2 \cos x) \, dx + (xe^y + \sin y) \, dy$

13. $\int_C e^x \sin y \, dx + e^x \cos y \, dy$

14. $\int_C (\sin y - y \sin x) \, dx + (\cos x + x \cos y) \, dy$

In exercises 15-16 approximate the area of the given region, R, bounded by the curve C using the line integral

$$A = (1/2) \int_C x \, dy - y \, dx$$

15. R: bounded by the quadrilateral with vertices (1,0), (4,8), (17,3) and (19,-5)

16. R: bounded by the hypocycloid $x^{2/3} + y^{2/3} = 1$

17. Modify program LINE INTEGRAL to use the trapezoidal rule in place of the midpoint rule in approximating the integral. Use the same test data as in the section.

18. Same as exercise 17 except use Simpson's rule.

19. Write a program which will approximate the value of a line integral given by

$$\int_C f(x,y) \, ds$$

As test data use:

$$\int_C xy \, ds$$

where C: y=4-x with x going from 0 to 2. The value is approximately 7.5425

21.2 SURFACE INTEGRALS--DIVERGENCE THEOREM

Programming Problem: Approximate the value of the surface integral

$$\iint_S g(x,y,z)\ dS$$

Output: The approximation of the surface integral.

Input: The function being integrated, the function which describes the surface, the x and y partial derivatives of the function which describes the surface, the limits of integration, and the number of subdivisions being used in the approximation.

Strategy: If the surface, S, over which the surface integral is taken is defined by the function z=f(x,y) above or below the region R in the xy-plane, then we know that

$$\iint_S g(x,y,z)\ dS= \iint_R g(x,y,f(x,y))\ \sqrt{1+f_x{}^2+f_y{}^2}\ dy\ dx$$

Since the integral on the right-hand side of the equation is a double integral, it can be approximated in the same manner as used in section 20.1 with the function being integrated

$$g(x,y,f(x,y))\ \sqrt{1+f_x{}^2+f_y{}^2}$$

However, instead of using the program for approximating double integrals from section 20.1 with the above function being integrated, we will modify that program so that the substitution of f(x,y) for z in g(x,y,z) and the computation of $\sqrt{1+f_x{}^2+f_y{}^2}$ is performed during the execution of the program.

Program:

```
100 REM ** SURFACE INTEGRAL **
110 REM ** FUNCTION TO BE INTEGRATED **
120 DEF FNG(X)=.....
130 REM ** FUNCTION OF SURFACE **
140 DEF FNF(X)=.....
150 REM ** X-PARTIAL SURFACE **
160 DEF FNA(X)=.....
170 REM ** Y-PARTIAL SURFACE **
180 DEF FNB(X)=.....
190 REM ** LOWER LIMIT-INNER INTEGRAL **
200 DEF FNS(X)=.....
210 REM ** UPPER LIMIT-INNER INTEGRAL **
220 DEF FNT(X)=.....
230 REM ** LIMITS-OUTER INTEGRAL **
240 LET L=.....
250 LET U=.....
260 PRINT "ENTER THE NUMBER OF SUBDIVISIONS."
270 INPUT N
280 PRINT"
290 REM ** MAIN PROGRAM **
300 LET S=0
310 LET H=(U-L)/N 320 REM ** LOOP-OUTER SUM **
330 FOR I=1 TO N
340 LET XL=L+(I-1)*H
350 LET XR=L+I*H
360 LET X=(XL+XR)/2
370 LET S1=0
380 REM ** FINDING DELTA-Y **
390 LET YL=FNS(X)
400 LET YU=FNT(X)
410 LET K=(YU-YL)/N
420 REM ** LOOP-INNER SUM **
430 FOR J=1 TO N
440 LET Y1=YL+(J-1)*K
450 LET Y2=YL+J*K
460 LET Y=(Y1+Y2)/2
470 LET Z=FNF(X)
480 REM ** FINDING G*SQR(1+FX 2+FY 2)*DX*DY **
490 LET S1=S1+FNG(X)*SQR(1+FNA(X)*FNA(X)+FNB(X)*FNB(X))*K
500 NEXT J
510 LET S=S+S1*H
520 NEXT I
530 PRINT "THE APPROXIMATION OF THE"
540 PRINT "INTEGRAL IS ";S
550 END
```

Test Data:

$$\iint_S 2x+y+3z \ dS$$

where S is the part of the plane 2x+4y+3z=12 in the first octant.

$$z=f(x,y)=(12-2x-4y)/3$$

$$f_x=-2/3 \qquad \text{and} \qquad f_y=-4/3$$

Thus, $\iint_S 2x+y+3z \ dS=$

$$\int_0^6 \int_0^{-.5x+3} (2x+4y+3(\frac{12-2x-4y}{3})) \sqrt{1+4/9+16/9} \ dy \ dx$$

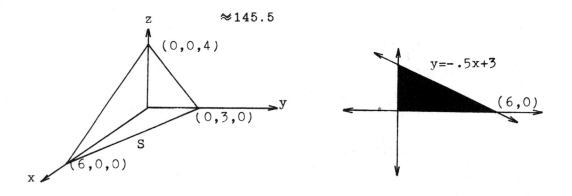

$$\approx 145.5$$

```
120 DEF FNG(X)=2*X+Y+3*Z
140 DEF FNF(X)=(12-2*X-4*Y)/3
160 DEF FNA(X)=-2/3
180 DEF FNB(X)=-4/3
200 DEF FNS(X)=0
220 DEF FNT(X)=-X/2+3
240 LET L=0
250 LET U=6
```

```
RUN
ENTER THE NUMBER OF SUBDIVISIONS.
?25

THE APPROXIMATION OF THE
INTEGRAL IS 145.418836
```

Example 1:

Approximate the surface area of the surface $z=x^2\ln(y+2)$ above the region $x^2+y^2-2x=0$ in the xy-plane using program SURFACE INTEGRAL.

To use the program SURFACE INTEGRAL to approximate surface area we let $g(x,y,z)=1$.

```
120 DEF FNG(X)=1
140 DEF FNF(X)=X*X*LOG(Y+2)
160 DEF FNA(X)=2*X*LOG(Y+2)
180 DEF FNB(X)=X*X/(Y+2)
200 DEF FNS(X)=-SQR(2*X-X*X)
220 DEF FNT(X)=SQR(2*X-X*X)
240 LET L=0
250 LET U=2
```

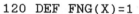

$$y=\sqrt{2x-x^2}$$

$$y=-\sqrt{2x-x^2}$$

```
RUN
ENTER THE NUMBER OF SUBDIVISIONS.
?25

THE APPROXIMATION OF THE
INTEGRAL IS 6.07383399
```

Thus, the surface area is approximately 6.074

Example 2: Approximate using program SURFACE INTEGRAL the mass of a thin sheet of material which has the shape of the surface $z=e^{xy}$ above $(x^2/4)+(y^2/9)=1$ if the material has a density function $p(x,y,z)=xyz$ gm/unit of area.

Since the density function is in grams per unit of area, the mass is given by the surface integral

$$\iint_S p(x,y,z)\ dS$$

```
120 DEF FNG(X)=X*Y*Z
140 DEF FNF(X)=EXP(X*Y)
160 DEF FNA(X)=Y*EXP(X*Y)
180 DEF FNB(X)=X*EXP(X*Y)
200 DEF FNS(X)=-SQR(9-9*X*X/4)
220 DEF FNT(X)=SQR(9-9*X*X/4)
240 LET L=-2
250 LET U=2

RUN
ENTER THE NUMBER OF SUBDIVISIONS.
?25

THE APPROXIMATION OF THE
INTEGRAL IS 1189.82516
```

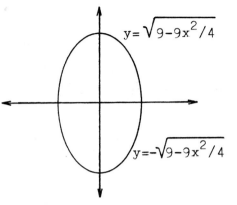

The mass is approximately 1189.8

Example 3: Use program SURFACE INTEGRAL to approximate the flux of $F(x,y,z)=3xy\mathbf{i}+4zy^2\mathbf{j}+xyz\mathbf{k}$ over the hemisphere $z=\sqrt{16-x^2-y^2}$.

By definition the flux of a vector function over a surface is given by

$$\iint_S F\, n\, dS$$ where n is the outward pointing unit normal

to the surface. Now, if $z=f(x,y)$ describes the surface, then the outward pointing unit normal is $n=\nabla F/|\nabla F|$ where $F(x,y,z)=z-f(x,y)$. Thus, for the hemisphere $z=\sqrt{16-x^2-y^2}$,

$$F=z-\sqrt{16-x^2-y^2}$$

$$\nabla F= \frac{x}{\sqrt{16-x^2-y^2}}\, \mathbf{i}+ \frac{y}{\sqrt{16-x^2-y^2}}\, \mathbf{j}+ \mathbf{k}$$

$$n= \frac{x}{4}\, \mathbf{i}+ \frac{y}{4}\, \mathbf{j}+ \frac{\sqrt{16-x^2-y^2}}{4}\, \mathbf{k}$$

Using this definition for flux we have

$$\text{flux} = \iint_S (3xyi + 4zy^2 j + xyzk)(\frac{x}{4} i + \frac{y}{4} j + \frac{\sqrt{16-x^2-y^2}}{4} k) \, dS$$

$$= \iint_S \frac{3x^2 y}{4} + zy^3 + xyz \frac{\sqrt{16-x^2-y^2}}{4} \, dS$$

```
120 DEF FNG(X)=3*X*X*Y/4+Y*Y*Y*Z+X*Y*Z*SQR(16-X*X-Y*Y)/4
140 DEF FNF(X)=SQR(16-X*X-Y*Y)
160 DEF FNA(X)=-X/SQR(16-X*X-Y*Y)
180 DEF FNB(X)=-Y/SQR(16-X*X-Y*Y)
200 DEF FNS(X)=-SQR(16-X*X)
220 DEF FNT(X)=SQR(16-X*X)
240 LET L=-4
250 LET U=4

RUN
ENTER THE NUMBER OF SUBDIVISIONS.
?25

THE APPROXIMATION OF THE
INTEGRAL IS -1.75116584E-07
```

We should approximate the value of the flux as 0. If the vector function represents the velocity vector of a fluid or gas, then the flux measures the amount of fluid flowing out of the surface per unit of time. A positive value for the flux means that fluid is flowing across the surface, and we say the fluid is expanding. A negative value for the flux means that the fluid is actually contracting, and we refer to the fluid as being compressed. Finally, a flux of zero means that the amount of fluid is stationary, and the fluid is called incompressible. This example shows that it is possible to have an incompressible fluid even though the velocity vector is not zero everywhere.

Example 4: Use programs SURFACE INTEGRAL and TRIPLE INTEGRATION to verify the Divergence Theorem:

Let Q be a solid region bounded by a closed surface S with outward normal n, and let F be a vector field with continuous partial derivatives in Q. Then,

$$\iint_S F \, n \, dS = \iiint_Q (\text{div } F) \, dV$$

483

for $F(x,y,z)=(x^2-1)i+(2y-x)j+zk$ where Q is the region in R^3 bounded by the surface S described by

$z=1-x^2-y^2$ if $z \geq 0$, and

$x^2+y^2 \leq 1$ if $z=0$.

Since the surface S is composed of two different pieces, paraboloid and disk, the surface integral on the left-hand side of the equation must be computed separately for each piece.

1. For the disk $x^2+y^2 \leq 1$, $z=0$, n=-k, and F n=-z. Thus,

$$\iint_S F\ n\ dS = \iint_S -z\ dS.$$ However, z=0 on the surface,

and we can conclude that the surface integral equals zero for the disk.

2.   For the paraboloid, $z=1-x^2-y^2$, $F=z+x^2+y^2-1$, $\nabla F=2xi+2yj+k$, and $n=(2xi+2yj+k)/\sqrt{1+4x^2+4y^2}$. Therefore,

$$\iint_S F\ n\ dS = \iint_S \frac{2x(x^2-1) + 2y(2y-x) +z}{\sqrt{1+4x^2+4y^2}}\ dS$$

with $S: z=1-x^2-y^2$, $z \geq 0$.

```
120 DEF FNG(X)=(2*X*X*X-2*X+4*Y*Y-2*X*Y+Z)/
              SQR(1+4*X*X+4*Y*Y)
140 DEF FNF(X)=1-X*X-Y*Y
160 DEF FNA(X)=-2*X
180 DEF FNB(X)=-2*Y
200 DEF FNS(X)=-SQR(1-X*X)
220 DEF FNT(X)=SQR(1-X*X)
240 LET L=-1
250 LET U=1

RUN
ENTER THE NUMBER OF SUBDIVISIONS.
?25

THE APPROXIMATION OF THE
INTEGRAL IS 4.70931046
```

Thus, $\iint_S$ F n dS for both surfaces is approximately 4.71

Now to approximate the triple integral $\iiint_Q$ (div F) dV

we will use program TRIPLE INTEGRATION with the function being integrated

$$(\text{div } F) = 2x+2+1 = 2x+3$$

```
120 DEF FNF(X)=2*X+3
140 DEF FNU(X)=0
160 DEF FNV(X)=1-X*X-Y*Y
170 DEF FNG(X)=-SQR(1-X*X)
180 DEF FNH(X)=SQR(1-X*X)
200 LET L=-1
210 LET U=1

RUN
ENTER THE NUMBER OF SUBDIVISIONS.
?15

THE APPROXIMATION OF THE
INTEGRAL IS 4.72538849
```

Taking into account round-off and discretization error, the volume can be considered equal, and the Divergence Theorem is verified in this instance.

EXERCISE SET 21.2

In exercises 1-5 use program SURFACE INTEGRAL to approximate the value of each integral.

1. $\iint_S xz \, dS$

   S is the portion of the cylinder $z=x^2$ bounded by $x=0$, $x=4$, $y=0$, and $y=4$

2. $\iint_S z(x^2+y^2) \, dS$

   S is the hemisphere $z=\sqrt{1-x^2-y^2}$

3. $\iint_S x^2+y^2+z^2 \, dS$

   S is the part of the plane $2x+4y+z=7$ in the first octant

4. $\iint_S x \, dS$

   S is the part of the cone $z^2=x^2+y^2$ between $z=2$ and $z=4$

5. $\iint_S xy \, dS$

   S is the ellipsoid $(x^2/9)+(y^2/4)+z^2=1$ with $z \geq 0$

In exercises 6-7 use program SURFACE INTEGRAL to approximate the surface area of the given surface.

6. S: the portion of the surface $z=\sin^2 x + \cos^2 y$ above the region $x^2+y^2=1$

7. S: the portion of the cylinder $z=x^3+1$ above the region $x^2+y^2-2Y=0$

In exercises 8-9 use program SURFACE INTEGRAL to approximate the mass of a thin sheet of material which has the shape of the given surface, S, and the given density function $p(x,y,z)$.

8. $p(x,y,z)=4x^2e^{yz}$; S is the part of the plane $x+y+z=3$ inside the cylinder $x^2+y^2=4$

9. $p(x,y,z)=\ln 16-xy + 3z$; S is the upper surface of the sphere $x^2+y^2+z^2=16$

In exercises 10-12 use program SURFACE INTEGRAL to approximate the flux of F over S.

10. $F=e^x i + e^y j + e^z k$; S the surface of the paraboloid $z=4-x^2-y^2$ with $z\geq0$

11. $F=2xi + 3y j+ 2xzk$; S the surface bounded by $z=x^2+y^2$ and $z=18-x^2-y^2$

12. $F=(x^2+z)i + (y^2+x)j + (z^2+y)k$; S is the surface $2z=y^2+x^2$ inside $x^2+y^2=2x$

In exercises 13-14 use programs SURFACE INTEGRAL and TRIPLE INTEGRATION to verify the Divergence Theorem for the given function and region.

13. $F=xi + yj + zk$; Q is the region bounded by the sphere $x^2+y^2+z^2=4$

14. $F=x^3 i + y^3 j + zk$; Q is the region inside $x^2+y^2+z^2=4$ and outside $x^2+y^2=1$

**
The div(grad f)= $\nabla \cdot \nabla f=\nabla^2 f$ is called the Laplacian. In exercises 15-18 use programs SURFACE INTEGRAL and TRIPLE INTEGRATION to verify the given expression for the given functions and regions. If possible prove that the expression holds in general.

15.   Green's First Identity:

$$\iiint_Q (f\nabla^2 g + \nabla f \cdot \nabla g)\, dV = \iint_S (f\nabla g) \cdot n\, dS$$

$f(x,y,z) = x^2 + y^2 + z^2$

$g(x,y,z) = x^2 y + y^2 z + z^2 x$

Q: region bounded by the sphere $x^2 + y^2 + z^2 = 4$

S: surface that bounds Q

n: outward unit normal to S

16.   Green's Second Identity:

$$\iiint_Q (f\nabla^2 g - g\nabla^2 f)\, dV = \iint_S (f\nabla g - g\nabla f) \cdot n\, dS$$

f,g,Q,S, and n as in exercise 15

17. $\iint_S \nabla f \cdot n\, dS = 0$

$f(x,y,z) = x^3 + 3xy^2 - 6z^2 x$

S: the hemisphere $x^2 + y^2 + z^2 = 1$ along with the unit disk in the xy-plane

Since this expression holds only if the Laplacian of $f(x,y,z)$ is identically zero, make the assumption in doing the general case. A function whose Laplacian is identically zero is called harmonic.

18. $$\iiint_Q \operatorname{grad} f \ ^2 \ dV = \iint_S (f \triangledown f) \ n \ dS$$

f,S as in exercise 17

Q: the region bounded by S

    As in exercise 17, this expression holds only for the harmonic functions f; make that assumption in doing the general case.

CHALLENGE ACTIVITY

Write a program which will approximate the value of the line integral

$$\int_C F(x,y,z) \, dR \text{ in space. As test data use}$$

$F(x,y,z)=(2x\sin z)i - (e^z \sin y)j + (x^2\cos z + e^z\cos y)k$

along the path $x=t$, $y=t^2$, $z=t$, $0 \leq t \leq 2$ which has a value of approximately -2.1926196

Use the program from above and program SURFACE INTEGRAL to verify Stokes' Theorem:

Let S be an oriented surface with boundary C, a simple closed path, and outward normal n. Let F be a vector field with continuous partial derivatives on S.   Then,

$$\oint_C F \cdot dR = \iint_S (\text{curl } F) \cdot n \, dS$$

for $F(x,y,z)=ye^x i + 2xe^z j + xye^y k$ and S the surface

$$x^2+z^2=1, \quad 0\leq y\leq 2, \quad z\geq 0$$

$$y=0, \quad x^2+z^2\leq 1, \quad z\geq 0$$

$$y=z, \quad x^2+z^2\leq 1, \quad z\geq 0$$

CHAPTER 22
Differential Equations

For centuries differential equations have been of fundamental importance in the application of mathematics to the physical sciences. In recent years, however, they have also gained importance in the biological and social sciences. In this chapter we will investigate numerical methods for solving differential equations. These methods do not yield a function which is a solution, but rather a numerical approximation to points on the curve of the solution.

22.1 EULER'S METHOD

In this section, as well as the next, we will be concerned with approximating a numerical solution to the initial value problem $y'(x)=f(x,y)$, $y(x_0)=y_0$, for a specific value $x=a$, that is, finding approximately the value $y(a)$. The simplest method is to make use of the fact that $y'(x)$ also gives the slope of the tangent line at any point on $y(x)$ and approximate $y(a)$ by y_a where

$$\frac{y_a - y_0}{a - x_0} = f(x_0, y_0)$$

or $y_a = y_0 + f(x_0, y_0)(a-x_0)$

However, using the value y_a to approximate $y(a)$ will generally result in a large error unless "a" is "quite close" to x_0.

Euler's Method attempts to eliminate the necessity of having "a quite close" to x_0 by using a sequence of line segments on the interval $[x_0, a]$ or $[a, x_0]$ instead of a single tangent line. For an illustration see the figure below. Suppose that $y(x)$ is the desired solution and that $[x_0, a]$ is subdivided into n subdivisions of equal length. Then y_1 can be determined by the tangent line method used above, namely

$$y_1 = y_0 + f(x_0, y_0)(x_1 - x_0)$$

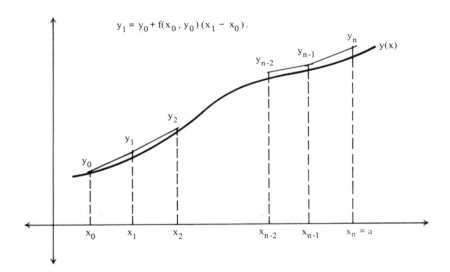

Now using the point (x_1, y_1) we can repeat the process to obtain y_2,

$$y_2 = y_1 + f(x_1, y_1)(x_2 - x_1)$$

and in general

$$y_{m+1} = y_m + f(x_m, y_m) * h \text{ where } h = (a - x_0)/n$$

which, after n iterations, will yield an approximation y_n for $y(a)$.

Example 1: Approximate the solution to $y'=2xy$, $y(1)=1$ at $x=1.5$ using Euler's Method with $h=.1$

Now $x_0=1$, $y_0=1$, $a=1.5$, $h=.1$
Finding y_1 : $y_1=y_0+hf(x_0,y_0)$
                    $=1+(.1)(2(1)(1))$
                    $=1.2$

Finding y_2 : $y_2=y_1+hf(x_1,y_1)$
                    $=1.2+(.1)(2(1.1)(1.2))$
                    $=1.464$

Continuing with the iteration we find:

$y_3=1.81536$
$y_4=2.2873536$
$y_5=2.92781260 \approx y(1.5)$

Programming Problem: Approximate the numerical solution to $y'(x)=f(x,y)$, $y(x_0)=y_0$ at a specific value for x using Euler's Method.

Output: A table listing the values of x and the corresponding approximations for $y(x)$.

Input: The function $f(x,y)$, the initial conditions, the value at which we are to approximate the solution, and the step size or length of a subdivision.

Strategy: Euler's Method employs an iteration formula which can be evaluated using a FOR-NEXT loop. Although the most natural FOR-NEXT loop would be to use x as the control variable starting at the initial value of x to the value at which we are approximating the solution incremented by the step size, this will not work properly on some computers if the values are decimals due to round-off error. Thus, it will be necessary to determine the number of times the iterative process is to be performed. This is performed in line 300 of the following program.

Program:

```
100 REM ** EULER **
110 REM ** FUNCTION **
120 DEF FNF(X)=.....
130 REM ** INPUT **
140 PRINT "ENTER THE INITIAL CONDITIONS."
150 PRINT "X VALUE THEN Y VALUE."
160 INPUT X1,Y
170 PRINT
180 PRINT "ENTER THE VALUE OF X AT WHICH"
190 PRINT "THE SOLUTION IS TO BE APPROXIMATED."
200 INPUT XN
210 PRINT
220 PRINT "ENTER THE STEP SIZE."
230 INPUT H
240 PRINT
250 REM ** HEADINGS **
260 PRINT "X-VALUE","APPROXIMATION OF Y"
270 PRINT X1,Y
280 REM ** MAIN PROGRAM **
290 REM ** DETERMINING NUMBER OF ITERATIONS **
300 LET N=ABS(INT((XN-X1)/H+.5))
310 FOR I=1 TO N
320 LET X=X1+(I-1)*H
330 LET Y=Y+H*FNF(X)
340 PRINT X+H,Y
350 NEXT I
360 END
```

Test Data: Approximate $y(1.5)$ for $y'(x)=2xy$, $y(1)=1$.

```
120 DEF FNF(X)=2*X*Y
```

```
RUN
ENTER THE INITIAL CONDITIONS.
X VALUE THEN Y VALUE.
?1,1

ENTER THE VALUE OF X AT WHICH
THE SOLUTION IS TO BE APPROXIMATED.
1.5

ENTER THE STEP SIZE.
?.1
```

X-VALUE	APPROXIMATION OF Y
1	1
1.1	1.2
1.2	1.464
1.3	1.81536
1.4	2.2873536
1.5	2.92781261

Example 2: For the differential equation $y'(x)=2xy$, $y(1)=1$ used in the test data, compare the results for $y(1.5)$ to the actual values for this differential equation which has solution $y=e^{x*x-1}$. Compute the percent of error using

$$\% \text{ error} = \frac{\text{approximation} - \text{actual value}}{\text{actual value}} \times 100$$

x	y-Euler	y-Calculator Value e^{x*x-1}	%-Error
1	1	1	0
1.1	1.2	1.23367806	2.730%
1.2	1.464	1.552707219	5.713%
1.3	1.81536	1.993715533	8.946%
1.4	2.2873536	2.611696473	12.419%
1.5	2.92781261	3.490342957	16.117%

As can be seen from the table the Euler Method for this problem has approximated a value which is in error 16.117%. In general, an error of this magnitude is unacceptable.

Although the Euler Method can yield a reasonable approximation in some problems (for example, it approximates $y'(x)=y$, $y(0)=1$ for $y(.5)$ to be 1.61051, an error of 2.318%), the inherent inaccuracies of the method led to the development of the modified Euler Method or Heun's Method.

One of a large class of what are called predictor-corrector methods, Heun's Method first uses Euler's Method to predict a value for y_{n+1} and then improves or corrects the prediction to obtain the final approximation for y_{n+1}. The general formula for this method is

$$y_{n+1} = y_n + (h/2)(f(x_n, y_n) + f(x_{n+1}, y^*_{n+1}))$$

where $y^*_{n+1} = y_n + f(x_n, y_n) * h$

Example 3: Approximate the solution to $y'=2xy$, $y(1)=1$ at $x=1.5$ using Heun's Method with $h=.1$, $x=1$, $y=1$, $a=1.5$

Finding y_1 :

$$y*_1 = y_0 + f(x_0, y_0)h$$
$$= 1 + (.1)(2(1)(1))$$
$$= 1.2$$

$$y_1 = y_0 + (h/2)(f(x_0, y_0) + f(x_1, y*_1))$$
$$= 1 + (.05)(2 + 2(1.1)(1.2))$$
$$= 1.232$$

Continuing with the iteration we find:

$$y_2 = 1.5478848$$
$$y_3 = 1.983150006$$
$$y_4 = 2.590787168$$
$$y_5 = 3.450928508 \approx y(1.5)$$

Comparing the approximation for $y(1.5)$ with the exact answer of $y(1.5)=3.490342957$ we have an error of only 1.129% as compared to 16.117% for Euler's Method.

Programming Problem: Approximate the numerical solution to $y'(x)=f(x,y)$, $y(x_0)=y_0$ at a specified value for x using Heun's Method.

Output: A table listing the values of x and the corresponding approximation of $y(x)$.

Input: The function $f(x,y)$, the initial conditions, the value at which we are to approximate the solution, and the step size.

Strategy: As with Euler's Method, Heun's Method also uses an iteration formula, and, thus, can be approximated in the same manner. The difficult part of the program involves having to compute $f(x_n, y_n)$ and $f(x_{n+1}, y*_{n+1})$. Since the DEF statement allows only one argument, the variable y must be assigned the proper value, y_n or $y*_{n+1}$ before the function is evaluated. However, we must also save the value y_n since it is used in the final computation of y_{n+1}.

Program:

```
100 REM ** HEUN **
110 REM ** FUNCTION **
120 DEF FNF(X)=.....
130 REM ** INPUT **
140 PRINT "ENTER THE INITIAL CONDITIONS."
150 PRINT "X VALUE THEN Y VALUE."
160 INPUT X1,Y
170 PRINT
180 PRINT "ENTER THE VALUE OF X AT WHICH"
190 PRINT "THE SOLUTION IS TO BE APPROXIMATED."
200 INPUT XN
210 PRINT
220 PRINT "ENTER THE STEP SIZE."
230 INPUT H
240 PRINT
250 REM ** HEADINGS **
260 PRINT "X-VALUE","APPROXIMATION OF Y"
270 PRINT X1,Y
280 REM ** MAIN PROGRAM **
290 REM ** DETERMINING NUMBER OF ITERATIONS **
300 LET N=ABS(INT((XN-X1)/H+.5))
310 FOR I=1 TO N
320 LET X=X1+(I-1)*H
330 LET D=FNF(X)
340 LET Y1=Y+H*D
350 LET Y2=Y
360 LET Y=Y1
370 LET Y=Y2+(H/2)*(D+FNF(X+H))
380 PRINT X+H,Y
390 NEXT I
400 END
```

Test Data: Approximate $y(1.5)$ for $y'(x)=2xy$, $y(1)=1$.

```
120 DEF FNF(X)=2*X*Y
```

```
RUN
```
ENTER THE INITIAL CONDITIONS.
X VALUE THEN Y VALUE.
?1,1

ENTER THE VALUE OF X AT WHICH
THE SOLUTION IS TO BE APPROXIMATED.
?1.5

ENTER THE STEP SIZE.
?.1

X-VALUE	APPROXIMATION OF Y
1	1
1.1	1.232
1.2	1.5478848
1.3	1.98315001
1.4	2.59078717
1.5	3.45092851

Example 4: Approximate $y(1)$ for the differential equation $y'(x)=y^2$, $y(0)=1$. Use program HEUN with h=.05

```
120 DEF FNF(X)=Y*Y

RUN
ENTER THE INITIAL CONDITIONS.
X VALUE THEN Y VALUE.
?0,1

ENTER THE VALUE OF X AT WHICH
THE SOLUTION IS TO BE APPROXIMATED.
?1

ENTER THE STEP SIZE.
?.05
```

X-VALUE	APPROXIMATION OF Y
0.0	1
0.05	1.0525625
0.1	1.11094891
0.15	1.17618234
0.2	1.24954004
0.25	1.33263734
0.3	1.42754732
0.35	1.53697431
0.4	1.66451453
0.45	1.81505402
0.5	1.99540229
0.55	2.21533702
0.6	2.48940891
0.65	2.84023518
0.7	3.30492918
0.75	3.94875911
0.8	4.89751933
0.85	6.42643723
0.9	9.26150806
0.95	15.9961517
1	43.1146689

For x=.95 and x=1 the corresponding approximations of y are erratic, making large jumps. To investigate the problem we will solve y'(x)=y , y(0)=1 analytically to obtain the solution y=1/(1-x) which is undefined for x=1. This not only explains the erratic behavior of the solution but also illustrates one of the problems that can arise when using a numerical method in approximating a solution to a differential equation. The solution may not exist for all values of x in the interval over which the approximation is being performed. Thus, it is important, as mentioned many times before, to carefully interpret output, questioning all results which appear inconsistent.

Ch. 22   DIFFERENTIAL EQUATIONS

EXERCISE SET 22.1

In exercises 1-4 use program EULER to approximate a solution to the given differential equation at the indicated value of a. If possible, determine the general solution and compute the percentage of error in the approximation by

$$\% \text{ error} = \frac{\text{approximation} - \text{actual value}}{\text{actual value}} \text{ X } 100$$

1. $y'=(2-y^2)/2xy$, $y(1)=2$, $a=1.5$, $h=.1$

2. $y'=(xy+x)/y^2$, $y(1)=1$, $a=1.5$, $h=.1$

3. $y'=e^{-2x}-2y$, $y(0)=4$, $a=.5$, $h=.05$

4. $y'=(3x^2+4xy)/(-2x^2-2y)$, $y(0)=1$, $a=.2$, $h=.01$

In exercises 5-8 redo exercises 1-4 using program HEUN. Compare the percentage of error in the two methods.

*
9.   A Bernoulli equation is given by $y'+P(x)y=Q(x)y^n$ where n is a nonzero real number. Use program HEUN to approximate $y(1.25)$ where $y(x)$ is the solution to the Bernoulli equation $y'+(1+x^2)y=4x^3y^4$, $y(.9)=.05$

10.   A Riccati equation is given by $y'=P(x)+Q(x)y+R(x)y^2$. Use program HEUN to approximate $y(2.6)$ where $y(x)$ is the solution to the Riccati equation $y'=-2x+4e^xy+(5x+1)y^2$, $y(1.9)=1$.

11.   Run program EULER to approximate $y(1)$ where $y(x)$ is the solution to $y'=e^{xy}$, $y(0)=1$. Use $h=.1$ Interpret your results carefully.

12.   Run program HEUN to approximate $y(1)$ where $y(x)$ is the solution to $y'=y+x^2$, $y(0)=1$. Use $h=.1$ Interpret your results carefully.

**
13.   Consider the differential equation $y' = \sqrt{|x|}$, $y(0)=0$.

     a.   Show that this differential equation has an infinite number of solutions by showing that for any real number $b>0$

$$y(x) = \begin{cases} 0, & x \leq b \\ (x-b)^2/4, & x>b \end{cases}$$

is a solution.

     b.   Decide which solution is found using Euler's Method. <u>Hint</u>: Run program EULER and determine b by setting the approximation for $y(1)$ equal to $(1-b)^2/4$.

     c.   Decide which solution is found using Heun's Method.

     d.   Is the solution of part b the same as part c?

***
     As with other numerical methods employed in this supplement the results are subject to error. The sources of possible error are:

     1.   Round-off error due to the machine.

     2.   Discretization or truncation error due to the method used.

     3.   Propagation error.

This last source of error, propagation error, has not been encountered before. To explain this error consider the iteration formula for Euler' Method, $y_n = y_{n-1} + hf(x_{n-1}, y_{n-1})$. In computing y_n only an approximation of $y(x_{n-1})$, y_{n-1} , is used. Thus, an error in the approximation of y_{n-1} will carry over and affect y_n. Since this type of error occurs at each step in the iterative process, a small error made in the beginning may grow into a significant error by the end of the process.

14.   If $f(x,y)$ is a function of x only, then the effect of propagation error is minimized for Euler's and Heun's Methods.

     a.   Both of the differential equations below have the same solution, $y(x)=e^x$.

$$y'(x)=y, \ y(0)=1 \qquad \text{and} \qquad y'(x)=e^x, \ y(0)=1$$

Run program EULER for both equations with h=.1 to approximate y(x) for x=.1,.2,...,.9,1 Using the exact value determine the percent of error in each approximation. Which equation has less error?

b. Same as part a but use Heun's Method with h=.1 to approximate y(x) for x=1.1,1.2,...,1.9,2 for the differential equations

$$y'(x)=2x, \ y(1)=1 \quad \text{and} \quad y'(x)=2y/x, \ y(1)=1$$

which both have solutions $y(x)=x^2$.

15. a. Run programs EULER and HEUN each five times for the differential equation $y'(x)=1+y$, $y(0)=0$ on the interval [0,.5] with h=.01,.02,.025,.05,.1

b. The solution to the differential equation of part a is $y(x)=e^x-1$. Determine the exact value for y(.1), y(.2), y(.3), y(.4), y(.5).

c. Using the results for Euler's Method complete a table similar to the one below for x=.1,.2,.3,.4,.5

x=

h	approximation y(x)	error
0.01		
0.02		
0.025		
0.05		
0.1		

d. Same as part c except use the results from Heun's Method.

e. For each table in part c plot the points on a separate graph using step size h on the horizontal axis and the error on the vertical axis. You should have five graphs, one for each of the values of x=.1,.2,.3,.4,.5 For each graph draw a smooth curve through the points

f. Same as part e except use the tables from part d.

g. What type of curves connected the points in part e? in part f?

h. What type of relationship (linear, quadratic, cubic, none, etc.) appears to exist between the error and step size for Euler's Method? for Heun's Method?

16. Modify program HEUN so that the process is repeated over again with a smaller step size until two successive approximations of the desired value are within a specified precision. You should include in the input the precision and for the output only the approximations of the desired value for the different step sizes. Be sure to include a procedure to terminate the program if it does not appear that successive approximations will ever get within the desired precision. As test data use $y'(x)=2xy$, $y(1)=1$ and approximate $y(1.5)$.

17. Run the program from exercise 16 using the differential equation in example 5. If your program did not conclude that $y(1)$ could not be approximated, make appropriate modification so that it will make this conclusion.

22.2 RUNGE-KUTTA METHODS

Although simple to use, the methods of Euler and Heun are not accurate enough in many applications unless the step size is very small. However, small step sizes require more iterations and, thus, increased computer (computation) time. Also, at some point the round-off and propagation error due to the smaller step size and the increased number of iterations will overcome any improvements in accuracy expected by using a smaller step size. To overcome these problems many other methods have been developed which are more efficient than either Euler's or Heun's. One of the most widely used of these is the fourth-order Runge-Kutta method.

Developed by Carl Runge and M. Wilhelm Kutta, two German applied mathematicians in the 1890's, this method, which is just one of many fourth-order methods, uses the iterative formula

$$y_{n+1} = y_n + (1/6)(k_1 + 2k_2 + 2k_3 + k_4)$$

with

$$k_1 = f(x_n, y_n) * h$$

$$k_2 = f(x_n + (1/2)h, y_n + (1/2)k_1) * h$$

$$k_3 = f(x_n + (1/2)h, y_n + (1/2)k_2) * h$$

$$k_4 = f(x_n + h, y_n + k_3) * h$$

where h is the step size, to approximate solutions to

$$y'(x) = f(x,y), \quad y(x_0) = y_0$$

Example 1: Use the fourth-order Runge-Kutta method to approximate $y(1.5)$ where y is the solution of $y'(x) = 2xy$, $y(1) = 1$ with $h = .1$

For y_1:

$$k_1 = f(x_0, y_0) * h$$
$$= (.1)(2(1)(1))$$
$$= .2$$

$$k_2 = f(x_0 + (1/2)h, y_0 + (1/2)k_1) * h$$
$$= (.1)(2(1.05)(1.1))$$
$$= .231$$

$$k_3 = f(x_0 + (1/2)h, y_0 + (1/2)k_2) * h$$
$$= (.1)(2(1.05)(1.1155))$$
$$= .234255$$

$$k_4 = f(x_0 + h, y_0 + k_3) * h$$
$$= (.1)(2(1.1)(1.234255))$$
$$= .2715361$$

Thus,

$$y_1 = y_0 + (1/6)(k_1 + 2k_2 + 2k_3 + k_4)$$
$$= 1 + (1/6)(.2 + 2(.231) + 2(.234255) + .2715361)$$
$$= 1.23367435$$

Similarly, we will find y, y_3, y_4, and y_5 as summarized below.

For y_2:

$k_1 = .271408357$
$k_2 = .3149570616$
$k_3 = .3199651626$
$k_4 = .372873483$

$y_2 = 1.552695398$

For y_3:

$k_1 = .3726468955$
$k_2 = .4347547115$
$k_3 = .4425181884$
$k_4 = .5187555325$

$y_3 = 1.993686769$

For y_4:

$k_1 = .51835856$
$k_2 = .6082738333$
$k_3 = .6204123953$
$k_4 = .7319477661$

$y_4 = 2.611633233$

For y_5:

$k_1 = .7312573053$
$k_2 = .8634059469$
$k_3 = .8825674999$
$k_4 = 1.04826022$

$y_5 = 3.490210636$

Thus, $y(1.5)$ is approximately 3.490210636

From section 22.1 we know that for this equation the exact answer is 3.490342957. The percent of error is .0038%, compared to 16.117% for Euler's Method and 1.129% for Heun's Method using the same step size.

Programming Problem: Approximate the numerical solution to $y'(x)=f(x,y)$, $y(x_0)=y_0$, at a specific value for x using the fourth-order Runge-Kutta method.

Output: A table listing the values of x and the corresponding approximations for $y(x)$.

Input: The function $f(x,y)$, the initial conditions, the value at which we want to approximate the solution, and the step size.

Strategy: Although more complicated, the iterative formula for the fourth-order Runge-Kutta method can be evaluated using a FOR-NEXT loop similar to the programs in section 22.1 Also, it will be necessary to continually change the value of y used in evaluating $f(x,y)$ to compute k_1 through k_4

Program:

```
100 REM ** RUNGE-KUTTA **
110 REM ** FUNCTION **
120 DEF FNF(X)=.....
130 REM ** INPUT **
140 PRINT "ENTER THE INITIAL CONDITIONS."
150 PRINT "X VALUE THEN Y VALUE."
160 INPUT X1,Y
170 PRINT
180 PRINT "ENTER THE VALUE OF X AT WHICH"
190 PRINT "THE SOLUTION IS TO BE APPROXIMATED."
200 INPUT XN
210 PRINT
220 PRINT "ENTER THE STEP SIZE."
230 INPUT H
240 PRINT
250 REM ** HEADINGS **
260 PRINT "X-VALUE","APPROXIMATION OF Y"
270 PRINT X1,Y
280 REM ** MAIN PROGRAM **
290 REM ** DETERMINING NUMBER OF ITERATIONS **
300 LET N=ABS(INT((XN-X1)/H+.5))
310 FOR I=1 TO N
320 LET X=X1+(I-1)*H
330 LET Y1=Y
340 LET K1=H*FNF(X)
350 LET Y=Y1+.5*K1
360 LET K2=H*FNF(X+.5*H)
370 LET Y=Y1+.5*K2
380 LET K3=H*FNF(X+.5*H)
390 LET Y=Y1+K3
400 LET K4=H*FNF(X+H)
```

```
410 LET Y=Y1+(1/6)*(K1+2*K2+2*K3+K4)
420 PRINT X+H,Y
430 NEXT I
440 END
```

Test Data: Approximate y(1.5) for y'(x)=2xy, y(1)=1 with h=.1

```
120 DEF FNF(X)=2*X*Y
```

RUN
ENTER THE INITIAL CONDITIONS.
X VALUE THEN Y VALUE.
?1,1

ENTER THE VALUE OF X AT WHICH
THE SOLUTION IS TO BE APPROXIMATED.
?1.5

ENTER THE STEP SIZE.
?.1

X-VALUE	APPROXIMATION OF Y
1	1
1.1	1.23367435
1.2	1.5526954
1.3	1.99368677
1.4	2.61163323
1.5	3.49021064

Example 2: The differential equation

$$y'(x)=(xy+y^2)/x^2 \quad , \quad y(1)=1$$

has a solution $y(x)=x/(1-\ln\ x)$. Use program RUNGE-KUTTA to approximate solutions of this differential equation for x=1.1, 1.2, 1.3, 1.4, and 1.5 Then determine the exact value for these values of x and determine the percent of error.

```
120 DEF FNF(X)=(X*Y+Y*Y)/(X*X)

RUN
ENTER THE INITIAL CONDITIONS.
X VALUE THEN Y VALUE.
?1,1

ENTER THE VALUE OF X AT WHICH
THE SOLUTION IS TO BE APPROXIMATED.
?1.5

ENTER THE STEP SIZE.
?.1

X-VALUE             APPROXIMATION OF Y
1                   1
1.1                 1.21587959
1.2                 1.46755338
1.3                 1.762358
1.4                 2.10988583
1.5                 2.52290466
```

x	y-Program	y-Exact	% Error
1.1	1.21587959	1.215886346	.00056%
1.2	1.46755338	1.467569568	.0011%
1.3	1.762358	1.7623875	.0017%
1.4	2.10988583	2.10993432	.0023%
1.5	2.52290466	2.522980603	.0030%

The extremely small percent of error indicates the reason for the popularity of this method.

Example 3:   In the 1840's the Belgian biologist-mathematician P.F. Verhulst developed the differential equation

$$dP/dt = AP-BP^2 \text{ , A and B constants,}$$

for determining population sizes. Referred to as the logistic equation, it has proven to be rather accurate in predicting patterns of growth for bacteria, protoza, and fruit flies in certain situations. For this equation A is the average birth rate, and B is a constant related to the average death rate of the population. Using this differential equation and program RUNGE-KUTTA, approximate the population at 250 hour intervals for a population in which A=.006, B=1.14×10^{-9}, and P(0)=2,500,000 for the next 5000 hours.

```
120 DEF FNF(X)=.006Y-1.14E-9*Y*Y

RUN
ENTER THE INITIAL CONDITIONS.
X VALUE THEN Y VALUE.
?0,2500000

ENTER THE VALUE OF X AT WHICH
THE SOLUTION IS TO BE APPROXIMATED.
?5000

ENTER THE STEP SIZE.
?250
```

X-VALUE	APPROXIMATION OF Y
0	2500000
250	4215863.28
500	4960766.73
750	5179416.6
1000	5240175.69
1250	5256867.22
1500	5261437.3
1750	5262687.39
2000	5263029.24
2250	5263122.72
2500	5263148.28
2750	5263155.27
3000	5263157.18
3250	5263157.7
3500	5263157.85
3750	5263157.88
4000	5263157.9
4250	5263157.9
4500	5263157.9
4750	5263157.9
5000	5263157.9

Notice that from about the 3000th hour on the population remains at about 5,263,157. This is the equilibrium point for this population where births and deaths will equal each other. For the logistic equation the value of the equilibrium point is given by A/B which for this example is A/B = .006/1.14x10^{-9} = 5,263,157.895

EXERCISE SET 22.2

In exercises 1-5 use program RUNGE-KUTTA to approximate the solution of the given differential equation at the indicated value of x using the given h.

1. $y'(x) = \dfrac{ye^{xy} + 4y^3}{2y - 12xy^2 + xe^{xy}}$, $y(0) = 2$, $h = .1$, $x = 1$

2. $y'(x) = \dfrac{-y-x}{x+2y}$, $y(2) = 3$, $h = .1$, $x = 3$

3. $y'(x) = \dfrac{8xy - 2x}{-4 - x^2}$, $y(0) = -1$, $h = .05$, $x = .5$

4. $y'(x) = \dfrac{2 \sin(y - \pi x)}{\pi \sin(2y - x)}$, $y(3) = 1$, $h = .05$, $x = 3.5$

5. $y'(x) = \dfrac{-2y + e^{-x}}{y}$, $y(1) = 1$, $h = .025$, $x = 2$

*
6. In example 3 the population grew from 2,500,000 to about 5,263,157 which was the equilibrium point. Investigate what would happen to the population if the initial population was 7,000,000 instead of 2,500,000 by running program RUNGE-KUTTA. Does the population reach an equilibrium point? If so, what is it?

7. A dog sees his owner jogging at a constant rate of 6 mph along a straight road and runs toward him at a constant rate of 10 mph. If the dog is originally 1/4 mile from his owner and always runs directly toward him, use the steps below to determine how long it will take for the dog to catch up to his owner. <u>Hint</u>: The path of the dog is determined by the equation $y'(x) = (1/2)\{(x/.25)^{6/10} - (x/.25)^{-6/10}\}$ with initial conditions $y = 0$ when $x = .25$

    a.   Use program RUNGE-KUTTA to approximate y when $x = 0$. What problem do you encounter?

b. To avoid this problem approximate the solution at x=.005 instead of x=0 using program RUNGE-KUTTA.

c. Find the actual solution to this problem at x=.005. What is the percent of error?

d. Why is the accuracy poor in this problem? <u>Hint</u>: See chapter 13

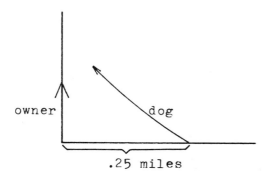

8. Suppose the dog and the owner in exercise 7 were both travelling at a speed of 7 mph. The equation of motion would then be $y'(x)=(1/2)(x/.25)-(x/.25)^{-1}$. Run program RUNGE-KUTTA for this equation to approximate the solution at x=.005   Should the dog ever reach his owner? Does the output support your conclusion about the dog reaching his owner?

**
9. Assuming only the force of gravity a body falling toward the earth satisfies

$$\frac{dy}{dt} = -\sqrt{2gR^2} * \frac{\sqrt{A-y}}{\sqrt{Ay}}$$

where g=32/5280, R=4000, and A is the initial distance from the center of the earth.

If the moon is about 240,000 miles from the center of the earth, how long would it take for the moon to fall to the earth? <u>Hint</u>: Use program RUNGE- KUTTA for different values of t until you have y=4000, as initial conditions use (0,239999).

10. Benjamin Gomperta (1779-1865), an English mathematician, modified the logistic equation to

$$\frac{dP}{dt} = P(A-B \ln P)$$

which today is used in actuarial predictions, population changes, and in forecasting revenues for commercial sales. Run program RUNGE-KUTTA and estimate the population after 5000 hours for a population which has 2,500,000 inhabitants at t=0, A=.005, and B=.00032258. Use h=250.

***
11. Modify program RUNGE-KUTTA so that the process is repeated over again with a smaller step size until two successive approximations of the desired value are within a specified precision.   You should include in the input the precision and for the output only the approximations of the desired value for the different step sizes. Be sure to include a procedure to terminate the program if it does not appear that successive approximations will ever get within the desired precision. As test data use y'(x)=2xy, y(1)=1 and approximate y(1.5).

12. Finding numerical solutions of a first order differential equation is similar to finding numerical solutions to definite integrals. In fact, if the function f(x,y) is a function of x only, then approximating y(a) for y'(x)=f(x), y(x0)=y0 is equivalent to approximating the integral

$$\int_{x0}^{a} f(x) \ dx \qquad since$$

$$y'(x)=f(x)$$

$$dy=f(x) \ dx$$

$$\int_{y0}^{y(a)} dy = \int_{x0}^{a} f(x) \ dx$$

$$y(a) - y_0 = \int_{x_0}^{a} f(x) \ dx$$

$$y(a) = y_0 + \int_{x_0}^{a} f(x) \ dx.$$

a.   Show that if $f(x,y)$ is a function of x only, then Heun's method for approximating y_{n+1} on the interval $[x_n, x_{n+1}]$ reduces to the trapezoidal rule for approximating definite integrals.

b.   Same as part a except show that the Runge-Kutta method reduces to Simpson's Rule.

c.   If $f(x,y)$ is a function of x only, then to which method for approximating definite integrals does Euler's method reduce?

CHALLENGE ACTIVITY

The methods of this unit can be extended to solve a system of first order differential equations, and, thus, differential equations of any order since any order differential equation can be expressed as a system of first order equations. A system of first order differential equations is given by

$$y'_1(x) = f_1(x, y_1, \ldots, y_n)$$

$$y'_2(x) = f_2(x, y_1, \ldots, y_n)$$

$$\vdots$$

$$y'_n(x) = f_n(x, y_1, \ldots, y_n)$$

To reduce the n-th order equation $y^{(n)}(x) = f(x, y(x), y'(x), \ldots y^{(n-1)}(x))$ to a first order system, let $y_1(x) = y(x)$ and

$$y'_1(x) = y_2(x)$$

$$y'_2(x) = y_3(x)$$

$$\vdots$$

$$y'_{n-1}(x) = y_n(x)$$

$$y'_n(x) = f(x, y_1(x), y_2(x), \ldots, y_n(x))$$

To illustrate, consider the differential equation $y''(x) = 4e^x + xy' + y^3$, $y'(x_0) = A$, $y(x_0) = B$. Let $y_1(x) = y(x)$. Then

$$y'_1(x) = y_2(x)$$

$$y'_2(x) = y''_1(x) = y''(x) = 4e^x + xy_2 + y_1^3$$

with $y_1(x_0) = B$ and $y_2(x_0) = A$.

Now the extension of the fourth-order Runge-Kutta method to the initial value problem $x'(t)=f(t,x,y)$, $y'(t)=y(t,x,y)$ with $x(t_0)=x_0$ and $y(t_0)=y_0$ is

$$x_{n+1}=x_n+(1/6)(k_1+2k_2+2k_3+k_4)$$

$$y_{n+1}=y_n+(1/6)(L_1+2L_2+2L_3+L_4)$$

where

$$k_1=f(t_n,x_n,y_n)*h$$

$$k_2=f(t_n+(1/2)h,x_n+(1/2)k_1,y_n+(1/2)L_1)*h$$

$$k_3=f(t_n+(1/2)h,x_n+(1/2)k_2,y_n+(1/2)L_2)*h$$

$$k_4=f(t_n+h,x_n+k_3,y_n+L_3)*h$$

$$L_1=g(t_n,x_n,y_n)*h$$

$$L_2=g(t_n+(1/2)h,x_n+(1/2)k_1,y_n+(1/2)L_1)*h$$

$$L_3=g(t_n+(1/2)h,x_n+(1/2)k_2,y_n+(1/2)L_2)*h$$

$$L_4=g(t_n+h,x_n+k_3,y_n+L_3)*h$$

and h is the step size.

Write a program which will approximate the solution to the first order system $x'(t)=f(t,x,y)$, $y'(t)=g(t,x,y)$, $x(t_0)=x_0$, and $y(t_0)=y_0$ at $t=a$ using the extension of the fourth order Runge-Kutta method given above. As test data convert the differential equation $y''-2y'+y=xe^x+4$, $y(0)=1$, $y'(0)=1$ to a first order system and approximate $y(1)$. Use $h=.1$ The analytic solution is $y=4xe^x-3e^x+(1/6)x^3e^x+4$.

APPENDIX A
A Brief Summary Of BASIC

This appendix summarizes the features of BASIC utilized in this book. Although only those statements and functions available in minimal BASIC have been used, it is possible that not all will work as described on all computer systems. When difficulties occur refer to the manual for the particular system being used.

SECTION I: NUMERICAL AND RELATIONAL OPERATORS

OPERATOR	SYMBOL	EXAMPLE
Addition	+	A+B
Subtraction	−	B−D
Multiplication	*	F1*F2
Division	/	R/T
Exponentiation	^	B^2
Equal To	=	A=B
Greater Than	>	A>B
Less Than	<	A<B
Greater Than Or Equal To	>=	A>=B
Less Than Or Equal To	<=	A<=B
Not Equal To	<>	A<>B

SECTION II: STATEMENTS

DEF FNG(X)=.....   Allows you to define your own functions. Here the function name is G and the argument is X.

DIM A( ),B$( )   Allocates storage in the computer's
                        memory for numerical or string arrays.
                        An array is similar to a matrix with
                        specific elements referred to by using
                        the array name in combination with
                        appropriate subscripts. One or more
                        integers are placed in the parentheses
                        to dimension the array.

END   Causes the program to halt execution.

FOR I=A TO B STEP C   Starts a loop that repeats a set of
                        instructions with the first value A
                        for I until the control variable I is
                        incremented to a value larger than B.
                        The increment is C which may be
                        negative. If   C   is   omitted,   the
                        increment   will   be   1.   Each   FOR
                        statement must be combined with
                        a NEXT statement.

GO TO N   Transfers control of the program to
                        the statement with line number N.

IF (expression) THEN N Allows for conditional transfer of
                        control of the program. If the
                        expression after the IF is true,
                        control is passed to the statement
                        with line number N. Otherwise,
                        control passes to the next line
                        in the program.

INPUT A,B$ Used to enter numeric or string values
                        into the program from the keyboard.

LET A=expression   Assigns the value of the expression to
                        the variable A.

LIST   Displays the program currently in memory
                        by listing all the lines of the program
                        on the monitor or printer

APPENDIX A

NEXT I   Indicates the end of a loop created by
                                a FOR/NEXT statement.

ON (expression) GO TO   Transfers control of the program to one
    N1,N2,...,etc.   of the statements with line number N1,N2
                                etc. The exact line to which control is
                                passed is determined by the value of the
                                expression following the ON.

PRINT   Outputs characters, calculated values,
                                and values of variables to the monitor
                                or printer. See SECTION IV for more
                                details.

RUN   Causes the computer to execute(run) the
                                the program currently in memory.

SECTION III: BUILT-IN FUNCTIONS

    In addition to the above statements, most computers provide a
certain number of "built-in" functions to aid in writing programs.
In this book the following functions will be used.

ABS(X)   Absolute value of X.

ATN(X)   Arctangent, answer is in radians.

COS(X)   Cosine of X; X expressed in radians.

EXP(X)   Value of e raised to the X power.

INT(X)   Greatest integer less than or
                                equal to X.

LOG(X)   Natural logarithm.

SIN(X)   Sine of X; X expressed in radians.

SGN(X)   Signum function; +1 if X is positive,
                          0 if X is 0, and -1 if X is negative.

SQR(X)   Square root of X.

TAB(N)   Used with the PRINT statement to
                          control output. See SECTION III.

TAN(X)   Tangent of X; X expressed in radians.

SECTION IV: FORMATTING FEATURES--PRINT AND TAB

The PRINT statement and TAB function are called formatting
features since they allow the programmer to control the appearance
of the output. For use with the PRINT statement we have a number
of options, each instructing the computer to arrange our output in
a specific manner.

A.   Quotation Marks

At times we want the computer to print out a message. To
alert the computer to this fact we use the PRINT statement and
enclose the message in quotation marks. Notice in the example
below that the quotation marks are not part of the output.

        100 PRINT "THIS IS AN EXAMPLE OF A MESSAGE."
        110 END

        RUN
        THIS IS AN EXAMPLE OF A MESSAGE.

The computer will print exactly what is between the quotation
marks, including any symbols, numbers, or blank spaces.

B.   Lack Of Punctuation

To instruct the computer to print the value of a variable, X,
we simply write PRINT X . With no punctuation to guide it, the
computer will print the value of X in a predetermined location on
the monitor or paper.

```
100 LET A=10
110 PRINT A
120 END

RUN
10
```

However, the machine will also calculate a value and print the final result from the PRINT statement. For example,

```
100 LET A=10
110 LET B=2
120 PRINT A*B
130 END

RUN
20
```

When dealing with more complex expressions the normal algebraic order of operations is followed. Finally, the PRINT statement alone will cause the computer to skip a line each time it is encountered.

```
100 LET A=10
110 PRINT A
120 PRINT
130 PRINT A
140 END

RUN
10

10
```

C.   Commas

Commas are used as separators to instruct the computer to print more than one quantity on a line. When the computer encounters a comma within a PRINT statement, it will divide the output line into a fixed number of print zones each of which contains a fixed number of print positions. Successive output values are place in successive print zones justified to the left.

```
100 PRINT "12345678901234567890"
110 LET A=10
120 LET B=20
130 PRINT A,B
140 END

RUN
12345678901234567890
10              20
```

If more data items are to be printed than the number of print zones on a line the output will be continued on the next line.

D. Semicolons

The use of semicolons packs the output more closely together as illustrated in the following example.

```
100 LET A=10
110 LET B=20
120 PRINT A;B
130 END

RUN
1020
```

Computers vary as to the exact location of the second value. Some computer will leave up to as many as three spaces between the values.

APPENDIX A

E.   TAB Function

The TAB function is used to set up positions similar to a
typewriter.   When the computer encounters the TAB(N) instruction
in a PRINT statement it will begin printing in column N.

```
100  PRINT "123456789012345678901234567890"
110  LET A=10
120  PRINT
130  PRINT TAB(10);A;TAB(20);A
140  END

RUN
123456789012345678901234567890
         10        10
```

The N may be a variable or a positive integer.

F.   Combinations

Commas, semicolons, quotation marks, and TAB may be combined
in a PRINT statement. Each punctuation mark will exert its
influence as explained above.

```
100  LET A=10
110  LET B=20
120  PRINT "THE VALUE OF A IS ";
130  PRINT A
140  PRINT TAB(10);"THE VALUE OF B IS ";B
150  END

RUN
THE VALUE OF A IS 10
         THE VALUE OF B IS 20
```

A comma or semicolon encountered at the end of a PRINT statement
will suppress line feed as illustrated in line 120.

APPENDIX B
READ And DATA Statements

The READ and DATA statements provide an alternate means for entering data into a BASIC program when working with a non-interactive computer system or when large amounts of data are needed to be entered. As with the FOR/NEXT statements every READ statement must have a DATA statement to go with if. However, more than one DATA statement may be accessed by a single READ statement. The general format of the READ and DATA statements is:

```
line# READ variable list
line# DATA data values
```

A valid example of the READ and DATA statements are given below.

```
100 READ X,Y,A$
       .
       .
       .
       .
400 DATA 10,20,BILL
410 DATA 15,40,"JAMES"
```

Using the READ statement for entering data works slightly different than the INPUT statement. When encountered during the execution of the program the READ statement instructs the computer to search the program for a DATA statement. Then the data values are assigned consecutively to the variables in the READ statement. Thus, for the example above, when line 100 is encountered the computer will assing 10 to the variable X, 20 to the variable Y, and the character string BILL to the variable A$.

The simplified description of the working of the READ/DATA statements presented above is accurate as long as there is only one READ statement which is executed one time. However, using multiple READ statements or repeating the same READ statement within a loop requires an understanding of how the computer processes DATA statements. DATA statements are non-executable statements and may be placed anywhere in the program. When the program is run, the computer will take all of the data items in all DATA statements and form one combined data list, ordering the data starting with the lowest line numbered DATA statement to the highest. Thus, for example, the following program segments produce the same data list.

```
100 DATA 10,20,BILL          100 DATA 10,20,BILL,15,40,"JAMES"
110 DATA 15,40,"JAMES"
```

Now when READ statements are encountered data is assigned to the variables in the order in which it occurs in the data list. Once assigned, however, the data values are deleted from the data list. Thus, returning to our original example, if the READ statement of line 100 is executed a second time 15 will be assigned to the variable X, 40 to the variable Y, and the character string JAMES to the variable A$. Notice that the inclusion of the quotation marks about character strings in DATA statements is optional.

Several common errors made when using READ/DATA statements are mixing data type, trying to assign a character string to a numeric variable, and not providing a sufficient number of data items for all the variables listed in the READ statements. Examples illustrating these errors as well as correct usage of the READ/DATA statements are presented below.

Example 1:

```
100 REM ** READ IN A LOOP **
110 FOR I=1 TO 4
120 READ X,Y,A$
130 PRINT X,Y,A$
140 NEXT I
150 DATA 45,27,BILL,12,13,FRED
160 DATA 83,72,JAMES,1,2,DAVE
170 END
```

```
RUN
45              27              BILL
12              13              FRED
83              72              JAMES
1               2               DAVE
```

Example 2:

```
100 REM ** SEVERAL READ STATEMENTS **
110 LET N=0
120 READ X
130 READ Y
140 PRINT X,Y,X+Y
150 DATA 34,56,72,89
160 LET N=N+1
170 IF (N<>2) THEN 120
180 END

RUN
34              56              90
72              89              161
```

Example 3:

```
100 REM ** MIXED DATA TYPES **
110 READ X,A$
120 PRINT X,A$
130 DATA BILL,45
140 END

RUN
?SYNTAX ERROR IN 130
```

Example 4:

```
100 REM ** INSUFFICIENT DATA **
110 READ X
120 PRINT X,X*X
130 GO TO 110
140 DATA 1,2,3,4,5
150 END

RUN
1               1
2               4
3               9
4               16
5               25

?OUT OF DATA ERROR IN 110
```

APPENDIX C
Derived Mathematical Functions

The following mathematical functions are defined using the DEF statement.

Trigonometric Functions

COSECANT(X)=1/SIN(X)

COTANGENT(X)=1/TAN(X)

SECANT(X)=1/COS(X)

Inverse Trigonometric Functions

ARCCOSINE(X)=-ATN(X/SQR(1-X*X))+1.5707633

ARCCOTANGENT(X)=1.5707633+ATN(-X)

ARCCOSECANT(X)=ATN(1/SQR(X*X-1))+(SGN(X)-1)*1.5707633

ARCSECANT(X)=ATN(SQR(X*X-1))+(SGN(X)-1)*1.5707633

ARCSINE(X)=ATN(X/SQR(1-X*X))

Logarithmic Functions

LOGA(X)=LOG(X)/LOG(A)

LOG10(X)=LOG(X)/2.30258509

Hyperbolic Functions

HYPERBOLIC COSECANT(X)=2/(EXP(X)-EXP(-X))

HYPERBOLIC COSINE(X)=(EXP(X)+EXP(-X))/2

HYPERBOLIC COTANGENT(X)=2*EXP(-X)/(EXP(X)-EXP(-X))+1

HYPERBOLIC SECANT(X)=2/(EXP(X)+EXP(-X))

HYPERBOLIC SINE(X)=(EXP(X)-EXP(-X))/2

HYPERBOLIC TANGENT(X)=-2*EXP(-X)/(EXP(X)+EXP(-X))+1

Inverse Hyperbolic Functions

INVERSE HYPERBOLIC COSECANT(X)=LOG((SGN(X)*SQR(X*X+1)+1)/X)

INVERSE HYPERBOLIC COSINE(X)=LOG(X+SQR(X*X-1))

INVERSE HYPERBOLIC COTANGENT(X)=LOG((X+1)/(X-1))/2

INVERSE HYPERBOLIC SECANT(X)=LOG((SQR(1-X*X)+1)/X)

INVERSE HYPERBOLIC SINE(X)=LOG(X+SQR(X*X+1))

INVERSE HYPERBOLIC TANGENT(X)=LOG((1+X)/(1-X))/2

APPENDIX D
Arrays

An array in BASIC is used to refer to a collection of variables conveniently. In this book we make use of one dimensional arrays, which can be thought of as a list, and two dimensional arrays, which can be thought of as a table or matrix. The individual elements of an array, called subscripted variables, act in the same manner as ordinary variables and thus can contain either numeric or character string values. However, all elements of the same array must be of the same type, numeric or character string.

To refer to a specific element in an array we do so by giving the array name, followed by the value of the subscript, enclosed in parentheses. For example, if A is a one dimensional array containing seven elements then the seven variable names associated with the array would be:

A(0), A(1), A(2), A(3), A(4), A(5), A(6) .

Notice that the first subscript is 0. However, for many programming problems it is more convenient to ignore the zero subscript and think of the array as starting with a subscript of 1. Elements in two dimensional arrays are referenced in a similar manner except that two subscripts seperated by commas are enclosed within the parentheses. Thus, B(4,7) is the variable name corresponding to the element in the fourth row, seventh column of the array named B.

Most implementations of BASIC will automatically assign 11 elements to a one dimensional array, subscripts 0 through 10, and 121 elements to a two dimensional array, subscripts (0,0) through (10,10). However, if arrays of larger size are needed, the largest subscript must be provided to the computer. This is accomplished through the use of the DIMension statement which has the general format

line# DIM array name(largest subscript),...
,array name(largest subscript)

For example, the statement

100 DIM A(40),B(10,11),C$(15)

reserves 41 variables for array A, 132 variables for array B, and 16 string variables for array C$. Although not always necessary it is good programming practice to dimension all arrays.

It is not necessary for the subscript to be a constant. In fact, variables, expressions, and functions can all be used as long as the value of the variable, expression, or function is an integer or a value which can be truncated to an integer which results in a valid subscript value. This ability to use variable expressions to determine the subscript value is why arrays are so useful in BASIC.

Example 1: Read 5 values into an array and print the values in reverse order.

```
100 DIM A(5)
110 REM ** READING VALUES **
120 FOR I=1 TO 5
130 PRINT "ENTER THE ";I;"-TH VALUE."
140 INPUT A(I)
150 NEXT I
160 REM ** OUTPUT **
170 PRINT
180 PRINT "THE ARRAY IN REVERSE ORDER IS:"
190 FOR I=5 TO 1 STEP -1
200 PRINT A(I);" ";
210 NEXT I
220 END

RUN
```

ENTER THE 1-TH VALUE.
?23
ENTER THE 2-TH VALUE.
?78
ENTER THE 3-TH VALUE.
?56
ENTER THE 4-TH VALUE.
?91
ENTER THE 5-TH VALUE.
?100

THE ARRAY IN REVERSE ORDER IS:
100 91 56 78 23

Example 2: Although not a problem involving calculus, an interesting problem involving two dimensional arrays is that of deciding if a two dimensional array with the same number of rows as columns is a magic square. A magic square has the same sum for each row, column, and both diagonals. The following program allows the user to enter any array up to an 8X8 and then determines if it is a magic square and prints out the array along with an appropriate message.

```
100 REM ** MAGIC SQUARE **
110 DIM A(8,8)
120 REM ** INPUT **
130 PRINT "ENTER THE NUMBER OF ROWS."
140 INPUT N
150 PRINT
160 REM ** ENTERING THE ARRAY **
170 FOR I=1 TO N
180 FOR J=1 TO N
190 PRINT "ENTER THE ELEMENT WITH SUBSCRIPT ";I;",";J
200 INPUT A(I,J)
210 NEXT J
220 NEXT I
230 PRINT
240 PRINT "THE ARRAY IS:"
250 PRINT
260 FOR I=1 TO N
270 FOR J=1 TO N
280 PRINT A(I,J);" ";
290 NEXT J
300 PRINT
310 NEXT I
320 PRINT
330 PRINT
340 REM ** TESTING **
350 REM ** MAIN DIAGONAL **
360 LET S=0
370 FOR I=1 TO N
380 LET S=S+A(I,I)
390 NEXT I
400 REM ** OTHER DIAGONAL **
410 LET S1=0
420 FOR I=0 TO N-1
430 LET S1=S1+A(N-I,1+I)
440 NEXT I
450 IF S<>S1 THEN 590
460 REM ** ROWS AND COLUMNS **
470 FOR I=1 TO N
480 LET S1=0
490 LET S2=0
500 FOR J=1 TO N
510 LET S1=S1+A(I,J)
520 LET S2=S2+A(J,I)
530 NEXT J
540 IF S<>S1 THEN 590
550 IF S<>S1 THEN 590
560 NEXT I
570 PRINT "THE ARRAY IS A MAGIC SQUARE."
580 GO TO 600
590 PRINT "THE ARRAY IS NOT A MAGIC SQUARE."
```

```
600 END

RUN
```

ENTER THE NUMBER OF ROWS.
?5

ENTER THE ELEMENT WITH SUBSCRIPT 1,1
?11
ENTER THE ELEMENT WITH SUBSCRIPT 1,2
?10
ENTER THE ELEMENT WITH SUBSCRIPT 1,3
?4
ENTER THE ELEMENT WITH SUBSCRIPT 1,4
?23
ENTER THE ELEMENT WITH SUBSCRIPT 1,5
?17
ENTER THE ELEMENT WITH SUBSCRIPT 2,1
?18
ENTER THE ELEMENT WITH SUBSCRIPT 2,2
?12
ENTER THE ELEMENT WITH SUBSCRIPT 2,3
?6
ENTER THE ELEMENT WITH SUBSCRIPT 2,4
?5
ENTER THE ELEMENT WITH SUBSCRIPT 2,5
?24
ENTER THE ELEMENT WITH SUBSCRIPT 3,1
?25
ENTER THE ELEMENT WITH SUBSCRIPT 3,2
?19
ENTER THE ELEMENT WITH SUBSCRIPT 3,3
?13
ENTER THE ELEMENT WITH SUBSCRIPT 3,4
?7
ENTER THE ELEMENT WITH SUBSCRIPT 3,5
?1
ENTER THE ELEMENT WITH SUBSCRIPT 4,1
?2
ENTER THE ELEMENT WITH SUBSCRIPT 4,2
?21
ENTER THE ELEMENT WITH SUBSCRIPT 4,3
?20
ENTER THE ELEMENT WITH SUBSCRIPT 4,4
?14
ENTER THE ELEMENT WITH SUBSCRIPT 4,5
?8
ENTER THE ELEMENT WITH SUBSCRIPT 5,1
?9
ENTER THE ELEMENT WITH SUBSCRIPT 5,2
?3

```
ENTER THE ELEMENT WITH SUBSCRIPT 5,3
?22
ENTER THE ELEMENT WITH SUBSCRIPT 5,4
?16
ENTER THE ELEMENT WITH SUBSCRIPT 5,5
?15

THE ARRAY IS:

11   10   4   23   17
18   12   6    5   24
25   19  13    7    1
2    21  20   14    8
9     3  22   16   15


THE ARRAY IS A MAGIC SQUARE.
```

APPENDIX E
PROGRAM FOR GRAPHING A FUNCTION
ON AN APPLE COMPUTER

Note: Apple is a registered trademark of APPLE COMPUTER INC.

This program is written so that the function to be plotted is entered by a DEF statement in line 130. Additional input required is the interval on the x-axis over which the graph is to be plotted.

The program itself is designed to compute all functional values over the entered interval so as to determine the range of values on the y-axis. Although this relieves the user of having to know the maximum and minimum functional values over the interval, it will result in the scales on the coordinate axes being different, causing a distortion in the graph. However, the alternative of having the same scale on both axes also has drawbacks. The main one being that for many functions only a small portion of the graph can be viewed at any one time and no "feel" of the total graph is obtained.

As with the other programs in this book, the user is encouraged to modify this program to suit individual needs.

```
100 REM ** FUNCTION GRAPHER **
110 TEXT
120 REM ** FUNCTION **
130 DEF FNF(X)=.....
140 REM ** INPUT **
150 PRINT "ENTER THE ENDPOINTS"
160 PRINT "OF THE INTERVAL ON THE"
170 PRINT "X-AXIS.   LEFT ENDPOINT"
180 PRINT "THEN RIGHT ENDPOINT, ALL"
190 PRINT "ON THE SAME LINE."
200 INPUT XL,XR
210 REM ** FINDING SCALE ON X-AXIS **
220 LET XS=(XR-XL)/279
230 REM ** FINDING FUNCTIONAL VALUES **
240 REM ** AND MAX AND MIN VALUES **
250 LET Y1=FNF(XL)
260 LET Y2=Y1
270 DIM Y(280)
280 LET Y(0)=Y1
290 FOR I=1 TO 279
300 LET Y(I)=FNF(XL+XS*I)
```

```
310 IF Y(I) > Y1 THEN Y1=Y(I)
320 IF Y(I) < Y2 THEN Y2=Y(I)
330 NEXT I
340 REM ** FINDING SCALE ON Y-AXIS **
350 LET YS=(Y1-Y2)/159
360 REM ** SETTING UP GRAPHICS PAGE **
370 HGR:HCOLOR=3
380 REM ** FINDING ORIGIN **
390 REM ** X-AXIS **
400 IF Y2 < 0 THEN 430
410 LET J=159+INT(Y2/YS)
420 GO TO 510
430 IF Y1 > 0 THEN 460
440 LET J=-INT(ABS(Y1)/YS)
450 GO TO 510
460 FOR J=0 TO 159
470 IF (Y1-J*YS) <= 0 THEN 500
480 NEXT J
490 GO TO 510
500 HPLOT 0,J TO 279,J
510 REM ** Y-AXIS **
520 IF XL > 0 THEN 590
530 IF XR < 0 THEN 590
540 FOR K=0 TO 279
550 IF (XL+K*XS) >= 0 THEN 580
560 NEXT K
570 GO TO 590
580 HPLOT K,0 TO K,159
590 REM ** PLOTTING GRAPH **
600 LET D=0
610 LET E=-SGN(Y(0))*INT(ABS(Y(0))/YS+.5)+J
620 FOR I=2 TO 278 STEP 2
630 REM ** FINDING COORDINATES **
640 LET R=-SGN(Y(I))*INT(ABS(Y(I))/YS+.5)+J
650 HPLOT D,E TO I,R
660 LET D=I
670 LET E=R
680 NEXT I
690 PRINT "X-AXIS:[";XL;",";XR;"] WITH SCALE ";XS
700 PRINT "Y-AXIS:[";Y2;",";Y1;"] WITH SCALE ";YS
710 END
```

Answers To Selected Exercises

Exercise Set 1.1

3.  ```
    100 PRINT "******************"
    110 PRINT "*                *"
    120 PRINT "* WILLIAM OBERLE *"
    130 PRINT "*                *"
    140 PRINT "******************"
    150 END
    ```

5.  ```
    100 PRINT 1/2
    110 PRINT 1/3,2/3
    120 PRINT 1/4,2/4,3/4
    130 END
    ```

The values are not exact and will cause loss of accuracy when used in computations.

Exercise Set 1.2

1. a. valid
    b. valid
    c. invalid, must start with a letter
    d. invalid, must start with a letter
    e. valid
    f. invalid, no special characters
    g. valid
    h. invalid, no special characters

3. a. $(N^3+3*N+1)/6$
    b. $((X+Y)/(Z+T))^4$
    c. $(A*X^3+B*X^2+C*X+D)/(E*X^2+F*X+G)$

7.
    ```
    100 PRINT "WHAT IS THE OBJECT?"
    110 INPUT A$
    120 PRINT "WHAT IS THE WEIGHT IN POUNDS?"
    130 INPUT P
    140 LET K=P*.454
    150 PRINT "THE WEIGHT OF ";A$
    160 PRINT "IS ";P;" POUNDS AND "
    170 PRINT K;" KILOGRAMS."
    180 END
    ```

9.
    ```
    100 PRINT "ENTER THE NUMBER OF MOLES, N."
    110 INPUT N
    120 PRINT "ENTER THE VALUE OF R, A, AND B."
    130 INPUT R,A,B
    140 PRINT "ENTER THE TEMPERATURE, T."
    ```

```
150 INPUT T
160 PRINT "ENTER THE VOLUME, V."
170 INPUT V
180 LET P=(N*R*T)/(V-N*B)-(A*N^2)/V^2
190 PRINT "THE PRESSURE FOR N=";N
200 PRINT ",R=";R;" ,T=";T;" ,V=";V
210 PRINT "A=";A;" , AND B=";B;" IS ";P
220 END
```

Exercise Set 2.1

1. a. valid
   b. invalid, need A$ not A
   c. invalid, 120.5 is not a valid line number
   d. valid
   e. valid
   f. invalid, only one line number in a GOTO statement
   g. invalid, B>7 is not a BASIC statement
   h. invalid, X*2 is not a relational expression

3. The numerical expression will be truncated or rounded depending on the computer being used.

7.
```
    IF A+2=1 THEN 100
    IF A+2=2 THEN 200
    IF A+2=3 THEN 300
```

9.
```
    100 PRINT "ENTER THE THREE SIDES OF THE TRIANGLE."
    110 INPUT A,B,C
    120 PRINT "FOR LINE SEGMENTS OF LENGTHS ";A;",",B;", AND ";C
    130 LET S=(A+B+C)/2
    140 LET A=S*(S-A)*(S-B)*(S-C)
    150 IF A<=0 THEN 180
    160 PRINT "A TRIANGLE OF AREA ";A^.5;" IS FORMED."
    170 GO TO 190
    180 PRINT "NO TRIANGLE IS FORMED."
    190 END
```

Exercise Set 2.2

1. Negative

3. One

9. (J-I)/K+1 truncated to the nearest integer

13. $D_N = F_{N+2}$

Exercise Set 3.1

1.   Any legal BASIC expression--numeric variable, constant, expression, or another function.

3.   -1,0, or 1

5.
```
100 PRINT "ENTER TWO SIDES AND THE INCLUDED ANGLE."
110 INPUT A,B,C
120 LET AREA=A*B*SIN(C/2)*COS(C/2)
130 PRINT "THE AREA OF THE TRIANGLE IS ";AREA
140 END
```

7.   Yes, given two angles of a triangle the third angle is known.

9.
```
100 PRINT "ENTER THE YEAR."
110 INPUT Y
120 LET A=Y-INT(Y/19)
130 LET B=Y-INT(Y/4)*4
140 LET C=Y-INT(Y/7)*7
150 LET D=(9*A+24)-INT((19*A+24)/30)
160 LET X=2*B+4*C+6*D+5
170 LET E=X-INT(X/7)*7
180 LET Z=22+D+E
190 IF Z<=31 THEN 230
200 PRINT "THE DATE OF EASTER SUNDAY IN THE YEAR ";Y
210 PRINT "IS APRIL ";31-Z;"."
220 GO TO 250
230 PRINT "THE DATE OF EASTER SUNDAY IN THE YEAR ";Y
240 PRINT "IS MARCH ";Z;"."
250 END
```

11.   Add the following lines:
```
124 LET N=2
125 IF (X/2)=INT(X-2) THEN 180
126 LET K=INT(SQR(X))
```
Change the following line:
```
130 FOR N=3 TO K STEP 2
```

Exercise Set 3.2

3.   In minimal BASIC the name of a user-defined function must be a single letter.

5. x^4

7.   DEF FNF(X)=X-SGN(X)*INT(ABS(X))

ANSWERS

9.

```
100 DEF FNX(T)=(3*T)/(1+T^3)
110 DEF FNY(T)=(3*T^2)/(1+T^3)
120 PRINT "T";TAB(10);"X";TAB(22);"Y"
130 FOR T=0 TO 10
140 PRINT T;TAB(10);FNX(T);TAB(22);FNY(T)
150 NEXT T
160 END
```

Exercise Set 4.1

1. Results will vary depending on the computer being used.

5. $f(x+h)-f(x)$ should never be zero. For small values of h, x and x+h will have the same computer value.

7. Results will vary depending on the computer being used. Try x=90 and y=45 degrees.

9. a. $1/11^7+1/11^6+1/11^5+1/11^4+1/11^3+1/11^2+1/11+1$
   b. $1+1/11+1/11^2+1/11^3+1/11^4+1/11^5+1/11^6+1/11^7$
   c. No, the sums are being added in reverse order.
   d. Program 9B. 9A requires 14 multiplications, 1 division, and 8 additions.
        9B requires 7 divisions and 8 additions.

Exercise Set 4.2

1. $x=.1, y=z=.00005$

5. a.
```
100 PRINT "ENTER THE COEFFICIENTS."
110 INPUT A,B,C
120 LET D=B*B-4*A*C
130 IF D<=0 THEN 170
140 PRINT "THE TWO REAL ROOTS OF THE EQUATION ARE:"
150 PRINT (-B+SQR(D))/(2*A);" AND ";(-B-SQR(D))/(2*A)
160 GO TO 210
170 IF D=0 THEN 200
180 PRINT "NO REAL ROOTS."
190 GO TO 210
200 PRINT "THE DOUBLE ROOT TO THE EQUATION IS ";-B/(2*A)
210 END
```

   b. $X=(-B-SQR(B^2-4*A*C))/(2*A)$
      $X=(-4*A*C)/(2*A*(B+SQR(D)))$

c. Modify line 150 to:
    150 PRINT (-4*A*C)/(2*A*(B+SQR(D)));" AND ";(-B-SQR(D))/(2*A)
d. X=(-B+SQR(B^2-4*A*C))/(2*A)
    X=(4*A*C)/(2*A*(-B+SQR(D)))

Exercise Set 5.1

1. Add the following lines:
    210 PRINT "THE TOTAL COST OF THE WHEAT IS"
    220 PRINT (S/550000)*2.5
    230 END

3.
    100 PRINT "ENTER A POSITIVE INTEGER."
    110 INPUT N
    120 IF (N/2)=INT(N/2) THEN 150
    130 LET N=3*N+1
    140 GO TO 160
    150 LET N=N/2
    160 PRINT N;" ";
    170 IF N<>1 THEN 120
    180 END

Exercise Set 6.1

1. Actual value: 22

3. Actual value: 1.2

5. Actual value: 1

7. Actual value: 0

9. Actual value: 5/3

11. Change line 230 to:
    230 LET X=A+(.5)^I

13. Change line 190 to:
    190 LET X=-(10^I)
The actual value of the limit is 2.

15. The value of the limit for any n and a is na.

Exercise Set 6.2

1. Removable, assign a value of 0

ANSWERS

3. Removable, assign a value of .5

5. Removable, assign a value of $-.91023923=-1/\ln(3)$

7. At x=3 a removable discontinuity, assign a value of .25
    At x=-1 a nonremovable discontinuity

9. At x=-3 a nonremovable discontinuity

Exercise Set 7.1

1. The actual values are:
    Example 1.   -.04
    Example 2.   .25
    Example 3.   -.5517767

3. Actual value: .2

5. Actual value: 108.074074074

7. Actual value: .011021668

9. Actual value: .020088976

11. b. No
    c. Yes

Exercise Set 7.2

1. Actual value: 12

3. Actual value: -2/27 or -.074074074

5. -3.9999797

7. Velocity: 67
    Acceleration: 32

Exercise Set 8.1

1. Actual values: at x=1, y'=7
                     x=-1, y'=7
                     x=3, y'=1327

3. Actual values: at x=0, y'=0
                     x=1, y'=-10.84
                     x=-.37, y'=-.9257314687

5.   Horizontal tangents will occur for x values of approximately 1, .2, and -.2

Exercise Set 8.2
1. f(-.12)=-.61618496   f'(-.12)=.974624
    f(.1)=-.50243   f'(.1)=.0871
    f(3)=-1.73   f'(3)=3.86

3. f(-.3)=2.932195   f'(-.3)=5.2275
    f(.1)=4.170235   f'(.1)=1.3275
    f(.6)=4.33336   f'(.6)=.300000001

5. Tangent Line: y=77.5372x-138.89168
    Normal Line: y=-.012897x+62.738572

7. a. 8207.36
    b. population: 2481290.11   growth rate:792830.976
    c. Yes, at a value of t about 26.7667406999

Exercise Set 9.1

1. 1.43287625

3. -1.24232973

5. One point of intersection (-7.73046003,-7.73046003)

7. -2.84548187

9. 2.14622498

11.   One disadvantage is that you must know an interval containing a root.

13.   No root exists

Exercise Set 9.2

1. Max:(4,636.8)   Min:(0,-2)

3. Max:(2,.263157895)   Min:(-2,-.15894737)

5. Maximum Height:1.71320804
    Minimum Height:-1.66375782

ANSWERS

Exercise Set 10.1

1. 3.14161393

3. .881382795

5. Actual Value: 77.25

7. Actual Value: -.13414323

9. .333365861

11. 87.1117964

13. .536224922

17. 5.09297862

Exercise Set 10.2

1. 81.5895353

3. 19.1569703

5. .650875183

7. 2.0569752

9. 10.5016997

11. 39.4784512

13. 66.3103388

15. 1516.923234

Exercise Set 11.1

1. Actual Value: .34657359

3. Actual Value: 553.481566

5. a. Let r=1
   b. .359057908
   c. .927581048

7. .166606995

Exercise Set 11.2

1. 9

3. 13

5. Actual Value: .076598722

7. Actual Value: 3.6340434

Exercise Set 12.1

1. -1.23372115

3. .925620367

7. -.00155216293

11. π if m=n, otherwise 0

13. a. 0
    b. 0

Exercise Set 12.2

5. a.   The   interpolating polynomial is only good for x between 1
and 3.1415 Maximum error at x=1.5 is 0.017099,   maximum   error   at
x=3 is 0.0058958
    b.   Effects of round-off error.

Exercise Set 13.1

1. Diverges

3. Not enough information with the upper limit 200.

5. .131811928

7. a.   Diverges
   b.   Pi
   c.   Diverges

9. 20000

11. Actual Value: .5

ANSWERS

Exercise Set 13.2

1. 1.24972217

3. Diverges

5. -1.08866264

7. Diverges

9. .40985853

11. Diverges

Exercise Set 14.1

1. Actual Value: .25

3. Actual Value: 0

5. Actual Value: 5

7. Actual Value: 0

9. Actual Value: 0

11. .641185744

13. 2

15. 3.14159266

17. -.349976485

Exercise Set 14.2

1. 5.00000001

3. Diverges

5. .421097686

7. 1.5

9. 1.99229477

11. -.189041926, N=10, Error=.0000888

13. -.249126806, N=126, Error=.0000984

15. Absolute Convergence

Exercise Set 15.1

1. Hyperbola, rotated 63.43 degrees with center (0,0), denominator of $x^2 = 1/9$, denominator of $y^2 = -1$

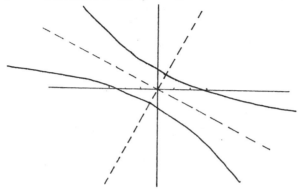

3. Ellipse, rotated 71.57 degrees with center (0,0), denominator of $x^2 = .323$, denominator of $y^2 = 10$

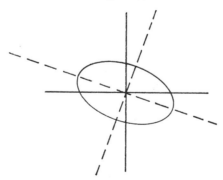

5. Ellipse, rotated 45 degrees with center (-1.41,0), denominator of $x^2 = 1.98$, denominator of $y^2 = .66$

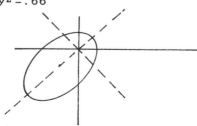

7. Hyperbola, rotated 76.72 degrees with center (.536,.33), denominator of $x^2 = .179$, denominator of $y^2 = -.179$

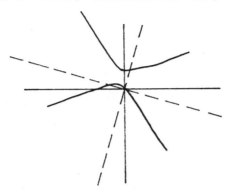

9. Hyperbola, rotated 78.69 degrees with center (5.1,5.1), denominator of $x^2 = 4$, denominator of $y^2 = -4$

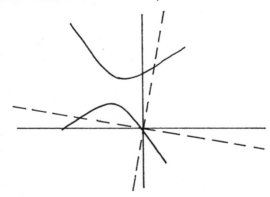

11. Ellipse, rotated 15.48 degrees with center (-2.1,.404), denominator of $x^2 = 11.512$, denominator of $y^2 = 2.45$

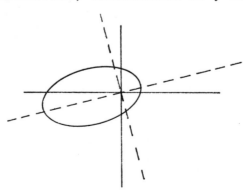

Exercise Set 16.1

1. 3.14160193

3. 7.0685775

5. .127820966

7. Center: $(\sqrt{2}/2, \sqrt{2}/2)$ Radius: 1

Exercise Set 17.1

1. Position: $2i+.693j+7.389k$
   Velocity: $1i+.5j+7.389k$
   Speed: 7.473
   Acceleration: $-.25j$
   Unit Tangent: $.134i+.067j+.989k$
   Unit Normal: $.009i-.998j+.066k$
   Binormal: $.991i-.134k$

3. Position: $.667i+.5j+3k$
   Velocity: $-.222i+.75j+3k$
   Speed: 3.1
   Acceleration: $.148i+.325j$
   Unit Tangent: $-.072i+.242j+.968k$
   Unit Normal: $.437i+.88j-.188k$
   Binormal: $-.897i+.409j-.169k$

5. Position: $-.301i+1.382j+k$
   Velocity: $-.841i+.54j+2k$
   Speed: 2.236
   Acceleration: $-1.382i-.302j+2k$
   Unit Tangent: $-.376i+.242j+.894k$
   Unit Normal: $-.54i-.842j$
   Binormal: $.753i-.483j+.447k$

7. Osculating Plane:
   $-.502(x+.707)+.707(y+.785)-.498(z-.707)=0$
   Normal Plane:
   $.5(x+.707)+.707(y+.785)+.5(z-.707)=0$
   Rectifying Plane:
   $.705(x+.707)+.002(y+.785)-.709(z-.707)=0$

9. Osculating Plane:
   $-.053(x-4.81)-.053(y-0)+.997(z-4.81)=0$
   Normal Plane:
   $.707(x-4.81)-.707(y-0)=0$
   Rectifying Plane:
   $.705(x-4.81)+.705(y-0)+.074(z-4.81)=0$

ANSWERS

Exercise Set 17.2

1. $F(0)=-.5$, Curvature=2, Radius of curvature=.5, and the center of the osculating circle is $(0,0)$.

3. $F(1)=1$, Curvature=.849906847, Radius of curvature=1.17659953, and the center of the osculating circle is $(1.83198229, 1.83198073)$.

5. .0192423394

7. .511976133

9. .235633463

Exercise set 18.1

1. Actual Value: does not exist

3. Actual Value: 0

5. Actual Value: 1

7. Actual Value: does not exist

9. Actual Value: does not exist

11. No possible value

13. C=0

Exercise Set 18.2

1. Actual Values: $f_x=-26$, $f_y=96$, Use a precision of .001

3. Actual Values: $f_x=-.2186589$, $f_y=.2099125$

5. Actual Values: $f_x=-.0034553554$, $f_y=-.0038872748$

7. Maximum Error=.0043261136

9. Tangent Plane: $3.52531624(x+1)+.168160604(y-2)-(z-.07)=0$
   Normal Line: $\dfrac{x+1}{3.525} = \dfrac{y-2}{.168} = \dfrac{z-.07}{-1}$

Exercise Set 19.1

1. (3,0)

3. (.768169157,.694819691)

5. (1.04542972,-.317364114)

7. b.   There is no point of intersection.

9. (0,0,3) Local Max
   $(2\pi/3,2\pi/3,-1.5)$ Local Min
   $(\pi,\pi,-1)$ Saddle Point
   $(4\pi/3,4\pi/3,-1.5)$ Local Min

Exercise Set 19.2

1. 20.3485678i+21.2577938j

3. 2.99999676i+1.84147374j+1.84147374k

5. 5.32487783i+3.01408179j-1342577.96k

7. Tangent Plane: .117(x-1.6)+3.75(y+1)=0
   Normal Line: 3.75(x-1.6)=.117(y+1),z=0

9. Direction: 3.637i+.25j+.584k, Magnitude: 3.6920624

Exercise Set 20.1

1. 18.9558365

3. .605914403

5. 16.9656451

7. .215199988

9. 1.14238807

11. 2.85404683

13. a.(0,0)
    b.(1.799989,0)

ANSWERS

Exercise Set 20.2

1. Actual Value: 2048/3

3. Actual Value: 4/3

5. Actual Value: 0

7. 25.202072

9. 0

11. 34.725221

13. 318.112936

15. 608365.148

17. 18824.1041

19. 171.768743

21. (3.1415927,.5698185,5.5399498)

Exercise Set 21.1

1. -.0500130112

3. .516199294

5. 0

7. 7.033012045

13. a. .205541757
    b. .205538091
       $f(x,y)=exp(x)*sin(y)$

15. Actual Value: 126.5

Exercise Set 21.2

1. 1657.32225

3. 150.305248

5. 0

7. 4.32248241

9. 861.981375

11. 1276.034421

Exercise Set 22.1

1. Actual Value: 1.8257419

3. Actual Value: 1.6554575

9. .023432312

11. Results are getting larger

13. b. b=-.599308
    c. b+-.6302718

Exercise Set 22.2

1. -.0892495073

3. -.730831166

5. .0702504029

9. Between 417000 and 418000 seconds.